Delivering Sustainable Competitiveness

Global trends such as climate change, digitalisation, enhanced concepts of democracy and the consequences of the 2008 financial crisis are changing the playing field of cities across the world. Urban development objectives are shifting away from being purely concerned with wealth creation and competitiveness, to increasingly combining social and environmental dimensions. In this context, how can cities influence and sustain their competitive position over time? Which new types of urban strategies are emerging, and which organising capacities are proving the most important?

This book provides insight into the complex issue of delivering sustainable competitiveness by analysing a number of innovative urban development strategies in context. Questions and topics addressed include: how can new legacies of city events be secured; how can clean technology industries be nurtured through urban regeneration initiatives; and how can the impact of urban safety strategies be enhanced? These and other pivotal questions are explored through close attention to the enabling factors linking ideas with results, such as distributed leadership, collaboration, communication and experimentation. Combining case studies from Europe, Africa, South America and Southeast Asia, the book provides a truly international perspective on the potentials and limitations of a new generation of urban development and competitiveness strategies.

Luís Carvalho is Senior Researcher at the European Institute for Comparative Urban Research (EURICUR) and at the Centre of Studies in Geography and Spatial Planning (CEGOT), based at the University of Porto. His research addresses local economic development dynamics, innovation and economic geography issues. He has been closely involved in a number of international comparative studies in several European, Asian and Latin American cities and in the coordination of academic and executive education in his fields of expertise. Current research focuses on the geography of smart city innovation and socio-technical transitions in cities.

Leo van den Berg is Emeritus Professor at the Erasmus University Rotterdam and has been active in the field of urban management and regional economics for more than 40 years. He has been Director of the European Institute for Comparative Research (EURICUR) for more than 25 years and has conducted urban research in more than 130 cities in all parts of the world. Since 1995 he has been an active member of the Global Urban Competitiveness Project (GUCP). He is the (co-)author of 27 books and has contributed to more than 100 articles, book chapters and other publications.

Hazem Galal is a PwC Partner and the global Leader for PwC's Cities and Local Government Sector. With 25 years of consulting experience, he has expertise in strategy formulation and implementation for cities and local government entities. He has supported several city projects in the United States, Middle East, Latin America, Asia, Africa and Europe. Galal is a frequent speaker at numerous global conferences and panels on cities organised by the World Cities Summit, World Bank, World Islamic Economic Forum, Asian Development Bank, United Nations and other international bodies as well as local and national governments around the world. He has also been hosted as a city expert on several occasions by leading media channels such as the BBC, CNBC, Russia Today and Channel News Asia.

Peter Teunisse works at the Public Sector practice within PwC Advisory, the Netherlands and is Senior Director for Cities and Local Government. Since 1990 he has led several consultancy projects for the Dutch government, as well as for local and regional authorities. Since 2004 he has been a core-team member of the international PwC network for Cities, Regions and Local Government. He has published various articles and books about the organisation of Dutch cities and is one of the lead authors of the international PwC publications on the future of cities.

EURICUR Series

The European Institute for Comparative Urban Research (EURICUR) was founded in 1988 and has its seat with Erasmus University Rotterdam. EURICUR is the heart and pulse of an extensive network of European cities and universities. EURICUR's principal objective is to stimulate fundamental international comparative research into matters that are of interest to cities. To that end, EURICUR coordinates, initiates and carries out studies of subjects of strategic value for urban management today and in the future. Through its network EURICUR has privileged access to crucial information regarding urban development in Europe and North America and to key persons at all levels, working in different public and private organisations active in metropolitan areas. EURICUR closely cooperates with the Eurocities Association, representing more than 100 large European cities.

As a scientific institution, one of EURICUR's core activities is to respond to the increasing need for information that broadens and deepens insight into the complex process of urban development, among others by disseminating the results of its investigations through international book publications. These publications are especially valuable for city governments, supra-national, national and regional authorities, chambers of commerce, real estate developers and investors, academics and students, and others with an interest in urban affairs.

EURICUR website: www.euricur.nl

This book is part of a series published by Routledge under the auspices of EURICUR, Erasmus University Rotterdam

Titles in the series include:

Empowering Metropolitan Regions Through New Forms of Cooperation
Alexander Otgaar, Leo van den Berg, Jan van der Meer and Carolien Speller

Industrial Tourism
Alexander H. J. Otgaar, Leo van den Berg, Christian Berger and Rachel Xiang Feng

Towards Healthy Cities
Alexander Otgaar, Jeroen Klijs and Leo van den Berg

Cities as Engines of Sustainable Competitiveness
Edited by Leo van den Berg, Jan van der Meer and Luís Carvalho

Delivering Sustainable Competitiveness
Edited by Luís Carvalho, Leo van den Berg, Hazem Galal and Peter Teunisse

Delivering Sustainable Competitiveness

Revisiting the organising capacity of cities

Edited by Luís Carvalho, Leo van den Berg, Hazem Galal and Peter Teunisse

Euricur
EUROPEAN INSTITUTE FOR COMPARATIVE URBAN RESEARCH

European Institute for Comparative Urban Research
Erasmus University Rotterdam
The Netherlands
www.euricur.nl

Routledge
Taylor & Francis Group
LONDON AND NEW YORK

First published 2017 by Routledge

2 Park Square, Milton Park, Abingdon, Oxfordshire OX14 4RN
52 Vanderbilt Avenue, New York, NY 10017

Routledge is an imprint of the Taylor & Francis Group, an informa business

First issued in paperback 2020

British Library Cataloguing in Publication Data
A catalogue record for this book is available from the British Library

Library of Congress Cataloging in Publication Data
Names: Carvalho, Luis Miguel da Costa Monteiro de, editor.
Title: Delivering sustainable competitiveness : revisiting the organising capacity of cities / [edited] by Luâis Carvalho, Leo van den Berg, Hazem Galal and Peter Teunisse.
Description: Farnham, Surrey, UK ; Burlington, VT : Ashgate, [2016] |
Series: EURICUR series (European Institute for Comparative Urban Research) |
Includes bibliographical references and index.
Identifiers: LCCN 2015043217| ISBN 9781472482679 (hardback : alk. paper) |
ISBN 9781472482693 (epub)
Subjects: LCSH: Urbanization. | City planning. | Regional planning. |
Strategic planning. | Sustainable urban development. | Urban policy. |
Urban geography.
Classification: LCC HT361 .D45 2016 | DDC 307.76–dc23
LC record available at http://lccn.loc.gov/2015043217

ISBN: 978-1-4724-8267-9 (hbk)
ISBN: 978-0-367-66831-0 (pbk)

Typeset in Times New Roman
by Cenveo Publisher Services

Contents

Figures

Tables

Boxes

Preface

As of today, the governance of metropolitan regions has been and remains a central concern for urban scholars and policymakers alike. During the 1990s, the European Institute for Comparative Urban Research (Euricur), in partnership with several European cities, heartily addressed this issue and studied different ways through which urban stakeholders influenced change in cities and metropolitan regions. These early studies resulted in a number of volumes, namely *Governing Metropolitan Regions* (Leo van den Berg, Arjan van Klink and Jan van der Meer, 1993) and *Metropolitan Organising Capacity* (Leo van den Berg, Erik Braun and Jan van der Meer, 1997), both published by Ashgate.

By that time, an important notion emerging from these studies was 'organising capacity', defined as the ability (of those in charge of urban development) to enlist all actors involved, and with their help generate new ideas and develop and implement policy designed to respond to fundamental developments and create conditions for sustainable development. Parallel to other conceptual developments in the urban planning literature, organising capacity called attention to the organisational factors behind successful urban transformation projects and strategies, namely leadership, vision, strategic networks, social and political support and communication.

Yet, since then, the playing field of cities has been changing substantially. New trends, such as climate change, digitalisation, ageing, state rescaling and new concepts of citizen participation, gained ground at the same time as urban development objectives shifted from wealth creation to combine social and environmental dimensions. In this new context, how can cities influence their competitive position over time? Which new types of urban development strategies are emerging and which organising capacities are proving important? Are they just 'old wine in new bottles'? What will be the critical issues for the coming years?

This book looks into these issues through the lens of a large international comparative study, carried out during 2012–14, in partnership between Euricur, PricewaterhouseCoopers (PwC) and the Institute for Housing and Urban Development Studies (IHS). The early idea for this study dates back to a joint discussion platform on the future of European urban policy, initiated by Euricur together with the City of Rotterdam and with city officials from Antwerp, Barcelona, Budapest, Dortmund, Dublin and Helsinki (2010–11). During 2012, this discussion platform was expanded to transform it into an international comparative study, broadening its European scope to include cities and developments in other parts of the world. It followed up and deepened previous Euricur and IHS studies, as well as recent global surveys by PwC that examined enabling factors for successful urban strategy execution.

This book is grounded in a number of in-depth case studies and could not exist without the enthusiasm and commitment of their authors. However, many other people were essential in making it happen. Alexander Otgaar and Egon de Haas were fundamental in managing the whole research project and, together with Jan van der Meer, Leon van den Dool and Saskia du Bois-Schutz, formed the core team and provided valuable inputs throughout the research process. Nick Jones reviewed previous versions of the introductory and concluding chapters and gave us many comments and a lot of 'food for thought' to improve them further. We thank Sarah Horsfield (for her collaboration in the case study of Rio de Janeiro), as well as Kees van Rooijen (IHS), Peter Kresl (Global Urban Competitiveness Project), Chris de Lange and Hans Verdonk (City of Rotterdam) for their support throughout the process. We are also grateful to Monique Valkenburg, Eva Smeding, Brenda Verberne, Stella Soekhlall and Ingeborg Haasbeek for their organisational support.

Finally, we are indebted to all the city officials directly involved in this study for their willingness to share their experiences and discuss insights and challenges with us. We are also thankful to all the interviewees and discussion partners for their time and willingness to contribute to this study, making the final result a true co-production.

The editors,
Luís Carvalho, Leo van den Berg,
Hazem Galal and Peter Teunisse

Contributors

Leo van den Berg is Emeritus Professor at the Erasmus University Rotterdam and has been active in the field of urban management and regional economics for more than 40 years. He has been director of the European Institute for Comparative Research (Euricur) for more than 25 years and has conducted urban research in more than 130 cities in all parts of the world. Since 1995 he has been an active member of the Global Urban Competitiveness Project (GUCP). He is the (co-) author of 27 books and has contributed to more than 100 articles, book chapters and other publications.

Luís Carvalho is Senior Researcher at the European Institute for Comparative Urban Research (Euricur) and at the Centre of Studies in Geography and Spatial Planning (CEGOT), based at the University of Porto. His research addresses local economic development dynamics, innovation and economic geography issues. He has been closely involved in a number of international comparative studies in several European, Asian and Latin American cities and in the coordination of academic and executive education in his fields of expertise. Current research focuses on the geography of smart city innovation and socio-technical transitions in cities.

Jan Fransen is Deputy Director of IHS, the international Institute of Housing and Urban Development Studies at Erasmus University Rotterdam. He is specialised in the field of local economic development, with a focus on innovation, governance and the role of small- and medium-sized enterprises. Fransen has considerable international experience and has conducted research, education and advisory work in over 15 different countries in Europe, Africa and Asia.

Hazem Galal is a PwC Partner and the global Leader for PwC's Cities and Local Government Sector. With 25 years of consulting experience, he has expertise in strategy formulation and implementation for cities and local government entities. He has supported several city projects in the United States, Middle East, Latin America, Asia, Africa and Europe. Since 2007, he has led PwC's global Thought Leadership for cities. He is a frequent speaker at numerous global conferences and panels on cities organised by the World Economic Forum, World Cities Summit, World Bank, World Islamic Economic Forum, Asian Development Bank, United Nations and other international bodies as well as local and national governments around the world. He has also been hosted as a city expert on several occasions by leading media channels such as the BBC, CNBC, Russia Today and Channel News Asia.

Alberto Gianoli, currently working at Erasmus University Rotterdam, is an economist and management specialist with 15 years' experience as a researcher and consultant. His main areas of expertise and interest include urban environment and infrastructure, project and infrastructure finance, urban and regional management and governance, and urban and regional economic development. Gianoli has carried out assignments for the World Bank, European Commission, United Nations (UNDP, UNEP, UN HABITAT, UNCDF), Organization for Security and Co-operation in Europe and several national, regional and local governments worldwide.

Sarah Lidé is a manager at PwC with a focus on strategic development and analysis. She has been actively involved in a variety of projects related to government, healthcare and life science, including national and regional innovation and development, and her work requires her to keep abreast of the latest trends and developments in a broad range of industries. She has led strategic processes in public sector organisations as well as conducted trend briefings at various conferences and seminars. She has also written a number of articles and leadership-related publications.

Jan van der Meer is retired Associate Professor of urban and regional economics at the Erasmus School of Economics (Erasmus University Rotterdam) and former Associate Director of the European Institute of Urban Research (Euricur). Since the early 1990s he was involved in international comparative urban studies. He is (co-) author and editor of more than 100 publications in the domains of urban economics and urban management.

Alexander Otgaar is Senior Researcher and Lecturer in urban economics and management at Erasmus University Rotterdam, the Netherlands. He has more than 15 years' experience with international (comparative) studies on the development of cities worldwide, addressing topics such as regional cooperation, cluster development, place marketing and urban attractiveness.

Peter Teunisse works at the Public Sector practice within PwC Advisory Netherlands and is Senior Director for Cities and Local Government. Since 1990 he has led several consultancy projects for the Dutch government, as well as for local and regional authorities. Since 2004 he has been a core-team member of the international PwC network for Cities, Regions and Local Government. He has published various articles and books about the organisation of Dutch cities and is one of the lead authors of the international PwC publications on the future of cities.

Erwin van Tuijl is a researcher and lecturer at Regional, Port and Transport Economics (RHV) and member of the European Institute of Comparative Urban Research (Euricur) at Erasmus University Rotterdam. He has been involved in several international comparative studies in Western and Eastern Europe, Latin America and Asia, on topics such as 'manufacturing in the new urban economy', 'events and urban development', 'energy transitions in cities' and 'urban agriculture in the knowledge economy'. His PhD thesis (2015) deals with the upgrading of the Chinese automotive industry. His main research interests include the geography of value chains, knowledge economy, firms' innovation strategies, upgrading, manufacturing, and creative industries.

1 Organising capacity for sustainable urban competitiveness

Luís Carvalho, Leo van den Berg, Hazem Galal and Peter Teunisse

1 Introduction

1.1 Setting the scene

Cities have always been engines of growth and development. It is widely stated that they boost people's productivity, wealth and creativity, but also energy efficiency and social mobility (Glaeser, 2011). Despite globalisation and the opportunities opened by information and communication technologies, it is still in cities that new ideas circulate faster and, therefore, where new productive combinations and innovation tend to emerge first. Cities are hotspots for ideas exchange among skilled individuals. Indeed, over the last few decades, as economies have become more knowledge intensive, the role of cities has clearly increased (van den Berg *et al*., 2005; McCann, 2008).

Naturally, and despite the much-heralded urban revolution and the growing urban population worldwide (United Nations, 2008), cities have rather heterogeneous capacities to grow and develop over time. Moreover, the fortune of cities is not perennial: they face cycles of growth and decline (van den Berg, 1987). Throughout history, many cities lost the edge they once had – just think of Renaissance Venice or industrial Detroit. Indeed, sometimes cities face problems to reinvent themselves in new global economic contexts, particularly after periods of strong economic specialisation. Physical assets become redundant and under used. But also their economic structures and institutions tend to become inflexible and inward looking (Grabher, 1993). As their innovation drive and position in the international labour division slips, the city's future is put at jeopardy (Storper, 2013).

In the medium and long term, a city's economy is driven by its capacity to innovate and diversify towards new economic activities, namely by nurturing and attracting new jobs, people and skills (Jacobs, 1969; Boschma, 2015). However, social and environmental dimensions are increasingly important for societal progress as well (Pike *et al*., 2010). Poverty, crime and widening divides between the 'haves' and 'have-nots' are critical issues in large metropolises as diverse as Paris, Rio de Janeiro or Cape Town. These are not only increasingly unacceptable from a moral perspective, but, over time, they endanger a city's very economic foundations and attractiveness (Hall and Pfeifer, 2000). The same goes for pollution, climate change and environmental problems, increasingly acute in rapidly growing cities such as Beijing or Shanghai. In the medium and long term, congestion and pollution negatively impact on a city's quality of life. Hence, social cohesion and environmental quality are no longer luxury alternatives or even 'restrictions' to growth and competitiveness – they are becoming integral to a city's economic development prospects.

The previous challenges have also been magnified by a number of currently unfolding megatrends. For example, climate change, digitalisation, population ageing, state rescaling, severe budget cuts and enhanced concepts of democracy-transparency (e.g. European Commission, 2011; PwC, 2013) are changing the playing field of cities. On the one hand, while many cities in emerging economies continue to thrive, it is uncertain whether such growth can be sustained over time. A pressing challenge is how to decouple growth from environmental dilapidation while making it more inclusive. On the other hand, in advanced economies a key challenge is how to re-ignite growth and create jobs while maintaining social welfare and quality of living. Even if much of a city's fortune has to do with external-to-the-city macroeconomic and political developments, cities are often not just passive receivers of a given context – they can also shape their own development.

1.2 The organising capacity of cities: issues and challenges

Naturally, a central question is how can cities (and their multiple actors) deal with those variegated trends and make urban change happen 'on the ground'. In the late 1990s it was argued that the success of urban transformation projects and strategies would be increasingly dependent on a city's organising capacity, defined as: 'The ability to enlist all actors involved, and with their help generate new ideas and develop and implement a policy designed to respond to fundamental developments and create conditions for sustainable development' (van den Berg *et al.*, 1997, p.1).

Organising capacity was defined as relying on a set of enablers or conditions, such as the quality of actor networks and stakeholder's collaboration, communication, leadership, social and political support. In other words, organising capacity was considered a fundamental mediator between a city's ability to identify present and future challenges and effectively act upon them. The organising capacity framework was relatively early combining notions of urban competitiveness and sustainability within a single conceptual model. Moreover, it called attention to relevant organisational features in urban management and to the role of agency in urban transformation. It has been tested and validated in several case studies, and provided important insights to understand the success (and failure) of urban development strategies during the 1990s, namely in Western Europe (e.g. van den Berg *et al.*, 1993, 1997, 2003, 2005, 2014).

Currently, the changing playing field of cities and evolving notions of urban competitiveness and sustainability around the globe call for a reappraisal of the role of organising capacity and its enablers in urban development projects. First, as a response to the aforementioned trends, a new generation of urban development and competitiveness-oriented projects and strategies has been emerging, whose visions, results and implementation modes are still to be assessed. Second, there is a strong need to understand development patterns and the success factors behind urban policies also outside western economies, namely in places where cities are growing at a very fast pace (e.g. Turok, 2013).

Overall, it is important to get a better understanding of how different types of contemporary urban development strategies are envisioning (and leading to) sustainable competitiveness, that is, 'the ability of cities to keep growing and developing over time while fostering social cohesion and environmental quality' (van den Berg *et al.*, 2014, p.4). How are cities currently dealing with new urban development challenges? And how (and to what extent) are cities moving beyond pure economic 'boosterism' strategies (Gold and Ward, 1994; Logan and Molotch, 2007) and combining economic competitiveness with higher order sustainability objectives?

1.3 Research questions and the organisation of this book

To deal with the aforementioned issues, this book offers a case-study-based survey of new types of urban development projects and strategies initiated, supported or carried out by local governments, with the ambition to improve a city's sustainable competitiveness. Its underlying research question is thus:

- How have cities been enhancing their sustainable competitiveness through new urban development projects and initiatives?

Each chapter in this book answers this broad question by focusing on a concrete urban development initiative and by answering to the following sub-questions:

- What contextual factors in the city (and external trends) explain the emergence of the initiative? How does it contribute to sustainable competitiveness?
- What are the key features of the initiative and what is most innovative about its content?
- Which enablers and organisational features (organising capacity) have been critical during implementation?
- What are the direct and indirect results of the initiative so far? What are the limitations and challenges ahead?

Before delving into each of the book's case studies, the remainder of this chapter provides a short review of the building blocks behind the notion of sustainable competitiveness, notably the complex and changing notions of sustainability and competitiveness (Section 2). Subsequently, we present the conceptual framework that guided the overall study, as well as the research setting and methods (Section 3).

Chapters 2–12 present concrete case studies, carried out by different authors. They analyse the emergence, implementation mechanisms and (early) results of contemporary types of urban development strategies. The cases come mostly from Europe, but include Asian, South American and African cities as well. Altogether, they bring to the fore a number of new (and not so new) urban development themes and strategies, as well as the organising capacities and enablers that mediated city ambitions and concrete achievements:

- *Chapter 2: Digital Strategy (Manchester)*, an integrated road-map to jointly tackle several dimensions of digitalisation in the city (by Luís Carvalho and Erwin van Tuijl);
- *Chapter 3: Dublinked (Dublin)*, an 'open data' initiative to foster innovation and entrepreneurship through the open release of data on urban transport, water, planning and energy (by Luís Carvalho and Alexander Otgaar);
- *Chapter 4: Integrated Housing Development Programme (Addis Ababa)*, an initiative to dramatically increase the city's housing supply and stabilise home ownership (by Alberto Gianoli and Alexander Otgaar);
- *Chapter 5: Kasongan pottery cluster (Yogyakarta)*, a combination of bottom-up and top-down initiatives to enhance and diversify one of Indonesia's most competitive and indigenous business clusters (by Jan Fransen and Erwin van Tuijl);
- *Chapter 6: Mobile World Capital (Barcelona)*, a new nexus of event legacy in the quintessential 'Olympic' city, this time focused on the relation between the Mobile World Congress and the city's ambitions to excel in digital innovation (by Alexander Otgaar and Luís Carvalho);

- *Chapter 7: Pacifying Police Units (Rio de Janeiro)*, an integrated type of security-social intervention to bring Rio's famous *favelas* 'back to formality' while enhancing the city's safety (by Hazem Galal);
- *Chapter 8: Park Railway North (Antwerp)*, the revitalisation of an old railway site towards a new urban green park (by Jan van der Meer and Alexander Otgaar);
- *Chapter 9: Phoenix project (Dortmund)*, the regeneration of a large brownfield site into a post-industrial location for leisure and new economic activity (by Erwin van Tuijl and Leo van den Berg);
- *Chapter 10: RDM Campus (Rotterdam)*, the redevelopment of deprived port areas for new industries, education and climate-proofing experimentation (by Erwin van Tuijl and Alexander Otgaar);
- *Chapter 11: Water management strategy (Singapore)*, a comprehensive strategy to enhance the link between water self-sufficiency, innovation and cluster development (by Sarah Lidé);
- *Chapter 12: World Design Capital (Helsinki)*, an 'umbrella' type of event with a focus on embedding design-thinking across the city as key legacy (by Alexander Otgaar and Luís Carvalho).

Chapter 13 wraps up and synthesises highlights from the various case studies across different themes and dimensions. It concludes with a number of conceptual reflections, policy implications and challenges ahead for urban managers.

2 The intricate notions of sustainability and urban competitiveness

2.1 Sustainability

It has become common sense that cities should develop in a sustainable way. Yet, the debate is easily undermined since there is not one single, unified meaning for sustainability. It is a very porous and elastic concept, which has been used to describe almost any kind of policy or intervention goal, even when radically different objectives are at stake. Sustainability can simply mean to sustain something over time; however, in many academic and policy spheres, it is popularly associated with 'green' visions and a plea to balance (or, sometimes, decouple) economic growth and its environmental consequences such as resource dilapidation and climate change.

Almost 30 years ago, the influential Bruntland report (WCED, 1987) imprinted an environmental notion to sustainability, also calling attention to the needs of future generations: since they cannot vote, we do not know their preferences and thus a cautionary principle (protecting environmental resources) should be followed. More recently, the sustainability debate gained a social dimension as well, a plea to balance growth and social equity. All in all, there is nowadays an awareness that 'sustainability' has somehow to do with balancing economic, social and environmental dimensions and goals, as they tend to be closely connected and influence each other (OECD, 2008).

Nevertheless, despite the general agreement that sustainability involves a sort of balance between economic, environmental and social dimensions, sustainability is still a highly contested and value-driven concept (e.g. Hopwood *et al.*, 2005). This becomes particularly evident when discussing urban planning and local development strategies (Campbell, 1996; Næss, 2001; Rees, 1995), for example in the integrated redevelopment of urban quarters and waterfronts. Although operating under the same 'sustainability banner', the involved stakeholders often have significantly different ideas about what sustainability means and use it according to their own agendas (e.g. pro-market vs pro-conservation vs pro-inclusion). Over time, those dissonances have been proving detrimental for the consistency of urban projects and strategies (Lombardi, 2011; Plácido Santos, 2011).

All in all, there is a relatively wide continuum between different notions of what sustainability is or should be. A standard distinction (see Figure 1.1) is often made between weak and strong sustainability (Williams and Millington, 2004). On the 'weakest' side, sustainability is broadly equated as continued economic growth. Here, the economy is the dominant dimension, and it is seen as practically independent from other environmental and societal spheres (which are perceived also as less relevant, as the size of the circles suggest). From this angle, it is implicitly assumed that economic growth is to be nurtured first since, in the end, new income and jobs will be created and wealth distributed in a trickle-down fashion: the 'Mickey Mouse' model. On the other extreme – 'strong sustainability' – the economy is considered as a minor subset of society and nature (the 'bullseye' model). From this perspective, the natural/environmental capital has precedence in relation to economic and social dimensions and has to be strictly maintained. According to this view, it is not legitimate or 'sustainable' to substitute natural resources with the help of new technology possibilities (Solow, 1986). This is an eco-centred model, associated with deep ecology and conservationist movements.

Naturally, there is a large continuum of sustainable (urban) development notions in-between these two extremes. In this study we start from a relatively more consensual approach. We consider that the economy, environment and society have a certain degree of independence, but recognise the close relations established between the different spheres. The intersections of the spheres (Figure 1.1) are associated with positive synergies (e.g. social inclusion through job creation, economic attractiveness due to the quality of the living environment), but also represent complex societal choice problems and negative externalities (e.g. pollution). All in all, this middle view assumes that

> To get caught up in an argument over whether sustainable development is more about the environment or about people is to miss the point: it is the connection of humans, their economies and societies to the ecosystems that support them which defines sustainable development. (OECD, 2008, p.33)

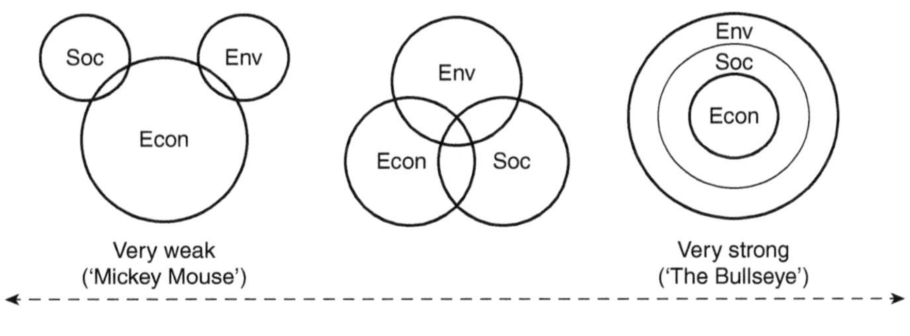

Figure 1.1 Sustainable urban development notions

Source: Own elaboration, inspired by Lombardi *et al.* (2011) and http://computingforsustainability.com/ (accessed 24 July 2012)

In line with Camagni *et al.* (1998), this notion suggests that sustainability implies a balanced and dynamic process – and not necessarily a concrete measurable outcome – in which the economy, society and environment are connected and interdependent. In practice, urban development strategies without economic growth and competitiveness dimensions have difficulties in succeeding – and making it through political arenas (e.g. Porter and Hunt, 2005). However, the capacity of such development strategies to deliver over time is also closely linked with the relations established with environmental and social dimensions (e.g. van Winden and Carvalho, 2015).

2.2 Urban competitiveness and its foundations

The interplays between economic, social and environmental spheres in urban development have also been permeating notions of urban competitiveness. In the late 1980s and 1990s, van den Berg and Braun (1999) suggested that globalisation, heightened factor mobility, digital accessibility and political integration were leading to growing competition between cities to attract jobs, people and investments. Since that time, there has been a long academic debate about whether notions of competitiveness, as used for firms, could be meaningfully applied for cities (e.g. Urban Studies, 1999; Regional Studies, 2004; OECD, 2006). There is nowadays some consensus that the answer is yes, but not in the same way. In this section we briefly elaborate on the changing notions of urban competitiveness to derive the perspective adopted in this book.

Urban competitiveness can ultimately be assessed by the performance of a city over time. In other words, it is about how well a city does in attracting and retaining jobs and people. As argued by Kitson *et al.* (2004), 'competitive regions and cities are places where both companies and people want to locate and invest in' (p.997). Thus, from this perspective, urban competitiveness is about two key issues: people development and business development. There are in principle two main channels for achieving this, which often work in combination: through nurturing existing, indigenous assets (e.g. upgrading of existing businesses and skills; creation of new start-ups) and through external-to-the-region attraction (e.g. of transnational companies and skilled migrants). In order to account for these outcomes, urban competitiveness is usually assessed through aggregated indicators such as per capita income, productivity, population growth and unemployment rates.

However, from a policy perspective, the idea that 'cities are in competition' tended to favour policies and urban strategies emphasising external-to-the-region attraction of footloose people, companies and investments (e.g. sports and cultural mega-events, multinational subsidiaries, creative workers). According to this view, cities can compete for quotas of investment, resources and markets in a certain moment of time, often by offering tax breaks, relocation incentive packages, good infrastructure and safe business environments. Yet, evidence suggests that this kind of policy has resulted in counterproductive and wasteful 'bidding wars' (e.g. Rodríguez-Pose and Arbix, 2001). In the absence of complementary policies, these types of competitiveness-boosting policies often led to zero and negative sum effects and temporary shifts of resources (and problems) from one city to the other (e.g. Pike *et al.*, 2006). Over time, in the absence of local anchoring capabilities – for example complementary skills, demanding local markets, adequate institutional settings – companies and investments tended to relocate again. Overall, there is plenty of evidence that those types of static competition policies based on external attraction seem to be incomplete (at best) or misleading (at worst), if the aim is to foster urban competitiveness and local economic development.

Therefore, there is a growing consensus that urban competitiveness must be deeply grounded in the structural drivers of a city's competitive advantage. Many authors have demonstrated that the long-term competitive advantage of places is created and sustained through highly localised 'assets', 'foundations' or 'capitals', such as the presence of specialised resources, skills, infrastructure and information flows (e.g. Porter, 1990; Saxenian, 1996; van Winden *et al.*, 2007; PwC, 2013). Those capitals change slowly as they are linked with industrial structures, socio-cultural bounds and actors' exchange routines (Storper, 1997). This is why it has proved so difficult to replicate 'Silicon Valleys somewhere' through subsidies and infrastructure (Hospers, 2006). In order to attract and retain jobs and people over time, the presence – but also the coherence – of such place-based capitals is of utmost importance: it is out of the match between external inflows of businesses and people and place-based capitals that regions diversify and remain competitive over time (Boschma and Frenken, 2011).

Urban competitiveness structural assets or 'capitals' result from a mix of historical processes, market forces and multi-level public interventions. They are not solely in the hands of local governments, but can be influenced by their action. Altogether, they form the socio-economic context of a city (van den Berg *et al.*, 2005; van Winden and Carvalho, 2008). There are many possible ways of clustering such capitals for analysis. In this study we considered the following: knowledge and economic capital; social conditions; environment and quality of life; image; and institutional conditions.

The *knowledge and economic capital* of a city is the most important capital. It is closely associated with its knowledge infrastructure (e.g. universities, R&D and higher education institutes) and the skills of the working population. There is a strong link between a city's knowledge capital and urban productivity increases (e.g. Glaeser, 2011). The skills of the working population tend to reflect the city's economic structure and its capacity to generate new innovative combinations. Highly specialised cities face problems in economically reinventing themselves; however, fully diversifying their industry and economic base is not the solution – if diversified urban economies tend to be more resilient to external shocks, they find fewer opportunities for productive recombination (Frenken *et al.*, 2007). Recent research suggests that, over time, the best performing and innovative local and regional economies are those which have a diversified yet related mix of industries and economic activities (Boschma and Frenken, 2011).

However, even in successful urban economies, social problems and inequalities can persist and hinder further development. For example, cities like Manchester have done remarkably well changing their economic structure and attracting new residents, but social deprivation and unemployment persists. Thus, *social conditions* are also considered by many as an important driver of urban competitiveness (e.g. widespread access to housing, basic skills, health and opportunity). A key issue has to do with the disparities between have and have-nots. Social disparities can be related to crime, decreasing a city's attractiveness for new residents and businesses (van den Berg *et al.*, 2005). However, beyond the pure economic and utilitarian rationale, disparities also raise many ethical considerations on urban justice since economic growth does not seem to trickle-down so easily into new jobs and opportunities (e.g. Storper, 2011; Fainstein, 2010). Overall, there is a well-known 'inverted-U' relationship between equality and economic performance in cities: over time, too high or too little social equality hinders growth (e.g. Castro and Jensen-Butler, 1999).

Environmental and quality of life conditions have been increasingly considered important capitals for urban competitiveness, namely when it comes to the attraction and retention of talented workers. Although in somewhat different fashions, both Florida (2002) and Glaeser

(2011) show positive relations between the environmental and amenity endowments of places and their competitiveness. The underlying logic is that a high premium has been placed on quality of living (van den Berg, 1987) and that talented workers chose to move or stay based not only on job opportunities but also on attributes of the built and natural environment, cultural endowments, advanced consumption opportunities and advanced health services. The market provides many of such amenities, but in most countries the public sector plays a key role as well. However, the opposite is also true: from a certain threshold of environmental degradation, a city's competitiveness is put at risk (even if jobs, consumption and cultural amenities are present).

A city's *image* and identity is also an important capital, and one that changes very slowly. Van den Berg and Braun (1999) stress the role of a city's image in its long-term competitiveness. When choosing a city, in a context of limited information and with other things being equal, its external image plays a role (Carvalho, 2013). Urban regeneration strategies, culture and the hosting of large events have been among the most common strategies deployed by cities to change their images to the outside world, but also to improve the self-esteem of their residents (Richards *et al.*, 2013). Former industrial cities provide vivid examples of the role of image in competitiveness: just think of the re-imagining of Barcelona and its increased post-Olympic attractiveness (Smith, 2005).

Finally, it is now widely acknowledged that place-specific *institutional conditions* are central in urban and regional development (e.g. Rodríguez-Pose, 2013). The capacity of a place to bring together stakeholders and get things done in the face of complexity, conflict and change are important factors behind a city's competitiveness (Stoker, 1995; Sotarauta, 2006). Rodríguez-Pose and Di Cataldo (2015) recently found a strong causal relation between the quality of the government and the capacity of regions to innovate and, on the flip side, confirm that ineffective and corrupt governments severely hinder the innovative capacity of regions, namely in more peripheral regions. Institutional conditions go beyond actors and administrative structures to include dimensions like social capital (Putnam, 2001) and the norms, cultures, exchange routines and regulations that guide and organise exchanges and interactions among different actors in place (e.g. North, 1990).

2.3 Organising capacity: old and new dimensions

Clearly, institutional conditions influence a city's overall capacity to identify and act upon its pressing development challenges and opportunities, reinforcing (or not) all the other 'capitals'. As mentioned earlier, van den Berg *et al.* (1997) speak about organising capacity as 'the ability to enlist all actors involved and, with their help, to generate new ideas and to develop and implement a policy designed to respond to fundamental developments and create conditions for sustainable development' (p.1). This capacity is pivotal to improve a city's aforementioned capitals over time and to mobilise new resources for a city's development. It refers to the capacity of places to pursue policy integration between different government layers (e.g. national, regional, local), but also between different stakeholders in a city (e.g. government, chamber of commerce, universities and other organisations) and to put their energy at the service of concrete urban development projects and strategies.

In this study, we revisit the original organising capacity framework (see introductory section; van den Berg *et al.*, 1997; van den Berg *et al.*, 2003) and suggest that the organising capacity required to implement increasingly complex urban development strategies – integrating competitiveness and sustainability goals – benefit from six key enablers: strategic networks and collaboration, leadership, communication, finance, prioritisation and piloting, and agile delivery. The first three enablers were present in the framework's original formulation, while the others are derived from recent studies and surveys on urban development strategies (e.g. Carvalho *et al.* 2014; van den Berg *et al.*, 2014; PwC, 2011, 2012, 2013).

Strategic networks and collaboration refers to the relations established between systems of urban actors whose strategies are influenced by (and may depend on) each other. As societal demands grow, the resources required to tackle integrated urban development strategies (e.g. knowledge, finance and legitimacy) are increasingly spread out (Teisman and Klijn, 2002) and hence the growing relevance of partnerships and 'co-production' in public governance and urban affairs (Pestoff *et al.*, 2012). Partnerships are time consuming (e.g. to align objectives and interests) but offer better guarantees of more sustainable outcomes, namely when 'birds of different feathers' are involved (van den Berg *et al.*, 1997; Carvalho, 2006). In this vein, the notion of 'triple helix' became popular to refer to the multitude of relations established between university-industry-government e.g. to promote local and regional innovation and economic development (Etzkowitz and Leydesdorff, 2000). More recently, notions of 'quadruple', 'penta' and 'N'-helix have been introduced to cater for the involvement of other actors in innovation processes, such as the civil society, not-for-profit organisations, etc. (Leydesdorff, 2012; Carvalho *et al.*, 2014).

Leadership has long been considered a critical element in urban development strategies. Good projects without good leaders can easily fail. Urban leadership is often associated with visionary and charismatic individuals that drive initiatives through the many obstacles on the way, linking actors and finding new solutions to old problems (van den Berg *et al.*, 1997). However, in urban and regional development studies, leadership is increasingly understood as a distributed capacity beyond hierarchical relationships (e.g. Beer and Clower, 2014), sometimes involving action in domains in which formal responsibilities are not attributed to concrete actors. Leadership is important to acquire social and political support for controversial urban projects and strategies (van den Berg *et al.*, 1997). Yet, the sources of power of urban leaders and urban development officials seem to be changing – more than exerting power through institutional and formal positions, the ability to connect visions and networks, inspire others to action and build convincing and appealing narratives is becoming ever more important (Sotarauta, 2006, 2009).

Communication is closely linked with the capacity to mobilise partnerships and involve citizens in urban projects, namely whose outcomes are hard to visualise beforehand. Many urban policies and projects increasingly have immaterial dimensions (e.g. involving investments in education or research), making it difficult to the lay citizen to understand its propositions and outcomes. Second, as physical urban projects often involve long time frames, communication is pivotal to keep enthusiasm and momentum over time. Over the last two decades, the role of communication has made strong inroads in planning theory (e.g. Innes, 1998; Healey, 1999). Innes (1998) reminds us that 'it is through communicative practice that they [planners] influence public action' and that 'being technically right is never enough to influence action' (p.52).

As new urban development strategies increasingly combine multiple dimensions and domains (economic, social, environmental, etc.), *prioritisation and piloting* is key. On the one hand, their ultimate objectives are too complex to be attained with single, short-term initiatives; on the other hand, they require concrete action to start and create momentum to achieve more demanding pursuits. Pilots and experiments allow for early visualisation of solutions and to see what works and what does not (e.g. Kemp *et al.*, 1998). Moreover, pilots support the formation of coalitions of stakeholders and can pave the ground for the legitimisation of initiatives that more strongly challenge the status quo (Bergek *et al.*, 2008; Carvalho *et al.*, 2012; Carvalho, 2015). Moreover, new visions, strategies and stakeholders' agendas will emerge along with changes during the long-span urban development initiatives (Teisman and Klijn, 2002), calling for flexibility and heterodox governance modes (Loorbach, 2010).

Due to their increasing complexity, it has been argued that successful urban development projects benefit from city administrations that are capable of incubating new ideas and accelerating their impact through rapid prototyping (Adams, 2015; PwC, 2013). Yet, ultimately, acting upon complex urban challenges requires *finance and agile delivery vehicles*. Sound finance is a necessary but not sufficient condition for the success of complex urban development strategies. In recent surveys (PwC, 2011, 2012), finance was portrayed as the most important barrier hindering the implementation of urban strategies, but not the most important enabler. At the same time, urban projects and strategies combining competitiveness and sustainability objectives call for new agile delivery vehicles. These are often important to challenge the silo-oriented mentality of many administrations, calling for action beyond 'comfort zones' and combinations of multiple actors and resources (Miles and Trott, 2011). Yet, the need for good finance and agility should not be associated with the financialisation of urban development or aggressive public service's privatisation that have been present in many urban development strategies over the last decades, clearly endangering the sustainable competitiveness of many cities (Sager, 2011; Weber, 2010).

2.4 Summing up

In this study, we primarily look at urban competitiveness as the dynamic capacity of a city to grow and develop over time, nurturing and attracting new jobs, people and skills. In a certain moment of time, cities can bid and 'compete' with each other to attract residents, investments, talent and companies. Yet, in the medium and long run, urban competitiveness and a city's capacity to reinvent itself are driven by a number of structural assets or 'capitals': knowledge and economic capital; social conditions; environment and quality of life, accessibility; image; and institutional conditions. Such capitals change slowly, but they can be influenced by hands-on policy initiatives and local development strategies.

Different and changing notions of sustainability provide additional twists to urban competitiveness strategies. There are different, often value-driven viewpoints of what sustainability means or should be. Yet, it is more consensual that in order to remain competitive, urban economic development dimensions have to be holistically combined with environmental and societal dimensions. Hence one can speak about sustainable competitiveness as a city's ability to keep growing and developing over time while fostering social cohesion and environmental quality. Yet, a key question is how can cities and their actors make such a complex objective happen, and which processes and enablers are necessary. Under a changing playing field in which urban development strategies become increasingly multi-dimensional and multi-stakeholder, urban development projects will likely rely on more (and new) types of designs, organisation features and implementation processes.

3 Research framework and case studies

3.1 Research framework

The case studies in this book analyse how different types of contemporary urban development strategies are contributing to sustainable competitiveness, namely by changing/reinforcing a city's structural foundations and 'capitals'. They analyse in depth the rationale behind those strategies and projects, their design and early observable results, but also, importantly, how they have been implemented 'on the ground'.

To this effect, a research framework is used that considers two main levels of analysis (city level and project level) and five spheres/dimensions, which interact with each other (Figure 1.2).

The *city level* is defined at the level of a functional urban region (vs a single administrative unit) and includes two spheres:

- the structural context of the city and its recent dynamics, assessed through the city's foundations, structural assets or 'capitals' as set out earlier (knowledge and economic; social conditions; environment and quality of life; image; and institutional conditions);
- the city's overall development vision and strategies, as expressed in formal plans and documents, but also in stakeholder narratives.

As suggested in Figure 1.2, the city's capitals are the starting point of the framework. There is a relationship between a city's development strategy and its capitals: different cities face different challenges and imbalances in their capitals, resulting (on average) in differentiated urban development strategies. Naturally, a city's overall vision and the progress of its capitals are also impacted by external influences and trends (e.g. new stages of globalisation, climate change pressures, macroeconomic context, national policies, changing worldviews,

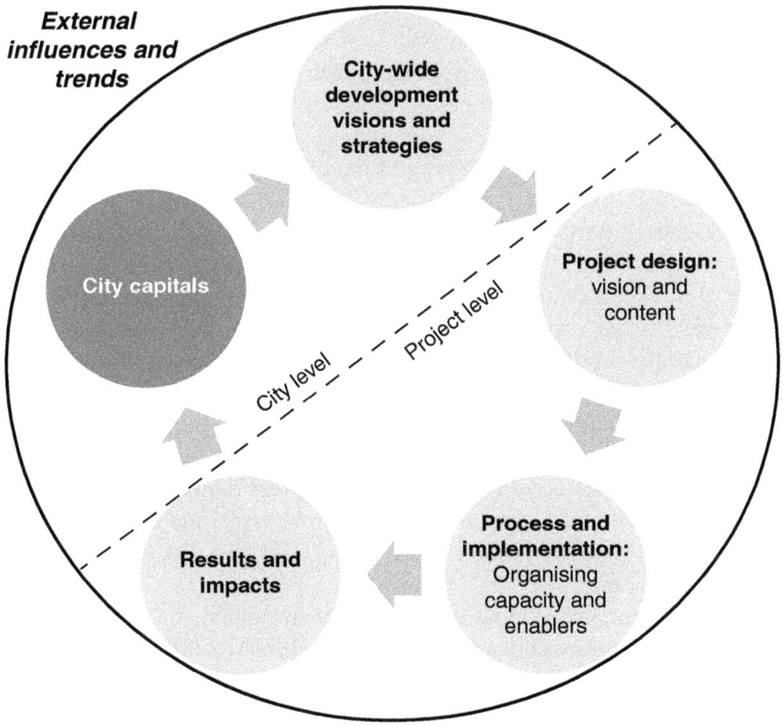

Figure 1.2 Research framework

Source: Own elaboration

etc.). All in all, the city level of the framework sets the context to better understand the ways through which different urban development projects and strategies are designed (i.e. their vision and content) and unfold (i.e. how cities manage to implement it).

The *project level* is influenced by the context of the city and includes three main spheres of analysis (Figure 1.2):

- project design: including a project's vision and content, namely its history and rationales, objectives, strategy, concepts and features;
- observable results and impacts ascribed to the project (across different dimensions of the city's capitals);
- process and implementation, including the organising capacities, enablers and critical processes that mediated between a project's design and its results. As framed earlier, we look at different dimensions of strategic networks and collaboration, leadership, communication, finance, prioritisation and piloting, and agile delivery.

It is assumed that the design and features of different projects are closely linked with a city's dynamics and development vision. Over time, the outcomes and results achieved feedback to the city's general context and profile, closing the loop and contributing (to a larger or lesser extent) to sustainable competitiveness. As urban development strategies increasingly target multiple dimensions, it is posited that more (and better) enablers and organising capabilities are required.

3.2 Case studies and research setting

This book presents 11 cases of contemporary urban development strategies throughout the world, envisioned to improve a city's sustainable competitiveness. Such projects and strategies are strongly embedded in the context of a city and are often deployed in partnership by a number of urban stakeholders. Figure 1.3 illustrates the spatial distribution of the case studies in this book.

Cases studies were chosen partly on the basis of theoretical sampling, focusing on the dimensions of innovativeness and impact. In terms of innovativeness, cases had to be innovative in at least one of two dimensions: 1) content and features, for example projects whose design and features are new or differentiating *vis-à-vis* conventional portfolios or urban development strategies (e.g. van den Berg *et al.*, 2014); and/or 2) process and implementation, for example projects whose design and purposes are not necessarily 'new' but whose processes of implementation and enablers revealed innovative features. In terms of the impact potential, the cases had to have strong impact (already revealed or potential) on the city's capitals and urban dynamics. Two types of initiatives were considered: 1) transformational, i.e. large-scale programmes, composed of multiple hard and soft projects; and 2) transitional, i.e. smaller yet path-breaking initiatives, with the aim to pave change towards new ways of doing things in cities.

The case studies touch upon one or more dimensions of sustainable urban development – for example, economic, environmental and social – resulting in different 'families' of projects (Figure 1.4). Some projects in the book focus more intensively on one of the dimensions, while others are more integrated. Nevertheless, to a larger or lesser extent, all the projects/strategies under analysis consider the relationships established with other development dimensions as an important condition towards the success of the project and urban sustainable competitiveness.

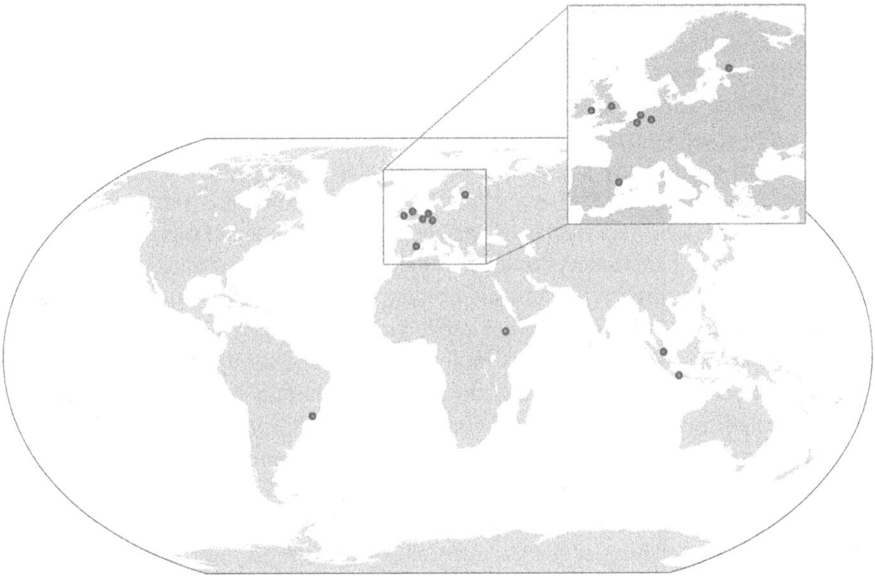

Figure 1.3 Spatial distribution of the case studies

Source: Own elaboration

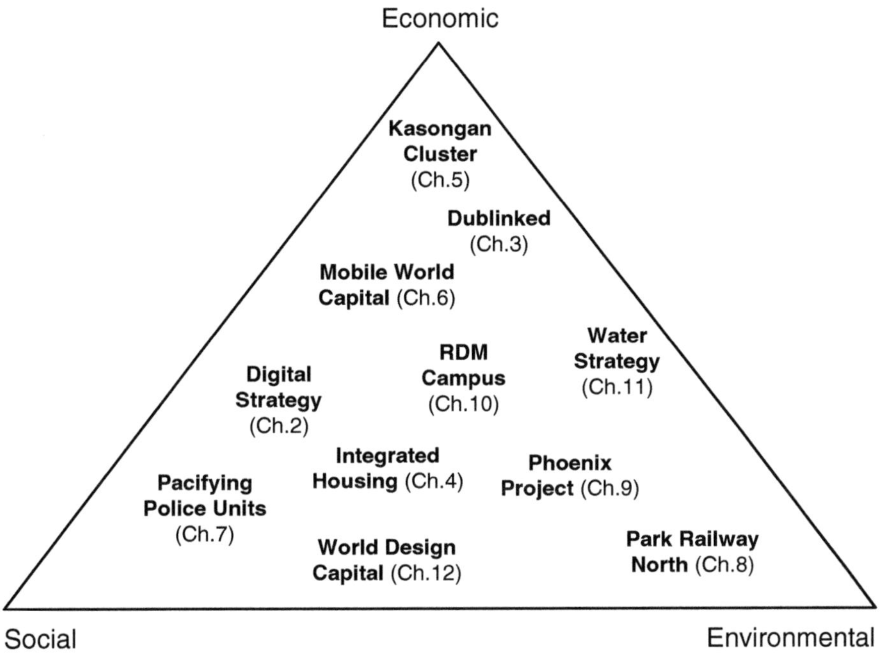

Figure 1.4 The projects under analysis along three axes

Source: Own elaboration

Case studies were developed by triangulating a number of research methods and data sources. Quantitative and qualitative information were used to answer the research questions and assess the different building blocks of the research framework. As secondary information, the research made use of descriptive statistics, city strategies, project documentation and previous studies, implementation reports, press releases, industry and government reports, online newspapers, weblogs, among others. Beyond desk research, the evidence base was complemented with primary information, collected through face-to-face interviews in each city with elected representatives, planners and government officials (from different departments and layers), project managers, private entrepreneurs and representatives of large companies, universities, non-governmental and not-for-profit organisations, grassroots movements, experts and other relevant stakeholders.

On average, a dozen of interviews were carried out in each city between March 2012 and February 2014, for the different projects under analysis, adding up to 124 interviews in 11 cities (Table 1.1). During the field visit, the first insights and impressions from the desk research were tested, further explored and largely improved. Some interviews had an ethnographic nature (Spradley, 1979; Kincaid and Bright, 1957), and were critical to grasp tacit knowledge, unwritten features and enablers associated with the projects, as well as to 'visualise' the project's development – including visits to development sites, but also attendance of project's events, networking sessions, informal discussion groups, among others. While

Table 1.1 Interviews conducted in the case study cities (count)

	Government (local, regional, national)	*Semi-public companies and delivery vehicles*	*Private companies*	*Chamber of commerce*	*Not-for profits and grassroots movements*	*Knowledge institutes*	*Total*
Digital Strategy	2	1	1	1	4	1	10
Dublinked	5	1	4	1	1	1	13
Integrated Housing Development Programme	1	—	—	—	2	2	5
Kasongan Pottery Cluster	4	3	4	4	1	1	17
Mobile World Capital	4	1	1	1	1	3	11
Pacifying Police Units	6	2	2	—	1	—	11
Park Railway North	4	4	4	—	—	—	12
Phoenix Project	4	3	1	—	1	—	9
RDM Campus	—	3	3	—	—	3	9
Water Management Strategy	—	1	—	—	—	7	8
World Design Capital	10	4	1	—	2	2	19
Total	40	23	21	7	13	20	124

Source: Own elaboration

some projects had been studied before (yet from other angles), others are very recent and their features and early results were being studied for the first time. Beyond 'success stories' and narratives, the interviews allowed for a clearer grasp of the project's challenges and tensions in context. They proved fundamental to deeply understand the nuts and bolts of the project, what is 'behind the scenes', the contours of the partnership, the 'how and why' of the decisions taken along the process, the critical and sensitive dimensions of its implementation, and many other relevant dimensions not written about anywhere. This allowed for rebuilding causal chains and understanding the critical success dimensions in a more nuanced way.

The following case study chapters opt for a narrative approach to present findings, as this offers more scope to do justice to the context specificity and helps to preserve most of the richness of each case (Yin, 1984). Yet, each case follows the same structure and looked into the same variables of interest, in order to allow for cross case comparisons, which are presented in the last chapter of the book.

References

Adams, E. (2015). Creating space for experimentation. In F. Jegou and M. Bonneau (eds), *Social Innovation in Cities*. Paris: URBACT.

Beer, A. and Clower, T. (2014). Mobilizing leadership in cities and regions. *Regional Studies, Regional Science*, 1(1), 5–20.

Bergek, A., Jacobsson, S., Carlsson, B., Lindmark, S. and Rickne, A. (2008). Analyzing the functional dynamics of technological innovation systems: A scheme of analysis. *Research Policy*, 37(3), 407–29.

Boschma, R. (2015). Towards an evolutionary perspective on regional resilience. *Regional Studies*, 49(5), 733–51.

Boschma, R. and Frenken, K. (2011). Technological relatedness and regional branching. In H. Bathelt, M. Feldman and D. Kogler (eds), *Dynamic Geographies of Knowledge Creation and Innovation*. Abingdon: Routledge.

Camagni, R., Capello, R. and Nijkamp, P. (1998). Towards sustainable city policy: an economy-environment technology nexus. *Ecological Economics*, 24(1), 103–18.

Campbell, S. (1996). Green cities, growing cities, just cities: Urban planning and the contradictions of sustainable development. *Journal of the American Planning Association*, 62(3), 296–312.

Carvalho, L. (2006). Governance challenges towards a knowledge economy–the case of the Metropolitan Area of Porto. Unpublished working paper. Rotterdam: European Institute for Comparative Urban Research, Erasmus University Rotterdam.

Carvalho, L. (2013). Knowledge locations in cities: emergence and development dynamics. *PhD series in Research in Management*, 274. Rotterdam: Erasmus Research Institute of Management.

Carvalho, L. (2015). Smart cities from scratch? a socio-technical perspective. *Cambridge Journal of Regions, Economy and Society*, 8(1), 43–60.

Carvalho, L., Mingardo, G. and van Haaren, J. (2012). Green urban transport policies and cleantech innovations: evidence from Curitiba, Goteborg and Hamburg. *European Planning Studies*, 20(3), 375–96.

Carvalho, L., de Haas, E., Galal, H., Jones, N., Otgaar, A., Teunisse, P., van den Berg, L., van den Dool, L., van der Meer, J. and van Tuijl, E. (2014). *Innovative City Strategies for Delivering Sustainable Competitiveness*. Rotterdam: Euricur, PwC Public Sector Research Centre and IHS.

Castro, E. A. and Jensen-Butler, C. (1999). Regional economic inequality, growth theory and technological change (No. 9903). Centre for Research into Industry, Enterprise, Finance and the Firm, School of Economics and Finance, University of St Andrews.

Etzkowitz, H. and Leydesdorff, L. (2000). The dynamics of innovation: from National Systems and 'Mode 2' to a Triple Helix of university–industry–government relations. *Research Policy*, 29(2), 109–23.

European Commission (2011). *Cities of tomorrow: Challenges, Visions, Ways Forward.* Brussels: European Commission, Directorate General for Regional Policy.

Fainstein, S. (2010). *The Just City*. Ithaca, NY: Cornell University Press.

Florida, R. L. (2002). *The Rise of the Creative Class: And How It's Transforming Work, Leisure, Community and Everyday Life*. New York: Basic Civitas Books.

Frenken, K., van Oort, F. and Verburg, T. (2007). Related variety, unrelated variety and regional economic growth. *Regional Studies*, 41(5), 685–97.

Glaeser, E. (2011). *Triumph of the City: How Our Greatest Invention Makes Us Richer, Smarter, Greener, Healthier and Happier*. New York: Penguin.

Gold, J. and Ward, S. (1994). *Place Promotion: The Use of Publicity and Marketing to Sell Towns and Regions*. New Jersey: John Wiley & Sons Ltd.

Grabher, G. (1993). The weakness of strong ties: the lock-in of regional development in the Ruhr area. In G. Grabher (ed.), *The Embedded Firm: On the Socioeconomics of Industrial Networks*. London: Routledge.

Hall, P. and Pfeiffer, U. (2000). *Urban Future 21: A Global Agenda for Twenty-first Century Cities*. Abingdon: Taylor & Francis.

Healey, P. (1999). Institutionalist analysis, communicative planning, and shaping places. *Journal of Planning Education and Research*, 19(2), 111–21.

Hopwood, B., Mellor, M. and O'Brien, G. (2005). Sustainable development: mapping different approaches. *Sustainable Development*, 13, 38–52.

Hospers, G. J. (2006). Silicon somewhere? Assessing the usefulness of best practices in regional policy. *Policy Studies*, 27(1), 1–15.

Innes, J. E. (1998). Information in communicative planning. *Journal of the American Planning Association*, 64(1), 52–63.

Jacobs, J. (1969). *The Economies of Cities*. New York, NY: Random House.

Kemp, R., Schot, J. and Hoogma, R. (1998). Regime shifts to sustainability through processes of niche formation: the approach of strategic niche management. *Technology Analysis & Strategic Management*, 10(2), 175–98.

Kincaid, H. V. and Bright, M. (1957). Interviewing the business elite. *The American Journal of Sociology*, 63(3), 304–11.

Kitson, M., Martin, R. and Tyler, P. (2004). Regional competitiveness: an elusive yet key concept? *Regional Studies*, 38(9), 991–9.

Leydesdorff, L. (2012). The Triple Helix, Quadruple Helix, …, and an *N*-tuple of Helices: explanatory models for analyzing the knowledge-based economy? *Journal of the Knowledge Economy*, 3(1), 25–35.

Logan, J. R. and Molotch, H. L. (2007). *Urban fortunes: the political economy of place*. Oakland, CA: University of California Press.

Lombardi, D. R., Porter, L., Barber, A. and Rogers, C. (2011). Conceptualising sustainability in UK urban regeneration: a discursive formation. *Urban Studies*, 48(2), 273–96.

Loorbach, D. (2010). Transition management for sustainable development: a prescriptive, complexity-based governance framework. *Governance*, 23(1), 161–83.

McCann, P. (2008). Globalization and economic geography: the world is curved, not flat. *Cambridge Journal of Regions, Economy and Society*, 1(3), 351–70.

Miles, E. and Trott, W. (2011). *Collaborative Working: How Publicly Funded Services Can Take a Whole Systems Approach*. London: Institute for Government.

Næss, P. (2001). Urban planning and sustainable development. European Planning Studies, 9(4), 503–24.

North, D. (1990). *Institutions, Institutional Change, and Economic Performance*. Cambridge: Cambridge University Press.

OECD. (2006). *Competitive cities in the global economy*. Paris: OECD Publishing.

OECD. (2008). *Sustainable Development. Linking Economy, Society and Environment*. Paris: OECD Publishing.

Pestoff, V., Brandsen, T. and Verschuere, B. (2012). *New Public Governance, the Third sector, and Co-production*. Abingdon: Routledge.

Pike, A., Rodríguez-Pose, A. and Tomaney, J. (2006). *Local and Regional Development*. Abingdon: Routledge.

Pike, A., Rodríguez-Pose, A. and Tomaney, J. (2010). *Handbook of Local and Regional Development*. Abingdon: Routledge.

Plácido Santos, I. (2011). Organising Capacity for Sustainable Urban Planning. The case of the riverside project in Coimbra. Master thesis, University of Porto.

Porter, L. and Hunt, D. (2005). Birmingham's Eastside story: making steps towards sustainability? *Local Environment*, 10, 525–42.

Porter, M. (1990). *The Comparative Advantage of Nations*. New York, NY: Free Press.

Putnam, R. (2001). *Bowling Alone: The Collapse and Revival of American Community*. New York, NY: Simon & Schuster.

PwC (2011). *Making It Happen: A Roadmap for Cities and Local Public Services to Achieve Outcomes*. London: PwC Public Sector Research Centre.

PwC (2012). *The Agile Council: Creating the Change Ready Organization*. London: PwC Public Sector Research Centre.

PwC (2013). *Future of Government*. London: PwC Public Sector Research Centre.

Rees, W. (1995). Achieving sustainability: reform or transformation? *Journal of Planning Literature*, 9(4), 343–61.

Regional Studies. (2004). Special Issue on Regional Competitiveness. *Regional Studies*, 38/39.

Richards, G., De Brito, M. and Wilks, L. (2013). *Exploring the Social Impacts of Events*. Abingdon: Routledge.

Rodríguez-Pose, A. (2013). Do institutions matter for regional development? *Regional Studies*, 47(7), 1034–47.

Rodríguez-Pose, A. and Arbix, G. (2001). Strategies of waste: bidding wars in the Brazilian automobile sector. *International Journal of Urban and Regional Research*, 25(1), 134–54.

Rodríguez-Pose, A. and Di Cataldo, M. (2015). Quality of government and innovative performance in the regions of Europe. *Journal of Economic Geography*, 15(4), 673–06.

Sager, T. (2011). Neo-liberal urban planning policies: A literature survey 1990–2010. *Progress in Planning*, 76(4), 147–99.

Saxenian, A. L. (1996). *Regional Advantage: Culture and Competition in Silicon Valley and Route 128*. Cambridge, MA: Harvard University Press.

Smith, A. (2005). Conceptualizing city image change: the 're-imaging' of Barcelona. *Tourism Geographies*, 7(4), 398–423.

Solow, R. M. (1986). On the intergenerational allocation of natural resources. *Scandinavian Journal of Economics*, 88, 141–9.

Sotarauta, M. (2006). Where have all the people gone? *Leadership in the Fields of Regional Development*. Sente Working Papers 9/2006, University of Tampere.

Sotarauta, M. (2009). Power and influence tactics in the promotion of regional development: an empirical analysis of the work of Finnish regional development officers. *Geoforum*, 40(5), 895–905.

Spradley, J. (1979). *The Ethnographic Interview*. New York, NY: Holt, Rinehart and Winston.

Stoker, G. (1995). Regime theory and urban politics. In D. Judge, G. Stoker and H. Wolman (eds), *Theories of Urban Politics*. London: Sage.

Storper, M. (1997). *The Regional World: Territorial Development in Global Economy*. New York, NY: Guilford Press.

Storper, M. (2011). Justice, efficiency and economic geography: should places help one another to develop? *European Urban and Regional Studies*, 18(1), 3–21.

Storper, M. (2013). *Keys to the City: How Economics, Institutions, Social Interaction, and Politics Shape Development*. Princeton, NJ: Princeton University Press.

Teisman, G. and Klijn, E. H. (2002). Partnership arrangements: governmental rhetoric or governance scheme? *Public Administration Review*, 62(2), 197–205.

Turok, I. (2013). Securing the resurgence of African cities. *Local Economy*, 28(2), 142–57.
United Nations. (2008). World Urbanization Prospects: The 2007 revision population database. Accessed at http://esa.un.org/unup/ (10 March 2011).
Urban Studies. (1999). Special Issue on Competitive Cities. *Urban Studies*, 36(5/6).
Van den Berg, L. (1987). *Urban systems in a dynamic society*. Aldershot: Gower.
Van den Berg, L. and Braun, E. (1999). Urban competitiveness, marketing and the need for organising capacity. *Urban Studies*, 36(5–6), 987–99.
Van den Berg, L., van Klink, A. and van der Meer, J. (1993). *Governing Metropolitan Regions*. Aldershot: Avebury.
Van den Berg, L., Braun, E. and van der Meer, J. (1997). *Metropolitan Organising Capacity: Experiences with Organising Major Projects in European Cities*. Aldershot: Ashgate.
Van den Berg, L., van der Meer, J. and Pol, P. (2003). Organising Capacity and Social Policies in European Cities. *Urban Studies*, 40, 1959–78.
Van den Berg, L., Pol, P., van Winden, W. and Woets, P. (2005). *European Cities in the Knowledge Economy: The Cases of Amsterdam, Dortmund, Eindhoven, Helsinki, Manchester, Munich, Münster, Rotterdam and Zaragoza*. Aldershot: Ashgate.
Van den Berg, L., van der Meer, J. and Carvalho, L. (eds) (2014). *Cities as Engines of Sustainable Competitiveness: European Urban Policy in Practice*. Aldershot: Ashgate.
Van Winden, W. and Carvalho, L. (2008). Urban competitiveness in the knowledge economy: evolution paths of the Portuguese metropolises. In T. Yigitcanlar, K. Velibeyoglu and S. Baum (eds), *Knowledge-based Urban Development: Planning and Applications in the Information Era*. Hershey, PA: IGI Global.
Van Winden, W. and Carvalho, L. (2015). *New Urban Economies: How Can Cities Foster Economic Development and Develop 'New Urban Economies'?* Paris: URBACT.
Van Winden, W., van den Berg, L. and Pol, P. (2007). European cities in the knowledge economy: towards a typology. *Urban Studies*, 44(3), 525–49.
WCED – World Commission on Environment and Development (1987). *Brundtland Report. Our Common Future*. World Commission on Environment and Development, United Nations.
Weber, R. (2010). Selling city futures: the financialization of urban redevelopment policy. *Economic Geography*, 86(3), 251–74.
Williams, C. and Millington, A. (2004). The diverse and contested meanings of sustainable development. *The Geographical Journal*, 170, 99–104.
Yin, R. (1984). *Case Study Research: Design and Methods*. Beverly Hills, CA: Sage.

2 Digital Strategy (Manchester)

Luís Carvalho and Erwin van Tuijl

1 Introduction

Located in the North West of England, Manchester was the cradle of the industrial revolution. Over the last decades, Manchester also became the quintessential example of a city in transition from manufacturing towards more knowledge-based industries. The city has managed to successfully bounce back and to diversify its economy; Information and Communication Technologies (IT) – or what has been recently dubbed the digital economy – is among the most important spearheads. However, despite all the progress, persistent skills mismatch, unemployment and social exclusion are still heavily felt.

This case study analyses Manchester's Digital Strategy (hereinafter, Digital Strategy). It consists in a transformative 'road map' of programmes of projects with the goal of making Manchester 'one of the world's leading digital cities by 2020 … the aim is to maintain and continue to grow the competitive advantage this will bring residents and businesses to promote economic growth' (MMC, 2012b). The Digital Strategy is currently championed by the City Council as an 'umbrella' or meta-level strategy, cutting across different domains of Manchester development strategies: economic attractiveness, skills and employment, social integration, access and governance. The first IT and digital-related initiatives in Manchester date back to 1989 and have been closely linked with the city's regeneration agenda (economic, social and physical). City priorities changed over time; yet, IT and digital initiatives have always been instruments to catalyse change. From 2008 onwards, the Digital Strategy has been recognised as a high-level priority in the city's administration, with the symbolic status of a 'master plan'. Among others, it aims to position Manchester in the global competition for digital media investment while tackling locally rooted and persistent social divides.

In this case study, we analyse the strategy's concept and its key priorities (e.g. widening the access to ultra-fast broadband to support new businesses' development, test-bed and digital skills enhancement). As the strategy became more demanding, so did the organisational features involved. Among others, distributed leadership, network organisation and communication become critical success factors. Moreover, we explore a number of outcomes and results linked with the Digital Strategy. The case study illustrates the benefits – and also some challenges ahead – involved in the development of such a comprehensive digital strategy.

The structure of this chapter is as follows. Section 2 analyses the structural context of Manchester across a number of 'capitals for sustainable competitiveness'; it briefly sketches Manchester's recent dynamics and how they influence the design of Manchester's Digital Strategy. Section 3 explores how the Digital Strategy fits within Manchester's City vision and other portfolios of local development initiatives. Sector 4 is the core of the case study and presents the strategic objectives of the Digital Strategy and its evolution over time (4.1), as well as critical implementation factors (4.2) and results so far (4.3). Section 5 concludes by presenting lessons and challenges that can be relevant for other cities as well.

2 City profile: capitals for sustainable competitiveness

2.1 Performance

Manchester is the largest city in England's North West and the biggest urban region outside London. Greater Manchester has a population of 2.6 million and a gross value added (GVA) of roughly £46 billion – representing 40 per cent of the GVA in the North West Region. During the second half of the twentieth century, Manchester went through a massive economic restructuring process in which the decline of manufacturing industries went hand-in-hand with unemployment and social deprivation. From the 1990s onwards, despite the persistence of structural unemployment, Manchester's economy bounced back as the city's economic structure steadily shifted towards more knowledge-based activities – notably based on information, knowledge and creativity.

During the last decade (2001–11), the city's population grew spectacularly (19 per cent), the highest growth rate within Greater Manchester and in the English core cities[1] (New Economy, 2012). After three decades of urban regeneration efforts, population growth has been primarily linked with the in-migration of young residents, students and families (New Economy, 2012), reflecting the increasing attractiveness of the city. Roughly 40 per cent of Manchester's population is between 16 and 34 years old (MCC, 2013a). In 2011, the number of start-ups and new business births outnumbered the deaths by 1,000, as well as the national rate of new business creation (New Economy, 2013, based on ONS – Office for National Statistics).

The prospects for Greater Manchester's economy are also rather positive. Recent studies (e.g. Greater Manchester Forecasting Model) estimate the creation of 85,000 extra jobs and a GVA annual average rise of 2.7 per cent until 2022. During 2007–12, and despite the global economic downturn, Greater Manchester attracted over 400 inward foreign investment projects, creating or maintaining more than 22,000 jobs, being the strongest FDI magnet outside London (MIDAS, quoted in New Economy, 2013). In Greater Manchester South (including the Municipalities of Salford, Stockport, Tameside and Trafford), GVA per head added to £22,000 in 2011, higher than the English and core cities average.

Nevertheless, due to the heavy toll inherited from the massive economic transition, the unemployment rate in Greater Manchester was 9.7 per cent, above the UK average of 7.9 in mid-2013 (NOMIS, quoted in New Economy, 2013). Beyond the impacts of the financial crisis, this number suggests the presence of structural unemployment.

2.2 Economic structure and knowledge

Over the last two decades, the recovery of Manchester's economy has been closely associated with its skills and knowledge base. Greater Manchester has one of the largest student populations in Europe: around 100,000 students are enrolled in one of Manchester's five higher education institutions; one-fifth of that figure is composed of international students (New Economy, 2013). The University of Manchester consistently ranks among the top 50 universities in the world – four Nobel Laureates are currently among the university's staff. In Greater Manchester, 31 per cent of the working age population is educated to degree level or higher (the UK level being 34 per cent: New Economy, 2013).

Nevertheless, at the same time, the number of residents with low qualifications is also substantial – in 2011, 12 per cent of the working-age population had no qualifications (vs 10 per cent in the UK). Hence, Manchester's (dual) urban economy still suffers from considerable skills mismatch; moreover, there is also evidence of misalignments between the industry's demands and skills of the graduates (MCC, 2013a). This is a transversal phenomenon in the UK, but strongly felt in Manchester. A concrete example is the lack of programming skills in the university's curriculum – as well as business skills among 'tech' graduates – which are essential for new media businesses.

As illustrated in Table 2.1, the city's economic structure changed considerably over the last decade. While manufacturing activities kept declining, the service sector expanded noticeably (namely information and telecommunication activities, financial and business services, health and education). One important sub-sector on the rise has been creative, IT and digital industries, representing 7.7 per cent of employment across Greater Manchester in 2010 (ONS, 2013, annual business inquiry). From 2009–10, 2,700 new jobs were created in activities associated with this sub-sector (sector code 34: culture, media and sports occupations), namely because of the relocation of BBC activities from London to Manchester/Salford (2009) and the reconfiguration of related value chains in the city.

2.2.1 *The digital economy*

Greater Manchester has currently the largest agglomeration of IT and digital media activities outside London (MCC, 2013a). As Figure 2.1 depicts, employment in the IT sector is substantial and primarily idea-driven – programming, telecommunications and advanced services are dominant while hardware manufacturing is residual. The employment in the top ten digital content sub-sectors is also illustrative of the local strengths and potential crossovers within the digital media field in Manchester (Figure 2.2).

The recent launch of BBC operations in Manchester/Salford (Media City) contributed to boost Manchester's digital economy and to anchor digital-related value chains (e.g. new local content providers). The presence of BBC, Sky1 and Channel 4 turned Manchester into a rather relevant media hub at European level. BBC's decision to decentralise had to do primarily with audience strategies, regional-level production quotas and national government

Table 2.1 Composition of the GVA by sector, Greater Manchester South

	1997	*2010*
Agriculture, forestry and fishing	0.2	0.1
Production	17.9	11.5
Construction	5.3	5.6
Distribution transport hospitality	23.1	19.5
Information and communication	4.9	5.7
Financial and insurance	7.7	11
Real estate	8.3	9.5
Business services	12	13.4
Public administration, education health	18.2	20.5
Other services	2.4	3.3

Source: ONS, 2013; New Economy, 2013

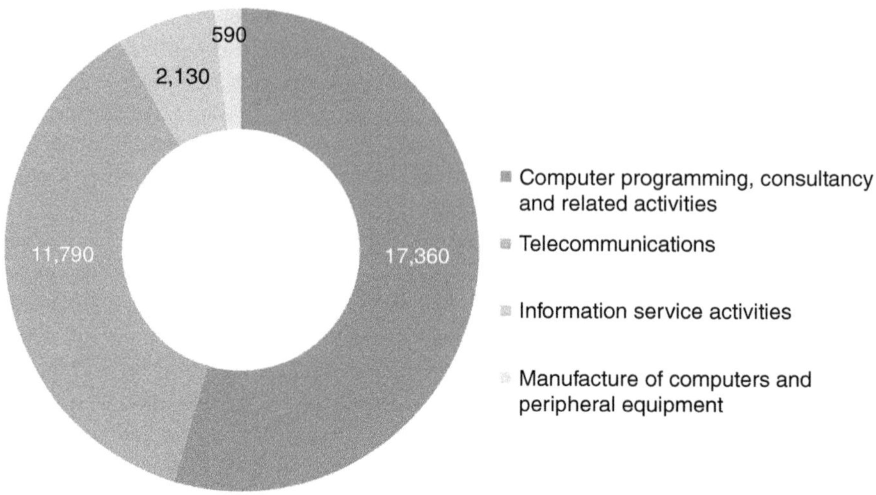

Figure 2.1 Employment in the ICT sector, Greater Manchester, 2011
Source: Business Register & Employment Survey, adapted from MCC (2013a)

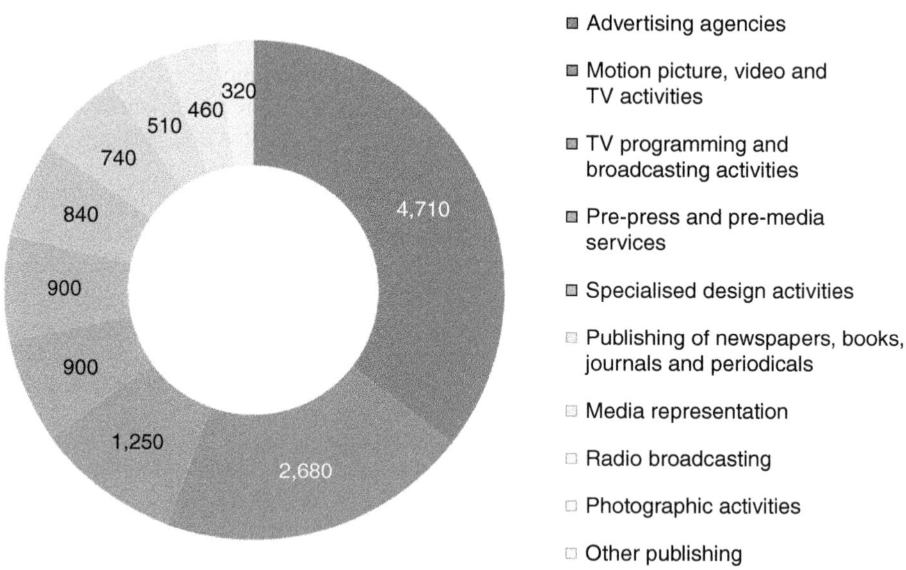

Figure 2.2 Employment in top ten digital content sub-sectors, 2011
Source: Business Register & Employment Survey, adapted from MCC (2013a)

decisions (de-concentration from London); however, the decision to locate in Manchester/Salford also had to do with the superior IT and media skills of the region.

Long before BBC, Manchester had a track record of digital and IT competence. Because of its industrial past, Manchester has universities and education institutes with tradition in the fields of engineering and technology. In 1948, the first centre working with computer-related technology outside London was founded in Manchester. In the 1960s, a reforming national government decided to locate the National Computing Centre (NCC) in Manchester, focusing on software and skill development. Moreover, despite the overall manufacturing decline, some older industries had been providing ample room to embed new digital and IT solutions (e.g. paper and publishing, industrial design and software, advertising, etc.).

Therefore, it was already possible during the 1980s to discuss the role of IT and technology in Manchester's policy agenda: as we explore in Section 4.1, the first digital initiatives of Manchester date from that period. Nowadays, besides BBC and other media companies, Manchester is home to important digital and IT industries such as cloud computing – Cisco, Oracle, EMC and other leading British companies have R&D operations in the city. Moreover, besides corporate activities, Manchester hosts relevant ecosystems of IT-related networks, informally organised through social proximity relations and local community groups. Owning start-ups and growing them up (vs selling to the highest bid) is often considered to be a cultural feature of the city's entrepreneurs (inherited from the city's industrial culture).

Manchester has excellent digital accessibility, counting with excellent fibre and broadband infrastructure. Private internet providers (BT, Virgin Media) have recently expanded the city's state-of-the-art fibre, giving Manchester an edge *vis-à-vis* other cities in the UK. Nevertheless, beyond the general access for the basic user, advanced new media companies require increasingly large capacity broadband solutions, which are still scarce and pricy (e.g. ultra-fast broadband) but essential for the development of new media businesses (e.g. viral connections, high capacity exchange of data). Recently, Manchester was awarded an Urban Broadband Fund of £12 million from the national government, in order to 1) help close the digital divide in some neighbourhoods and 2) to endow all businesses in the city with ultra-fast broadband in the near future.

All in all, Manchester's creative and digital industries are foreseen to keep growing over the next years, even if their contribution to the GVA is likely to grow stronger than the job creation rate until 2025: 72 per cent vs 16 per cent, respectively (MCC, 2013a). Many observers suggest that the supply of skills is one of the main obstacles to accomplish even stronger growth (MCC, 2013a). Misalignment of academic skills and industrial needs and the retention of talent are considered important issues to tackle in the city.

2.3 Social conditions

Despite the city's economic performance and changing economic structure, Manchester suffers from significant and persistent levels of deprivation and social exclusion. A recent National Index of Multiple Deprivation (health conditions, income, disability, skills, barriers to housing, living environment and crime) ranked Manchester as the fourth most deprived local authority in England and Wales (out of 236); Salford ranked eighteenth (Department for Communities and Local Government, 2010). Moreover, deprivation is spatially concentrated (Carter, 2013), namely in eastern and northern parts of the city (former industrial areas) and in the southernmost zone (social housing estate). Furthermore, 212 neighbourhoods in Greater Manchester are within the 5 per cent most deprived nationally (Department for Communities and Local Development, cited in New Economy, 2013).

Poor educational levels are at the root of the problem, as they are closely linked with social, economic and other forms of exclusion. As said, 12 per cent of the working age population in Greater Manchester have no qualifications (and 68,000 people in employment have no qualifications). Sixteen per cent of Manchester residents cannot access the internet; even if this phenomena is concentrated in the city's most deprived areas, digital exclusion still exists in all parts of the city (MCC, 2013a). Life expectancies at birth are lower than the regional and national average and over a quarter of all children (dependents under the age of 20) live in poverty. There were around 85,500 residents in Greater Manchester claiming for Jobseeker's Allowance in March 2013 (New Economy, 2013).

2.4 Image

Just like many other former industrial cities, Manchester's image was for many years closely associated with decline, unemployment and deprivation. In recent times, the overall perception seems to be changing (or at least diversifying). Manchester is now identified by many as a 'phoenix from the ashes' or as the 'original, modern' former industrial city (Carter, 2013; Marketing Manchester, 2009).

In the aftermath of the city centre's terrorist bombing in 1996, Manchester launched a large-scale rebuilding programme, leveraging public and private funds (£2 billion) and many city marketing initiatives. The reconstruction focused on improving Manchester's image and attractiveness as a cosmopolitan city (Young *et al.*, 2006), namely by placing a strong focus on the city centre, culture and flagship projects with an eye to attract professional and middle classes. Moreover, despite the persistence of visible deprivation and social exclusion in the city, crime levels dropped more than 50 per cent in Greater Manchester during the past ten years (New Economy, 2013). The hosting of the 2002 Commonwealth Games was also an important turning point in the 'cosmopolitanisation' process, as it contributed, among others, to attract further attention and investments to regenerate the city. Manchester is nowadays UK's third most internationally visited city (after London and Edinburgh). Its universities and new cultural-creative atmospheres (e.g. Northern Quarter) have contributing to the perceived vibrancy of the city.

2.5 Institutional conditions

Manchester is an independent local authority. There is no formal government tier between Manchester (and for that matter, the other nine local authorities) and the national government. In 2009, Greater Manchester was announced as a pilot 'city–region' – an effort to decentralise economic development competences from the national government to metropolitan regions. Hence, in April 2011, the ten local authorities of Greater Manchester were the first in the country to develop a so-called 'Combined Authority', with the aim of coordinating economic, regeneration, environment and transport policies. The new Authority has jointly developed a Local Enterprise Partnership (LEP) with an eye to pool joint financial resources (e.g. European Regional Development Fund (ERDF), national funding, pension funds) to tackle the financial constraints in the public sector and to 'finance growth'. Despite the latent inter-municipal competition to attract jobs and residents (e.g. between Manchester, Salford and Trafford), local authorities of Greater Manchester are working together on the development of a Manchester Investment Fund to support large-impact joint projects. Greater Manchester owns a number of think thanks and related agencies towards these aims, such as the New Economic, Marketing Manchester, Manchester's inward investment agency and Greater Manchester Growth Company.

As a leading former industrial region, local governments are traditionally dominated by the Labour party. National government's intervention is generally considered as small and hands-off, while the local cooperative movement is relatively strong. There is formally no mayor of Manchester, but a leader of a cabinet of elected politicians, which compose the executive political team. The leader of the cabinet has been stable for many years, facilitating the implementation of major reforms and forward-thinking initiatives (e.g. digital and IT strategies). The operational structure of the local authority is led by a general chief executive (not politically elected) and a number of appointed Assistant Chief Executives for different areas, such as 'Regeneration' (including, for example, the departments of planning, adult education, economic development), or, more recently, 'ICT & Communications'.

From the mid-1980s onwards, Manchester's regeneration policies started to more closely promote area-based development, a philosophy that remains to the current days. As explained by Carter (2013, p.2):

> [From the mid-1990s onwards,] Manchester City Council embarked on a radical new approach to regeneration. A new Economic Development Department was established and work was commissioned to bring together people with new and innovative ideas, from research bodies, business, trade unions, the voluntary sector and the wider community, to advise the City Council leadership on how to tackle economic restructuring ... Out of those discussions, which also included many 'heated debates' on priorities, a number of practical proposals emerged in the late 1980s (1987–89), including ... a focus on area and neighbourhood based working, devolving intervention to local areas and encouraging cross-sector working where the public sector would work on a more proactive basis with local communities, businesses and other public sector bodies ....

2.6 Conclusions: main issues

The economic structure of Manchester changed significantly over the last decades. While manufacturing activities declined, new knowledge-intensive activities became increasingly relevant. The city's more recent attractiveness for new residents, students, companies and investments is closely associated with this change. IT and digital media activities have been among the most important spearheads: they represent a growing slice of the city's economy. At the same time, the toll of deindustrialisation is still heavily felt. Manchester's new economy has been side-by-side with persistent skills mismatches, unemployment and social problems. Many observers speak about a 'tale of two cities' syndrome that can undermine growth and opportunities for citizens at large.

Thus, Manchester faces a clear sustainable competitiveness challenge: how to sustain economic growth (and the development of the city's most dynamic sectors) while making sure that citizens can benefit from it. Additionally, the ways of 'financing growth' and new development projects are increasingly under strain, requiring new types of partnerships across government levels and involving multiple stakeholders.

As we shall explore in the next sections, Manchester's new Digital Strategy illustrates how these dimensions have been simultaneously tackled with an eye to support Manchester's global 'digital city' ambitions.

3 City-wide vision and strategy

For more than three decades now the vision for Manchester's development has been centred on urban regeneration objectives (economic, social and physical). The re-development of the city centre after the 1996 bombing and the hosting of the 2002 Commonwealth Games were aligned to create conditions for economic growth and to increase the city's attractiveness for residents, investments, companies and visitors. As mentioned above (Section 2.4), improving arts and cultural supply has also been at the forefront of the city's development strategy; the same goes for facilitating networks between knowledge institutes and companies (see, for example, the large investments in cultural venues and science parks over the last decades). Moreover, substantial efforts have also been paid to tackle deprivation and alleviate poverty.

Despite the progress, the previously referred to objectives remain important for Manchester; yet, the city has reached a new development stage, with new ambitions. Currently, the development road map for Manchester is elaborated in the so-called 'Manchester's Local Development Framework' (adopted 11 July 2012) – core strategy for 2012–27 (MCC, 2012a).[2] This document combines strategic and spatial planning and follows up previous regeneration initiatives with a set of new-generation goals. The Core Strategy Vision has no single 'tagline', but is formed by ten strategic objectives (Box 2.1).

Box 2.1 Manchester's Core Strategy Vision

By 2027 Manchester will be:

- a successful sustainable and accessible City in the front rank of cities in Europe and the world
- a City with a growing economy driven by the strength of the Regional Centre and Manchester Airport which supports a successful City Region. The knowledge based economy flourishes within an entrepreneurial community, characterised by a fully skilled, inclusive working population
- meeting the challenge of climate change at the forefront of environmental initiatives and improvements and continuing to deliver sustainable development and a more effective Green Infrastructure
- a City with an increased and sustainable healthy population and a community which both benefits from and drives the City's productivity upwards
- a City with neighbourhoods where people choose to live all their lives because they offer a wide range of quality housing and an attractive environment where locally distinctive character is conserved and enhanced. Neighbourhoods in the North and East of the City will have benefited from a significant increase in the quantity and quality of housing
- a City with thriving district centres, which have a distinct local character and provide a good range of accessible public services, retail and local facilities
- a City which is supported by transport infrastructure which provides sustainable and efficient links locally, within the City, and more widely, between the Regional Centre and workers across the City Region. It will also have strong national and international connections made possible by its world class airport
- a City defined by excellence in urban design and environmental quality, where its distinctive historic environment is understood, valued, cared for and its potential fully realised
- a City where residents and visitors regularly enjoy a network of quality open spaces, parks and river valleys, enriched with biodiversity, which provide recreational routes and opportunities for sport across the City
- an international destination for tourism and culture with the Eastlands area of the Regional Centre a national attraction for sport, leisure and recreation

Source: MCC, 2012a, p.28

As we analyse in the next section, Manchester's more recent Digital Strategy cuts across many of the abovementioned objectives, namely the ones related to the city's global ambition, infrastructure development, business and innovation support and attractiveness.

Moreover, the Digital Strategy has linkages with many other on-going economic and social development projects in the city. These involve the development of mixed use and R&D-focused districts (to which broadband and social media interactions are to be part of the 'backbone'), the development of new affordable housing (IT connected) and many other skills enhancement initiatives, both associated with advanced IT skills and the closure of digital divides.

Moreover, one of the most emblematic projects is the Sharp Project, a recently developed new media location in the city, hosting Sky1 and Channel 4, among many other digital-based companies. It is currently a very dynamic location, to which many IT and media companies have relocated over the last two years. Ultra-fast broadband and 'digital' skills availability are of utmost importance for such companies. Sharp is currently one of the most visible faces of the Digital Strategy in Manchester.

4 Digital Strategy

4.1 Vision and content

Manchester's Digital Strategy is a holistic 'road map' of programmes of projects with the goal of making Manchester 'one of the world's leading digital cities by 2020. The aim is to maintain and continue to grow the competitive advantage this will bring residents and businesses to promote economic growth' (MCC, 2012b, p.3). The Digital Strategy was approved by Manchester City Council's Executive Committee on 12 March 2008, and reviewed and updated on October 2012 and July 2013 (MCC, 2012b, 2013b). It has a clear 'sustainable competitiveness' drive: sustaining (IT and digital-led) economic growth in Manchester, making sure that citizens at large can benefit from it.

The Digital Strategy has transformational goals but does not start from scratch. It is the follow-up of more than two decades of locally driven initiatives dealing with IT, broadband and new digital businesses (Section 4.1.1). Its vision is closely linked with the mission of Manchester Digital Development Agency (MDDA) the City Council's arm's length unit coordinating the Digital Strategy:

> To make Manchester a leading world class digital city, having one of the most competitive broadband infrastructures in Europe, attracting and sustaining investment in ICT and e-commerce across all sectors of the economy, generating new businesses, developing new learning cultures, promoting social inclusion and providing all residents with the skills and aspirations to play a full role in the information society.
>
> (Carter, 2013, p.178)

The Digital Strategy has strong ambitions and the character of a 'digital master plan'. It frames a number of actions by the local government towards 2020, but it is also intended as a 'guide' for other local stakeholders (e.g. industry, universities, community organisations, schools). It is implemented as a distributed effort – within and outside the City Council – in which digitalisation becomes a transversal field with the potential to catalyse other policy dimensions (e.g. economic growth, skills and social inclusion, connectivity, leadership, governance). All in all, the Digital Strategy is designed in a way that improving the city's 'digital competitive advantage' becomes everybody's business.

The Digital Strategy is organised as a ten-point plan, combining many hard and soft initiatives. Yet, before exploring these points and the concept in more depth, we briefly introduce its background and the track record of Manchester pioneering urban IT policies.

4.1.1 Brief story and background of the strategy

The first IT and digital strategies in Manchester date from the late 1980s and were closely linked with the Manchester's City Council Regeneration Agenda (economic, social and physical) – it provided ample space in which to anchor and gain legitimation for an emergent pipeline of urban IT policies. Hence, IT[3] and technology were first introduced as instruments to support better jobs, skills development and inclusion (van Winden, 2003). Back then, the efforts were coordinated by the so-called 'Technology Team' within the Economic Initiatives Group – EIG (Carter, 2013).

Manchester was one of the European pioneers in this field. The city established early cooperation links with other leading 'IT cities' (e.g. in Germany and Finland) through the EUROCITIES network (Carter, 2013). For example, the Telecities cooperation network contributed to gain inspiration, to widen the span of the local IT agenda and to further legitimate it inside the City Council. Moreover, Manchester's history with IT and computers (see Section 2) made it possible to include digital dimensions in urban policy in an early stage. As put by one interviewee, 'because of that [IT tradition], nobody thought it was too strange or too visionary'. Hence, the early development of digital ambitions and strategies in Manchester was more of an 'evolution' than a 'revolution'.[4]

The first milestones of Manchester's IT policy were the implementation of the Manchester Host (1991), the Network of Electronic Villages in local neighbourhoods (1992), Manchester City Council website (1993) and Manchester Community Information Network (1994). By the mid-1990s, an Internet Exchange was already established in Manchester's Science Park (Carter, 2013). Such a local 'digital awareness' contributed to attract Internet Service Providers to the city at an early stage, giving rise to a spiral of cumulative causation: new markets and opportunities were created for internet and hosting services, which in their turn attracted new companies (e.g. Telecity, one of Manchester's most successful start-ups); subsequently, an increasing number of companies became connected and the costs of service provision progressively felt.

Another milestone was the development of the 'Eastserve' initiative, in the deprived East Manchester (2002). The initiative was catalysed by the Commonwealth Games and besides improving the area's digital infrastructure, it was a test-bed for new ways of engaging with the local community (e.g. Smith and Fox, 2007). Eastserve turned into one of the biggest wireless community networks in Europe, championing public access in schools, buildings, etc. As a result, the area developed a large online community network (Carter, 2013).[5] Moreover, the games were also an opportunity for the city to re-organise its structure. The technology team within the EIG moved to a newly created municipal arm's length structure – Manchester Digital Development Agency (MDDA) – now exclusively focused on digital issues and with more leeway to start-up new initiatives and tender for external funding.[6]

In 2005, the UK government published a National Digital Strategy, encouraging cities to develop and scale up their own digital strategies: the so-called 'Digital Challenge'. Manchester took it and established the foundation for the creation of a Manchester Digital Strategy – which still grounds today's ambitions – with a vision of fostering the city-region as 'the most advanced "next generation" connectivity in the UK, providing a sustainable base for high growth business, innovation, transformational public services and an inclusive knowledge society' (Carter, 2013, p.178). Over time, new paradigm shifts (e.g. web 2.0, living labs, user involvement, smart cities, open data, internet of things) stared to permeate the strategy, adding up to the (now renewed) dimensions that formed the strategy for roughly 20 years: skills and inclusion, jobs and growth, access and infrastructure.

4.1.2 Digital Strategy: the concept and design

As it becomes clear from the previous section, the key tenets of Manchester's Digital Strategy have been inherited from the city's past initiatives and experience.[7] Yet, from 2012 onwards – with the appointment of an Assistant Chief Executive for 'ICT & Communication' within the City Council – the Digital Strategy gained a higher symbolic status of a 'master plan', or, as put by one interviewee, 'the digital now made serious'. Moreover, besides the inherited domains and approach, the concept of the Digital Strategy gained three new underlying rationales:

- A broader aspiration as a *meta-strategy*. IT and digital should become embedded everywhere. Hence, beyond the city's regeneration agenda, the strategy becomes 1) more holistic and with more objectives and 2) with higher global ambitions – attracting global companies and rolling out ultrafast broadband to outcompete other cities (inspired by New York's digital strategy and grounded in a recent analysis from Oliver & Ohlbaum Associates (2012)).
- *City stakeholders get more involved as co-producers*. Beyond the championing role of the City Council and MDDA, the strategy should be absorbed and implemented by many other agents in the city.
- The *mainstreaming of IT and digital* within the City Council's activities (from a relatively minor priority towards becoming a key instrument to increase efficiency, customer relationship manager, transparency, etc.)

Hence, the Digital Strategy is currently organised into ten points, under five underlying themes: *Access and Connectivity*, *Engagement*, *Industry*, *Place* and *Leadership* (Box 2.2). Naturally, fulfilling all the ten points requires the implementation of many projects and initiatives, both inside and outside the City Council. A detailed list of actions – and recent progress of each of them – can be found in MCC (2013b). For the sake of illustration, without claiming to be comprehensive, we provide examples of some of the most emblematic projects and initiatives in Box 2.3.

4.2 Implementation

Naturally, the implementation of such a large-scale strategy is a rather complex challenge. Section 4.3 explores some of the early successes and results that can be attributed to it. However, before that, this section explores a number of critical factors and the enablers that contributed to making it happen: leadership, finance, organisation and networking, and risk taking.

4.2.1 Leadership

The implementation of the Digital Strategy relies not on one but on a number of leaders and orchestrators. For this reason it is more adequate to speak about collective or distributed leadership, working at different levels within the City Council.

Box 2.2 Manchester Digital Strategy – 10-point plan

Access and Connectivity

1. **CONNECTED CITIZENS**
   100% Broadband Coverage:
   - Wi-Fi in all public spaces and high speed wireless across Manchester
   - Superfast and where available Ultrafast Broadband
2. **CONNECTED BUSINESSES**
   All businesses have access to fast broadband services and are supported with advice to best exploit it and/or financial assistance in accessing even better/faster services.
3. **CONNECTED MANCHESTER**
   Public and private sector better connected. Lead and deliver a more coordinated partnership to drive improvements and market the city as a successful digital place to attract new business.

Engagement

4. **DIGITAL SKILLS**
   From officers and members; primary school children; Academies and Universities; to adult education we need excellence at all ages to provide the best digital education and further education to support people today to get the skills needed for future jobs. Digital Championing of knowledge transfer about the benefits of 'Going Digital'. This includes use of digital technology but more importantly design, research and development.
5. **SOCIAL CITY**
   An open and transparent, digital embracing City Council and Marketing Board. Accelerate social media and engagement, advance our website and web advertising focus, enrich content, maximise offline digital advertising, fully utilise e-opportunities (e.g. e-zines/e-bulletins). Embrace growth in mobile by increasing mobile content, mobile advertising and mobile apps. The development of a Manchester super app (Mapp) is one opportunity, the delivery of many consolidated into a New York-style hub is another. Work with the private sector to support fast progress and crowd sourcing with a hackathon, fully exploit GIS mapping utilisation etc.
6. **DIGITAL REFORM**
   Encourage and enable more people to become self-service customers, use CRM integration to deliver reform and efficiencies. By providing continued focus on key digital platform excellence, openness, tools, fast accessible connectivity and skills to citizens and business we will allow people to become self-sufficient and help them achieve their full potential.

Industry

7. **ECONOMIC GROWTH**
   Stimulate and grow the creative and digital sector. As one of our three most important areas of job creation we have to support, encourage and instigate growth from skill development from an early age, inward investment, procurement to maximising global opportunities of our home-grown talent.

Place

8. **DIGITAL PLACE**
   The City's digital landscape supports the vision of Digital Manchester – from advertising options to wayfinding, real time tram information to digital tourism – Microsoft surface maps, interactive tours etc. Digital Master planning should be considered at the same time as physical infrastructure and should fundamentally be led by the needs of the digital resident. Providing a joined up voice to drive coordinated development is vital.

Leadership

9. **DIGITAL INVESTMENT / RESOURCES**
   Exploit new investment opportunities, public and private, that can help accelerate the delivery of our ambitions.
10. **DIGITAL LEADERSHIP**
    Digital at the heart of Manchester strategies to promote a truly digital city in all we do from transport to health, education to public services. Consolidate and coordinate approaches to development based on need – not multiple people trying to achieve the same end via different methods. Think digital first and lead the way.

Source: MCC, 2012b, 2013a, 2013b

Box 2.3 Digital Strategy: examples of on-going projects and initiatives

Rolling out super and ultra fast broadband access across Manchester

One of the key objectives is to roll out super (>30Mb/s) and ultra (>80Mb/s) fast broadband across the city, with the goal of creating an open access network and a new internet hub exchange, linking homes and offices. It intends to challenge the limited and pricy market supply in order to support the city's digital businesses to compete globally. Due to potential market distortions and conflicts with EU state-aid regulations, partnerships with private players are being considered.

The corridor

The corridor project involves installing high-capacity, open access wi-fi infrastructure along Oxford Street – an area that concentrates many higher education and R&D institutions, medical facilities, companies, etc. The aim is to provide a digital backbone to support the test-bedding of new businesses and digital solutions, for example through the deployment of living labs and viral interactions.

Sharp project
Sharp is a recent and successful hub for digital media companies (see Section 3), specially designed to foster interactions and synergies between tenants (media companies). Among others, one key initiative is the delivery of education, apprentices and training schemes to nurture 'digital talent' and raise awareness for the creative and digital sector. It is organised and supported through different public-private partnerships.

Digital skills and training
In order to counter the skills mismatch and to assure the supply of digital skills, several initiatives are being carried out in partnership with schools, education institutions and private parties. A 'digital skills strategy' is being defined that considers different educational pathways, apprentice opportunities, etc. Among other examples, there is a partnership between the City Council and the city's universities for the organisation of the Digital Skills Summit – a forum for digital and graduate hiring and training workshops.

GO ON Manchester
The initiative intends to teach digital skills to the population at large through the recruitment of voluntary 'digital champions' as tutors. It is a follow-up of a similar national-level initiative and of the IT-inclusion work championed by MDDA over the past years. It has been carried out together with the Regeneration unit at the City Council, libraries and other external parties from the community-voluntary sector.

Others
Living labs (for new digital technologies, 'green' test-bedding, smart metering and energy, health, etc.)
Open data diffusion; hackathons and open data events
App development and the deployment of many social media tools

Source: Adapted from MCC, 2012b, 2013a, 2013b; and interviewees

First, the Digital Strategy – and before that, other urban IT policies – have been strongly supported at higher echelons of policy making in Manchester. The leader of the political cabinet considered it an important road map. As put by one interviewee, 'he is an industrial chemist and knows little about digital but understands very well the role of knowledge and technology'. Moreover, the council's Chief Executive (see Section 2) has been personally involved in the city's regeneration agenda and major projects, thus providing visibility for the IT and digital efforts of the city over time. Again, it can be said that the regeneration agenda of the City acted as an 'incubator' for the Digital Strategy to reach its current stage of maturity.

Second, an appointed Assistant Chief Executive for the new field of ICT and Communications currently spearheads the Digital Strategy. This contributed to bringing the

Digital Strategy to a high level within the Council's managing structure. As recognised by the Assistant Chief Executive, 'it is unfair to say that it [the Digital Strategy] moved higher and only now is paid attention in the City Council – it was always very respected and dynamic'.

Third, and most importantly, the Digital Strategy is coordinated and led on a daily basis by MDDA – Manchester Digital Development Agency (see previous sections).[8] As explained, MDDA inherited and expanded the competences of the former Technology Group (Economic division). Currently, it is staffed with both 'creative and business development' and 'tech/IT programming' people (ten persons in total), allowing tackling and monitoring of the different dimensions of the Digital Strategy. For many less 'institutional' groups (e.g. social media activists and grassroots organisations), MDDA acts a 'route' to access the City Council and raise awareness for new developments and challenges of the digital and IT world (e.g. open data); MDDA supports new groups by facilitating their own (MDDA) services but also by providing them legitimacy and visibility within policy spheres.

The Head of MDDA is a highly respected, experienced and passionate officer who has been involved in the development of IT strategies in Manchester since the early beginning. His role has been essential to assure stability, while simultaneously allowing for smooth transitions and new variety to permeate the Digital Agenda over time (e.g. living labs, user involvement, new global orientation, etc.). Moreover, as the Digital Agenda becomes everybody's business, his role is pivotal as a boundary spanner who can connect, bring together and understand the 'languages' of different 'worlds' (public administration, universities, companies, community groups, IT activists, among others). As many stakeholders are co-responsible for the implementation of the Strategy (yet not compulsorily), this type of leadership has been of utmost importance to pool and leverage financial and organisational resources.

4.2.2 Finance

As in many other local governments throughout the world, Manchester City Council has also been under financial strain and budget cuts, with the pressure 'to do more with less'. As one interviewee puts it,

> The pressure to cut budgets has perhaps not been so strong here as in other European cities, but it does exist – I think we can speak of cuts of around 25 per cent in many fields .... I don't believe it is possible to do more with less, but I believe we can do better with less – we hope the Digital Strategy can also help with that [improving resources efficiency within the City Council]. I see the crisis as a good opportunity to provoke change in how we finance growth

The Digital Strategy (being a road map) has no specifically allocated or comprehensive budget. MDDA's funding – the Digital Strategy's main orchestrator – comes one-third from the City budget and two-thirds from external funding (e.g. European funding). However, the resources to implement and enable their many actions have been coming from many different streams. This illustrates the widening of the 'partnership space' to finance economic development initiatives, namely the Digital Strategy:

- *national funding* (e.g. for big investments and hard incentives, such as the £12 million recently raised for the Urban Broadband Fund, a funding programme from the national government);

- *European funding* (e.g. for the piloting and implementation of new initiatives, such as Living Labs, social and skills programmes, R&D pilots, some infrastructure, etc);
- *small seed funding* from the City Council (e.g. to leverage other organisations' projects, such as open data and app development contests, the 'Digital Skills Summit', etc.);
- *volunteer work* (e.g. the GO ON Manchester project);
- the internal funding and resources of the many organisations involved in IT and digital fields in the city, 'responsible' for implementing projects. This also includes private companies who co-support the development of, for example, 'coding schools' for programmers.

4.2.3 Organisation, networking and communication

As the Digital Strategy became an 'umbrella' strategy, the need for close cooperation, partnerships and networking increases. In a simplified way, the operationalisation of the Digital Strategy has been twofold:

1. direct implementation (of some parts of the Strategy), by the City administration
2. indirect implementation, by 'showing the way', lobbying and nudging other organisations in the city working with IT and digital.

As suggested above, distributed leadership has been pivotal for both; moreover, the Digital Strategy has required internal human resources and close collaborations across departments in the City. Again, the boundary-spanning role of the MMDA has been of utmost importance. On the one hand, contrary to the more conservative mindset of the IT department, MDDA is involved in out-of-the-box and more innovative projects and initiatives (e.g. open data, living labs, preparing bids and participating in city networks, etc.). The organisation and identification of new topics of the Digital Strategy thus required some 'slack' and innovative thinking. The IT department of the City is currently involved in some initiatives of the Digital Strategy but, as said by one interviewee, 'it took some time to break [the mindset]; the IT department is used to get all complaints [e.g. fighting viruses] from the City's staff; … their tasks are not very amenable to open-ended project'. On the other hand, initiating and carrying out some large-scale projects of the Digital Strategy have been facilitated by the history of close cooperation between MDDA, the Regeneration unit and other agencies dealing with, for example, education, economic and social affairs in the City.

Yet, the involvement and engagement with other actors beyond the City's structure has also proved to be an essential enabler. The main rationale has been to devolve the Strategy to the stakeholders and organisations in the city so that they can implement it without overlapping efforts. The stakeholders interviewed are of the opinion that there is currently strong complementarity and synergies between the action of the City (MDDA) and their own, which didn't happen in the past with other government organisations. As one organisation involved in the field of improving digital skills said:

> In the past other governmental organisations [Regional Development Agencies] launched projects and initiatives overlapping with ours, with public funding and thus difficult to outcompete. [Why? A:] That was the State's 'logic' of decentralising competences, but not really grassroots or articulated with the institutional fabric and on-going initiatives in the region.

MDDA has close formal and informal links and fluid communication channels with many other businesses and organisations within the city. The logic is that 'they know better what they [and Manchester] need, [for example] which types of digital skills are missing and what can the Digital Strategy do for them'. One example is the cooperation of MDDA with four universities in the city to organise the yearly 'Digital Skills Summit', a training, networking and match-making event between companies and graduates. Box 2.4 and Box 2.5 provide illustrations of other types of partnerships with organisations closely involved in providing input to the Digital Strategy.

Box 2.4 Mad Lab – Manchester Digital Laboratory

Mad Lab is an NGO providing hosting and working space for community groups interested in diverse types of creative and technical innovations, or as put by the organisers, for 'geeks, artists, designers, illustrators, hackers, innovators and idle dreamers' – Mad Lab (2013). It opened in 2009, in the Northern Quarter (old industrial quarter turned into 'creative district'), with the support of a small start-up grant from the UK government for supporting adult informal learning. By this time, MDDA was one of the official supporters officially recommending and 'legitimating' the organisation so that it could formally apply for the tender.

Four years later, Mad Lab already had a substantial turnover (from workshop fees, group fees, some cross funding, etc.). It is run by four managers and all the money is re-invested in the organisation. Back in 2009, there was the perceived need to develop a place for the meetings and workshops of community user groups in many domains. Thus, the objective was providing a place for informal learning and exchange, complementing meetings in pubs which were not the best place for workshops. So far Mad Lab has already hosted 50 community groups. It has grown substantially and is now connected with many similar initiatives throughout the world.

As explained, 'in the beginning it was very much about very geeky and male groups – e.g. sci-fi groups; Google groups, etc. – now there are more diverse ones such as women programmer groups, poetry, film … Some groups are diversifying into more cross-sector platforms, bringing people with related interests together, from different backgrounds'. One such group that started in Mad Lab was the Manchester Open Data Group. Mad Lab provided room for experimentation and for the prototyping of solutions. The first open data 'hackathons' took place there, with the support of the City Council and MDDA. Now the City Council is becoming involved as well in the provision of (open) datasets. As one interviewee puts it, 'this is not the place where many City officials would come in the first place; the MDDA helped to raise awareness about the open data movement'.

Box 2.5 Chamber of Commerce

Manchester's Chamber of Commerce has been working and lobbying to improve the conditions of companies involved with digital businesses in the city. There is inclusively a Digital Infrastructure group at the Chamber: one of their roles is to lobby with others for better digital connectivity. Currently, super and ultrafast broadband access is still constrained and very expensive for many advanced digital businesses (namely due to evidence of cartelisation between internet providers in the city). The City

Council and MMDA are also involved and closely cooperate in the Chamber's lobbying efforts, jointly championing the need for better access and digital infrastructure.

One illustration of the connectivity issue is illustrated by the recent UK regulation on the procurement for new buildings and infrastructure. In order to participate in such tenders, developers and construction companies need to exchange complex and very heavy information through software of building Info Design Modelling (3D) – this includes the design of the components plus other databases on embedded components, so that there is fluid integration between the constructor and the suppliers. Manchester has leading companies in the field, but having fast broadband and good IT infrastructure is essential to tender and to keep ahead of competition.

The Chamber is also involved in digital skills enhancement initiatives, namely for advanced and highly qualified technicians. As it is not easy to find workers combining tech-digital-creative-business skills (university curriculums are lagging behind), job poaching and competition push salaries higher. The Chamber is also involved in demand-driven education schemes in which it delivers tailor-made modules and training according to the needs of the firms ('beyond coding heroes').

4.2.4 Risk taking

The Digital Strategy is a combination of quick wins and long-term ambitions. So far, its orchestrators haven't been afraid to take a number of risks in unchartered territories. Examples are the promotion of relatively immature experiments related to open data, living labs, social media, etc. Moreover, taking the Digital Strategy to the next level – global ambitions, distributed efforts – entails risks of image (in case of failure) and of difficulties in monitoring and controlling the development of the Strategy (e.g. the impossibility of controlling other stakeholders' actions and budgets).

Furthermore, one important enabler in this respect has been the capacity to challenge established European and national regulations, e.g. for the rolling out of dedicated wi-fi infrastructure so that new business models and digital solutions can be developed by companies (e.g. in the corridor project).

4.3 Results

Over time, the deployment of IT and digital-related initiatives in Manchester has proved its value. Section 4.2 illustrated some of its already associated results, for example:

- The attraction of internet providers and IT companies to the city (mid-1990s), at least partly ignited by the pioneer involvement of the City in urban IT policies. Currently, BBC and many other digital companies still benefit from the competences nurtured in the region since this period.
- The results of the Eastserve initiative, increasing connectivity, IT abilities, the propensity to look for jobs and for being economically active in deprived communities (East Manchester).

As explored above (Section 4.1), IT and digital-related policies have been running for many years in Manchester, and it is thus difficult to establish a clear division between the past and the current (more ambitious) version of the Digital Strategy under analysis and the underlying results. Moreover, it is still early to assess the full scope of results of this more ambitious Digital Strategy. Nevertheless, having said that, the following points present a (non-exhaustive) number of outcomes and results that can be at least partly connected with the more recent version of Manchester's Digital Strategy (for more detail, see MCC, 2013b):

- The award of an Urban Broadband Fund (£12 million) from the National Department of Culture, Media and Sports (DMCS) to roll out superfast and ultrafast broadband towards covering 100 per cent of the city. (However, due to state aid regulations, the programme is being redesigned to be deployed through demand-led connection vouchers, wi-fi connectivity and several other R&D schemes.)
- The securing of £1.5 million from ERDF and a Regional Growth Fund (RGF) to stimulate business demand for connectivity in Greater Manchester.
- The recruitment of 800 volunteer 'digital champions' for the GO ON Manchester initiative, with progressive engagement of the population and business parties.
- Increasing social connectivity through the facilitation of a free public wi-fi network in public transport hubs. More than 16,000 customers accessed the network during the first six months (December 2012 – May 2013) in 14 different sites.
- Several new partnerships across the public and private sector have been developed with an eye to improving digital skills competences in the region (e.g. Digital Skills Strategic Framework).
- The joint organisation of several highly popular joint events (e.g. Digital Skills Summit, Future Everything innovation festival).
- Training initiatives within the City Council to champion social media and IT programmes in different neighbourhoods by City's staff.
- The establishment of an open data philosophy at the City Council – only by exception will data not be openly available – and acceleration of local open data ambitions.
- Improvements in the customer relationship management (CRM) services in the City Council towards increased digitalisation.

Moreover, a number of soft impacts could also be identified – associated with the meta-level dimension of the Digital Strategy:

- *Image building*. Together with the installation of the BBC offices, the Digital Strategy created considerable awareness about the relevance of the 'digital' for Manchester's development prospects. The city's digital skills, aspirations and the Digital Strategy were promoted in national and international media. There is evidence of at least one company that openly recognised the role of the Digital Strategy (and the possibility to find local business partners) in their decision to settle European headquarters in Manchester: EON Reality, from Los Angeles, one of the world's leading providers of digital and 3D solutions for education and edutainment. The company has created a 'school for coding' in the city and currently advertises the City's Digital Strategy as their own.
- *Advocacy and legitimation*. The Digital Strategy is contributing to legitimate the role of other organisation-level digital strategies in the city. Naturally, the stakeholders have their own agendas and will not change them just because of the Digital Strategy, but it helps to legitimate visions within large and fragmented organisations. Some interviewees explained how this process works:

> It helps that Manchester wants to be top-ten city in Digital … when I talk to my boss I can say 'look, this is the direction in which the City is moving' … it is important to make others recognise that [e.g. need to foster new creative-digital skills].

> I don't know the agenda's ten points, but I know the overall direction and it helps to say that the city is moving in a certain direction. … I know there are things related with educational media, and I have things on-going with that as well. It is comforting to have the agenda [even if not strictly enforced].

- *Uniting stakeholders around a single vision.* The Digital Strategy is contributing to bringing unusual stakeholders together that would hardly meet otherwise (e.g. high-level city officials, social media activists and data hackers), enabling them to share similar visions. There is evidence that the Digital Strategy is supporting the creation of a joint 'digital narrative' for the city, while inspiring other organisations. As one interviewee explains, 'we have our own strategies, but we have been drawing inspiration from the City's digital agenda as well. … It helps to make it our own.'

5 Conclusions

5.1 Broad assessment

Over the last decades, Manchester championed a large-scale urban regeneration strategy and the city's attractiveness increased substantially. Nowadays, despite the persistence of social deprivation and unemployment, the economic performance of the city has largely improved and growth has become very much idea-driven, with IT and digital media industries as important spearheads. Thus, one of the most pressing sustainable competitiveness challenges in Manchester is how to steer (and finance) new economic growth while making sure that citizens can benefit from it.

This case study looked into one of Manchester's strategies to tackle these issues: the Digital Strategy. IT and digital-related strategies are far from new in Manchester – they have been linked to urban regeneration for more than two decades now. However, as the city's priorities change – and new competitive advantages have to be nurtured – the current Digital Strategy envisions turning Manchester into 'one of the world's leading digital cities by 2020'. This requires, among other things, improving digital accessibility, skills, business and innovation opportunities, as well as diffusing more IT and digital opportunities throughout the city. The Digital Strategy has a global ambition – to better position Manchester in the international competition for digital businesses and investments – and the status of a digital 'master plan' within the administration. Its implementation is distributed and co-produced by different stakeholders in the city, in a combination of quick wins and long-term goals. It represents the attempt to mainstream IT and digital across the city, making it a rooted competitive advantage.

It is perhaps too early to assess the results of such a meta-level strategy. Yet, some promising outcomes can be already identified, namely the securing of significant broadband investment funds, the increase in digital social interaction, the attraction of new businesses, the development of new partnerships and of varied skill enhancement programmes, the recruitment of several 'digital champions' and the acceleration of new digital attitudes in the city (e.g. open data). Moreover, there is evidence that the deployment of the Digital Strategy as a meta-level 'road map' is contributing to strengthen Manchester's image as a digital city, to legitimate new digital endeavours within other city organisations and to create a joint 'digital narrative' for Manchester.

5.2 What have been critical factors?

Since the early 1990s, the deployment of IT and digital policies in Manchester benefited from the link with other important urban strategies in the city. Namely, as discussed, the urban regeneration agenda has somehow acted as an 'incubator' supporting the digital strategies to reach a higher degree of maturity.

However, a number of additional factors emerged as critical enablers behind the successes of the more recent Digital Strategy.

First, the presence of strong leadership distributed across multiple levels. The Digital Strategy benefited from the support of stable, leading officials in the City Council; moreover, the appointment of an assistant chief executive as directly responsible for the strategy contributed to giving it more attention and a higher 'symbolic status' within the administration. On the top of that, the operational leadership of MDDA – Manchester's Digital Development Agency – has been essential 1) to assure the smooth transition from the regeneration-driven agenda towards its new ambitions; 2) to provide innovative thinking and permeate the strategy with new variety and new digital paradigms; and 3) to act as a boundary spanner, listening, understanding and connecting new organisations and digital innovators to each other and to the City's Digital Strategy.

Second, the mobilisation of new partnerships and funding was pivotal. The Digital Strategy is anchored in varied funding schemes, such as European and national funding, combined with small seed money from the City (to leverage innovative projects), volunteer work and the own internal resources of the different stakeholders (including private companies). This combination of sources has been key in leveraging different projects in a context of growth financing constraints.

Third, some of the early achievements of the Digital Strategy relied on the capacity of its promoters to take risks. Examples are the challenging of national and European State Aid regulations to deploy new infrastructure and piloting services, as well as the involvement of unusual suspects (e.g. community groups) in the implementation of the Strategy.

Notes

1 The so-called 'core cities' is an advocacy group of large cities in England outside Greater London: Manchester, Birmingham, Bristol, Leeds, Liverpool, Newcastle, Nottingham and Sheffield.
2 This strategic Framework has been recently complemented by the 'Greater Manchester City Deal'. This primarily focuses on economic development and business support initiatives to support investment and unlock/optimise growth at the metropolitan level.
3 Actually referred to as 'telematics' at the time (early convergence of telephony and informatics).
4 The same did not happen, for example, in other English formerly industrial cities (e.g. Leeds, Birmingham), whose early regeneration plans were much more conventional. Also compared with other European cities like Helsinki – whose digital strategies were to a large extent nudged by the severe crisis of the late 1980s – the implementation of IT strategies was somehow smoother in Manchester.
5 The many evaluation studies about the initiative report that Eastserve beneficiaries became more likely to be aware of job opportunities and training, to seek work and to look for new challenges; 40 per cent of the residents in the area got IT training and were more likely to look for complementary skills afterwards (Carter, 2013).

6 MDDA was by this time composed of ten people. That number remained more or less stable over time (see Section 4.2). MDDA became responsible for dealing with digital-related issues and strategies in the City such as 1) inclusion, skills and jobs; 2) digital business sector development; and 3) innovative projects and new initiatives (e.g. emerging out of EU projects).

7 Moreover, today as in the past, it draws inspiration from the city networks in which Manchester long participates (e.g. EUROCITIES) and benefits from the experience of MDDA (see Section 4.2).

8 MDDA is tasked, among other things, with 1) organising training initiatives and raising digital awareness (e.g. with housing agencies, schools, associations, NGOs); 2) nudging and supporting digital-related actions with other stakeholders; and 3) strategic and practical work for the development of small IT infrastructure and programming (e.g. innovative IT pilots in the city).

References

Carter, D. (2013). Urban regeneration, digital development strategies and the knowledge economy: Manchester case study. *Journal of the Knowledge Economy*, 4(2), 169–89.

Department for Communities and Local Government (2011). *English Indices of Deprivation 2010.* London: Department for Communities and Local Government.

Mad Lab (2013). Accessed at http://madlab.org.uk/about/ (13 February 2013).

Marketing Manchester (2009). Original, modern. Marketing Manchester publications. Accessed at http://issuu.com/manchestermagazine/docs/original_modern?viewMode0presentation&mode0embed (13 February 2013).

MCC – Manchester City Council (2012a). *Manchester Core Strategy Development Plan Document* (adopted 11 July). Manchester: Manchester City Council.

MCC – Manchester City Council (2012b). *Report for resolution – Digital Manchester Update*, from the Chief Executive (24 November). Manchester: Manchester City Council.

MCC – Manchester City Council (2013a). *Digital Skills*, from the Chief Executive (6 March). Manchester: Manchester City Council.

MCC – Manchester City Council (2013b). *Economy Scrutiny Committee – 17 July 2013, Digital Strategy Update*, from Sara Tomkins, Assistant Chief Executive Communications and ICT. Manchester: Manchester City Council.

New Economy (2012). *Briefing 24 – Greater Manchester Growing Population – Census 2011.* Manchester: Greater Manchester Combined Authority.

New Economy (2013). *Greater Manchester Key Facts*, May. Manchester: Greater Manchester Combined Authority.

Oliver & Ohlbaum Associates Ltd (2012). Manchester: a truly global digital content hub by 2025. Brief report, 22 October.

ONS – Office for National Statistics (2013). Accessed at http://www.ons.gov.uk/ons/index.html (13 February 2013).

Smith, A. and Fox, T. (2007). From ‘event-led’ to ‘event-themed’ regeneration: the 2002 Commonwealth Games legacy programme. *Urban Studies*, 44(5–6), 1125–43.

Van Winden, W. (2003). Essays on urban ICT policies. PhD thesis, Erasmus University Rotterdam.

Young, C., Diep, M. and Drabble, S. (2006). Living with difference? The ‘cosmopolitan city’ and urban reimaging in Manchester, UK. *Urban Studies*, 43(10), 1687–714.

3 Dublinked (Dublin)

Luís Carvalho and Alexander Otgaar

1 Introduction

This case study explores the development of Dublinked – an open data and innovation network initiative aiming to make available to the wider public (businesses, researchers, governments, citizens) data about public provisions (e.g. water, transport, environment and planning) in the greater Dublin region.

Releasing public data to wider audiences has been gaining strong momentum during since 2008, with world leaders like Barack Obama calling for more transparency and accountability in the public sector. Going beyond such objectives, the European Commission and some countries recently started to explore the possibilities for open data to generate jobs and growth. Actually, some analysts argue that national governments are presently 'sitting on billions of euro of data' that could be transformed through innovations and monetised.

Yet, before being conceived at the national level, publishing data showed up in the radars of cities and municipalities. Leading North-American (e.g. Chicago, Vancouver) and British (e.g. London) cities pioneered the first open data experiments, but there are already many followers: a recent count identified more than 400 cities and towns making their data available on the internet under the name of 'open data'. So far, most cities have been using open data as instrumental for citizen empowerment, transparency, democratisation and participation. Yet others have the ambition to release data to help solve local problems and increase the efficiency of public service provision (e.g. transport, planning and water). The idea is that when data on such services is open, many 'innovative minds' beyond the local government (users, researchers, businesses) may find good solutions for the problem, update the available information and develop valuable products and services for the city, such as mobile apps, among others.

Dublin is not an open data world pioneer (design began on Dublinked in 2009 and implementation in 2011) but it is an innovative and ambitious one, pushing forward the state-of-the-art of such initiatives. Beyond citizen empowerment and transparency, Dublinked's main ambition is to spur economic and technological development in the city, promoting R&D and innovation, collaboration and new 'smart city'-related investments. To this effect, Dublinked started by providing a 'regular' open data zone plus a so-called 'R&D zone', where partners can experiment with more complex and valuable urban data towards the development of marketable solutions. Yet, releasing such data raises many technical, legal, cultural and organisational challenges thus providing critical tests and lessons for other cities wanting to develop more demanding open data initiatives.

In the remainder of this chapter, Section 2 analyses the structure (and recent dynamics) of Dublin across a number of dimensions or 'capitals' for sustainable competitiveness. Section

3 sketches the city region's economic vision and strategy, clarifying the fit of Dublinked in such a strategy. Section 4 is the core of the chapter and explores Dublinked in large depth, namely its start-up history, vision and content (4.1), the key features and enablers that 'made it happen' (4.2) and a number of preliminary results of the project (4.3). Section 5 concludes by raising lessons and challenges ahead.

2 City profile: capitals for sustainable competitiveness

2.1 Performance

With roughly 1.3 million inhabitants in the metropolitan area, Dublin is Ireland's largest urban agglomeration and national capital. It represents 40 per cent of the Irish economy, concentrating the European headquarters of many North American corporations. Since the early 1990s until recently, Dublin's demographics went hand-in-hand with the well-known Celtic Tiger phenomenon, during which the Irish economy grew at 6–8 per cent rates per annum.

Benefiting from the strong momentum of the world economy, the foundations of the Celtic Tiger are well documented: presence of a highly educated population, favourable business climate leading to the attraction of (knowledge-intensive) investments, labour and social reforms, public finance stability and access to the EU single market. As a result, the population grew steadily and the city attracted many migrants, making Dublin an increasingly diverse metropolis.[1] Population and jobs expanded fast from the core city to other areas, which became independent municipalities of the greater Dublin region.

From 2008 onwards, Ireland and Dublin's growth came to a halt. Catalysed by the North-American subprime crisis, the Irish economy faced a severe economic bust, linked with the explosion of a real estate property bubble, a financial and banking crisis and the collapse of the construction sector. The overall economy contracted fast, unemployment soared and financial markets crumbled, leading to an EU/IMF bailout in 2010. As the main Irish economic and financial hub, Dublin severely felt the effects of the economic contraction. Many of the young and highly educated left the city for better opportunities elsewhere. The mood turned grim in Ireland and in Dublin in general.

However, despite the economic slump and collapse of the construction and financial sector, the Irish economy recently started to bounce back. This has largely been due to export's growth and to the continued attractiveness of the city for foreign direct investment. Roughly two-thirds of the exports originate from Irish-based foreign companies, and more than 50 per cent of such companies and headquarters are located in Dublin (Cudden, 2011). This indicates that Dublin is playing a pivotal role as a 'recovery engine' for the Irish economy.

2.2 Economic structure and knowledge

In spite of the gloomy mood and economic slowdown, the foundations of Dublin's competitiveness didn't wash away, namely its innovative economic and knowledge base. Just like other European capital regions, the economic structure of Dublin is rather diversified; it is service intensive with relevant shares of advanced manufacturing and high-end industrial activities. Although unemployment soared, many companies are increasingly hiring and investing, especially foreign corporations, which represent an important share of Dublin's economy. The Irish Industrial and FDI attraction agency (IDA Ireland, 2012) reported that in

2011 foreign companies created 13,000 new jobs and invested €700 million in R&D,[2] and the lion's share of this is concentrated in Dublin (over 50 per cent).

Among other dynamic sectors (e.g. 'green' business, biosciences and financial services), IT and internet-related activities have been pivotal in the economic recovery. Since 2000, Dublin has been a magnet for global companies in software, animation, digital media and cloud computing. Moreover, many of Dublin's indigenous companies prospered. In 2011/12, leading companies such as Intel, Microsoft, IBM, Dell, Google, Facebook, LinkedIn, Zynga, Amazon and PayPal, among others, started or expanded operations in Dublin, making the city a leading world capital of digital media and the new 'app-economy'.

As a result, Dublin hosts a very dynamic IT-related innovation ecosystem, with relevant formal and informal exchange among multinational and indigenous companies, business associations, brokers, universities and venture capitalists (Irish and North American). As an illustration, the Science Foundation Ireland (2010) reported that the number of formal R&D projects between software companies and Irish researchers more than doubled between 2008 and 2010 (totalling more than 500 projects in 2010 alone).

Such developments are largely due to the quality of Dublin's knowledge base. Dublin hosts large universities and many well-ranked R&D institutes; its population is, on average, highly qualified with a large percentage of tertiary and college degrees. However, just like other places, Dublin faces a structural shortage in IT-related graduates, which manifests itself during times of economic growth (most graduates are still in the fields of business and financial services, and the number of internships are limited, making it hard for recent graduates to acquire working experience). This has resulted in some Irish companies attracting graduates with some experience from abroad, as is happening today as multinationals (e.g. IBM) expand R&D operations in Dublin.

2.3 Environment and accessibility

Dublin's infrastructure is the best in Ireland. It is well endowed with an international airport; moreover, through its multinational networks and diaspora, Dublin is one of the most internationally open cities in the world. As a national capital, Dublin has a relevant number of amenities and a buoyant creative and arts scene, including theatre, music and literature (Dublin has been a designated UNESCO City of Literature since 2010). Dublin is a seaside location and is surrounded by low mountains and farmland, which rank high in the overall quality of living (Cudden, 2011).

Yet the transport and energy infrastructure faces relevant challenges. Public transport systems are poorly integrated and have room to become more efficiently connected; this has contributed to a recognised overdependence on the car, urban sprawl and, hence, increasing pressure on public service provision.

2.4 Image

To the regular visitor, Dublin is famous for its music and pubs, and for being home of the well-known Guinness brewery (Cudden, 2011).

During recent decades, Dublin became associated with the image of the Celtic Tiger, of a fast-growing successful city. The recent economic downturn does not seem to have endangered the structural image of the city, but internally it has unleashed a grim mood. Presently the city actively works to further promote Dublin as an excellent place for business and investment, tapping into the top results achieved in many international rankings. Dublin City

Council has relevant on-going initiatives to promote its brand and identity dimensions, namely in the setting of global competition for talent, tourism and investment.

2.5 Institutional conditions

Traditionally, Irish local authorities are responsible for urban planning, infrastructure, housing, health, safety, environment and leisure. Economic development policies rested almost exclusively under the responsibility of the central government and national agencies. Despite its size and relevance, Dublin has not been an exception. Dublin City Council administers the metropolitan area of Dublin and the Dun Laoghaire/Rathdown, Fingal and South Dublin the county or more suburban areas of the region. Each local authority has a democratically elected city or county councils. Each council is headed by a Lord Mayor or chairperson. The elected councillors adopt key strategic plans for the region including the city or county Development Plans. Although policy frameworks are decided by the councillors, day-to-day management of the city, policy implementation and coordination are carried out by the city or county managers ('CEO style') and their staff.

Dublin officials have long called for more power and competence to intervene in competitiveness, innovation and economic development issues. This is more so as many national policies have tended to favour de-concentration of economic activity from Dublin to the countryside. One recent step towards a more active stance in economic-innovation issues was the creation of the Creative Dublin Alliance (CDA). Founded in 2007/8 it envisages tackling institutional fragmentation within Dublin towards the development of joint solutions to common challenges at the metropolitan level (NYC Global Partners, 2011). The CDA was designed to organise cooperation between key stakeholders, represented at the highest level of decision-making (university deans, company CEOs, President of the Chamber of Commerce, etc., chaired by the City Manager). The CDA embodies a new model of local economic governance, where the City leads and facilitates cooperation and alignment of other stakeholders' actions (including dynamic leading firms).

The action of the CDA and the development of competitiveness-related projects gained ground and legitimation in the aftermath of the economic slump. In a context of significant budget cuts and public debt, the key objective was to catalyse local economic and innovation engines towards a faster and sustainable recovery (in line with national 'smart economy' and national jobs recovery plans).

2.6 Conclusions: main issues

After more than one decade and a half of fast growth, the Irish economy and the city of Dublin are still experiencing the effects of the financial meltdown and economic downturn. Together with unemployment and budget cuts, the mood in the city turned grim. Yet, the key assets and 'capitals' beyond the Celtic Tiger phenomenon didn't wash away. The business and living environment together with the presence of a highly qualified labour force keep attracting global corporations, namely in the field of IT and new media, which have played a role in exports' growth.

A newly established economic governance mode in the City (whose 'face' is the CDA) has been catalysing further these assets towards faster and more sustained economic recovery. Among others, Dublinked started as one of the flagship efforts of the CDA, implemented through soft and relatively inexpensive measures, aiming to unleash new innovation networks in the city and the development of globally marketable technologies and solutions.

3 City-wide vision and strategy

3.1 Vision

The most recent city-wide vision for Dublin is the *City Development Plan 2011–2017*. It is innovative to the extent that it explicitly identifies a vision for 'Dublin 2030', considering 2017 an 'intermediate stage':

> Dublin 2030 is a compact city, with a distinct character, a vibrant culture and a smart, innovative economy. It's a city of urban neighbourhoods, connected by good public transport and green spaces. In 2030, Dublin is one of the most sustainable, dynamic and resourceful cities in Europe.
>
> (Dublin City Council, 2011)

This vision also explicitly considers a necessary balance between six elements: 'People', 'Culture', 'Places', 'Enterprise', 'Movement' and the 'Environment'. Therefore, as a broad reference, Dublin 2030 1) is clean, green, compact and connected; 2) has an innovative, smart economy; and 3) contains sustainable neighbourhoods and communities.

Yet, for the purpose of this case study, it is important to consider a complementary and more focused development plan: the *Economic Development Action Plan* (CDA, 2009), released in July 2009 (the bleakest period of the Irish economic downturn) and supported by the recently created CDA (see Section 2.6). Focusing on the greater Dublin region as a whole, this action plan inspired the above-mentioned city-wide vision. It intends to give it a strong economic drive by proposing a vision and number of strategies towards economic recovery. However, it hardly proposes quick fixes; it is much more focused on medium and long-term initiatives leading Dublin towards more sustainable competitiveness. The vision tag is the following:

> Positioning the Dublin City Region, the engine of Ireland's economy, as a significant hub in the European knowledge economy through a network of thriving spatial and sectoral clusters providing a magnet for creative talent and investment.
>
> (CDA, 2009)

3.2 Strategies

The Economic Development Action Plan paved the way for subsequent actions of the recently formed CDA, backed at the highest level by Dublin's City Manager, John Tierney. The strategic framework of the plan includes three key drivers/strategies and a number of actions, to be largely implemented under a 'triple helix' framework (government, university, privates/industry). These drivers are:

- 'to develop strong city leadership': including making the case for Dublin as an engine for recovery in national policy; developing further the CDA; working the city's brand and international profile; *establishing a central regional data point across the city authorities*;
- 'to create a vibrant place': including lobbying for better infrastructure; embedding economic policies in the city planning; agreeing on clusters and economic corridors to be supported; identifying economic opportunities in urban regeneration; linking environmental assets with new clean-tech industries;

- 'to nurture, retain and attract creative people': including, among others, identifying enterprise supports; developing policy thinking on diversity and equality; supporting cultural and creative industries; creating conditions for the attraction of talent through the improvement of Dublin as a place to live and visit for workers and students.

The genesis of Dublinked is explicitly mentioned under the first driver – *establishing a central regional data point across the city authorities* – as a less ambitious data collection initiative. However, when developed further and aligned with other stakeholders' visions, its design started to cut across more of the plan's actions, such as talent attraction through place quality and, mainly, (sustainable) economic recovery. Despite Dublinked's 'on-the-go' adaptations, its vision and features have always been aligned with different strategies of Dublin's Economic Development Action Plan. These dimensions are explored in the next section.

4 The project: Dublinked

4.1 Vision and content

Dublinked is a rather recent type of urban development initiative – open data – and, therefore, its vision and content are not easy to grasp at a first glance. Its promotional tagline ('sharing data, sharing ideas, connecting a city') doesn't get us much closer. Yet the reading of the memorandum of understanding signed between its proponents is more helpful. It clearly states Dublinked's vision as to 'promote economic and technological development in the Dublin region, based on open innovation principles, and with the aim of facilitating research and development regarding the use of data in the greater Dublin region' (Dublinked, 2010).

As we will explore further on, Dublinked is carried out in an association between the four local authorities of greater Dublin region (Dublin City, South Dublin, Fingal and Dún Laoghaire Rathdown) and the National University of Ireland Maynooth (NUIM). More concretely, it consists of a broad-scope, coordinated initiative that collects, standardises and shares data with the wider public on local services and provisions in fields such as energy, environment, waste collection, water, traffic, planning, among others. This data is released through an internet-accessible platform, with the ambition that users can 1) develop new innovative and commercial solutions on the basis of it, while 2) contributing to tackling Dublin's challenges in the service provision realm.[3] Still, before exploring further Dublinked's 'nuts and bolts', it is useful to understand how and why it emerged.

4.1.1 Brief story of the initiative

Dublinked has many 'parents', whose interests converged around 2008/9. Before this time, the City Manager John Tierney already had the idea of developing a data repository/observatory for the greater Dublin region, yet in a less ambitious way. This intention was formalised in the Dublin's Economic Development Action Plan, developed by the (recently formed) CDA (see Section 3).

Yet, the idea for Dublinked as such only matured during 2009. Specifically, this happened as IT global corporations with interest in urban technology solutions (e.g. Siemens, IBM, Accenture, Intel) were settling and expanding operations in Dublin (see Section 2). Such firms approached DCC's City Manager, showing interest in accessing data related to the city's service provision, namely in the fields of energy, transport and water. In particular, IBM was starting to explore possibilities for using Dublin as a 'smart city' test-bed for new urban solutions (see Box 3.1).

However, at this time, the first open releases of such (public service) data were starting up in the USA and UK, and there was very limited knowledge of how to do it. One of the first ideas in Dublin was to develop a 'shared data platform', in partnership between DCC and large IT companies. Yet this was soon abandoned since it would have violated EU state-aid regulations by giving privilege to such big players. Full outsourcing of such a platform would also not be a solution, introducing delays and complications due to the need to comply with rules of procurement and future commercial advantage.

Therefore, a more consensual solution emerged in the shape of a partnership between DCC and three other local authorities, with the technical coordination of NUIM (a local experienced university in data analytics, geo-location and innovation eco-systems).[4] Before 2009, NUIM had already conducted an exploratory study on open data strategies and their requirements. In 2009, DCC, NUIM and IBM met to discuss the design of Dublinked, now with the ambition of making it more than a data repository: a tool to spur innovation and growth.

With this view in mind, achieving the right urban scale was an issue (larger amounts of data at the relevant metropolitan level), and thus the strategy of making Dublinked a City Region initiative, with the participation of DCC plus three contiguous city councils. However, scale was not the only reason for regional collaboration – Fingal County Council (Box 3.2) already had a more established open data initiative by that time, from which all the others could learn.

Box 3.1 IBM, smart cities and Dublinked

In March 2010, IBM announced the establishment of a leading Research Lab in Dublin, focusing on the development and prototyping of 'smart city' solutions for global markets (with on-going international recruitment of 200 researchers). Yet, this location decision was not made from scratch. IBM and Dublin City Council (DCC) had worked together since 2009 on the development of smart city projects on energy, traffic and water. This partnership led to a greater interest by DCC and IBM in the potential of Dublin's data. During 2009, the talks between IBM and DCC progressed towards finding new ways to collaborate on those fields. The genesis of Dublinked started here, with an eye on involving many players interested in that data (public parties, universities, firms, entrepreneurs) by releasing it on a web platform.

As stressed by IBM representatives, the idea was not to 'technologyze the way out of a problem', but to foster the involvement of new people and organisations towards idea exchange, in an 'open-innovation' fashion. According to IBM, Dublin provides an excellent context for this, namely due to its optimum urban scale (not too small, not too big), innovation culture and the excellent relationship between DCC and IBM (important for face-to-face interaction, fast feedback, etc.). Since then, IBM has been instrumental in networking Dublinked with other relevant open data initiatives in Europe and the USA. Moreover, among others, IBM has been supporting Dublinked in dealing with metadata, publishing standards and the mechanics of publishing dynamic data.

Box 3.2 Fingal's open data initiative

In 1994, in order to cope with fast population growth, Dublin was split into four councils (Dublin City plus Fingal, South Dublin and Dún Laoghaire), each with an elected local authority. Due to this administrative reorganisation, fast-growing counties such as Fingal (where the airport and new industrial and residential areas are located) felt the need to have more information about its territory in order to better plan/provide public infrastructure and services. As a new council, few datasets had been broken down at this scale.

To cope with it, in the early 2000s Fingal's IT department started to consider the development of standardised data platforms; in 2006 it started to internally report the newly collected and standardised data about the county, from multiple departments, through the so-called Fingal Data Hub (e.g. geo-referenced data on health, education, policing, employment, planning). During the summer 2010 the first visualisations of the data were tried out (e.g. paths, circuits, location of infrastructure, etc.); in November 2010 the website Fingal Open Data was launched together with the 'hack-day' event – 'apps4fingal' – within the Innovation Dublin festival, to explore which types of apps could emerge out of the data. At this time, other counties and DCC were starting to think of similar initiatives and thus the momentum was created for cooperation, although the Fingal Open Data initiative developed alone and was the first in Ireland.

Fingal jointly led the Dublinked initiative but has maintained its own open data hub for the county's internal affairs and service information. Here rests one important difference between Dublinked and the Fingal Open Data initiative: while economic development, job creation and nurturing new innovation networks has always been the prime drive of Dublinked as such, in Fingal economic development is only one of the objectives among other primary objectives such as transparency, participation and collaboration (as in other more conventional open data initiatives).

Some important success factors for Fingal's initiative are 1) the fact that 'homework' was done before (standardising data, participation of different departments), nudged by a new administrative context and a proactive IT department; 2) the support of a dynamic county manager, interested in pioneering open data world initiatives; and 3) the technical support from a former communication manager of Fingal, subsequently (she has since moved) responsible for the London Data Store, one of the world's open data pioneers – this helped to prevent mistakes, show what works and what does not, and provide inspiration.

Hence, Dublinked (and its platform) was formally launched on 18 October 2011, during the yearly Innovation Dublin festival, in order to test the usability of the data and the feedback of potential users. The initial data collected and released (around 100 datasets) was primarily in the domains of traffic, environment, planning and water – the initial 'priority requests' of foreign IT companies operating in Dublin.

4.1.2 Dublinked: vision, features and design

To formalise Dublinked's partnership, the four city councils and NUIM signed Dublinked's Memorandum of Understanding, in which they explicitly state the aims of the initiative. Overall, there are three broad aims, ordered as follows:

- *support economic activity, job creation and innovation*: spurring innovation and attraction of 'smart city'-related R&D investment, encouraging the development of global marketable solutions by using Dublin as a laboratory, city data as 'raw material', and, in the process, making Dublin a high-profile R&D and testing hub in this field;
- *improve efficiency in local service provision*: providing solutions to specific problems faced by the local authorities – e.g. traffic, environment – while increasing the participation of Dublin's citizens in the planning and provision of such services;
- 'support other initiatives in the area of sustainability, eGovernment, health and employment through data access' (Dublinked, 2010).

In December 2013 – more than two and a half years after the start of the initiative – the vision for Dublinked was formally stated as follows:

> To participate actively in making the greater Dublin region a leader in innovation in the urban environment – focused on the global challenges arising from increasingly populated, urbanised living-spaces – through collaboration between private, public and research partners – with the city region as a proving ground
>
> (Dublinked, 2013)

Such aims reveal that the notion of sustainable competitiveness of Dublinked's proponents is clearly economic and innovation-oriented; yet, it is coupled with its potential contribution for the improvement of other urban-regional capitals such as accessibility, quality of life or the environment.[5] These aims have had natural implications in the design and features of Dublinked as an open data initiative. But what are those features and which main innovations can be identified?

In order to answer this question it is easier to refer first to the more conventional open data platforms so far established. Most of these initiatives welcome new commercial applications out of the released data (e.g. smartphone apps), but have a prime focus on transparency and citizen empowerment; therefore, the key challenge is mainly to polish and gather existing data, often not very 'polemic'[6] sets, and let the public use them as they wish. Published data is often static or only partially updated (e.g. when a new master plan is approved, or when new counts of a certain public service take place), resembling public repositories of data on public services and the built environment. Dublinked has these features as well (most of the datasets fit this profile), but it simultaneously tries to provide more advanced features.

First, Dublinked is working to complement its more conventional open data features with a so-called 'Research Zone': a specific area in the platform with more 'polemic' and technically complex datasets, licensed and with potentially higher commercial value (such as data streams of traffic, energy and water flows). The access to this area is membership-based, meaning that researchers, businesses, entrepreneurs and governmental organisations can access it through the payment of a fee.[7] This *open data 'plus'* feature of Dublinked recognises that big data from critical infrastructures (e.g. water, energy, transport) has strong operational sensitivity and can never be fully open, but can be shared (e.g. through the release of sample data streams). Larger innovations and commercial opportunities rest on this sort of data, and thus the Research Zone becomes a virtual area where innovators can adventure, draw inspiration, develop projects and motivate new research collaborations. Figure 3.1 illustrates the functioning of Dublinked and its two zones: conventional open data plus the Research Zone.

Second, and notwithstanding the membership fee for the Research Zone, Dublinked's data is released on a *neutral and equity basis*. Even if the initiative was motivated by the

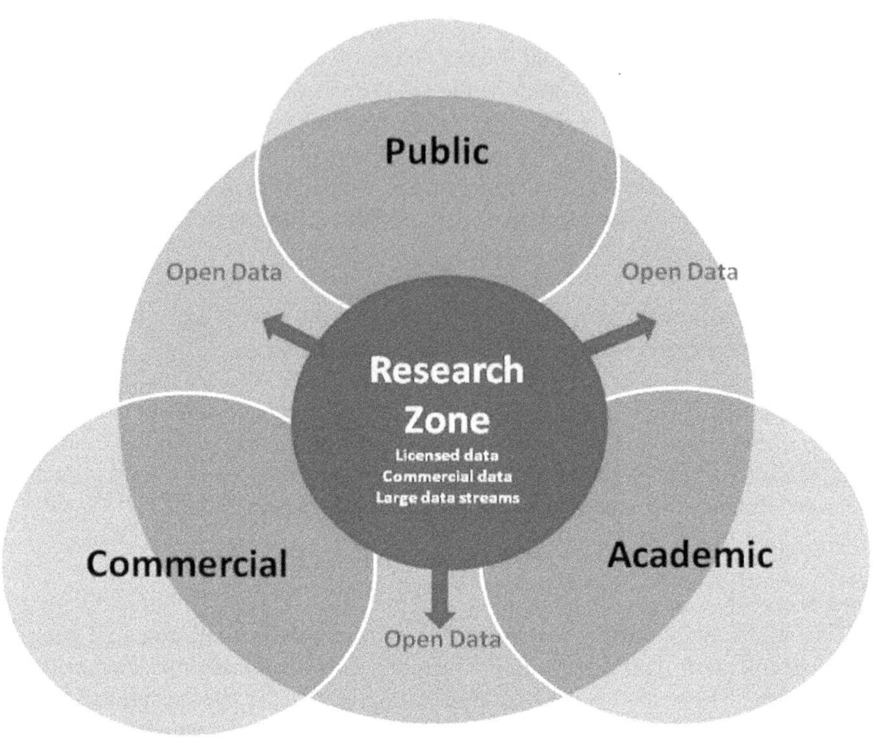

Figure 3.1 Dublinked, open data and the Research Zone.

Source: Dublin City Council – The Studio (2012)

demands of large companies, Dublinked avoids the risk of privileged treatment by making data open to everyone interested. Moreover, Dublinked is not involved with the monetisation or 'royaltisation' resulting from the development of private intellectual property (IP). Dublinked only and exclusively prepares, hosts and shares data, which can (or not) be used for innovations. Dublinked is also not about procurement: there are no specific requests directly made to companies, even if public bodies can indirectly learn about available capabilities through Dublinked; moreover, eventual access to more complete data streams or sell-out/commercialisations of applications has to be made directly between the developer and the original data owner (e.g. traffic authority). The Dublinked platform also has tight privacy features – in the Research Zone users can see who other registered users are, but not which data they visualised or downloaded, in order to avoid competition-sensitive issues.

Third, Dublinked is not only about the technical platform (software) and the data. Dublinked's proponents regularly host diffusion and networking events for potential data users and providers, as an integral part of the strategy. They do so in order to 1) raise awareness; 2) get feedback from the data-users (e.g. which data they want to see, in which formats); and, importantly, 3) to facilitate the development of new networks of innovators, communities and R&D projects around the data and the city's challenges. Dublinked events and workshops provide good opportunities to informally nudge issues and organisations and broker new partnerships.

4.1.3 Start-up challenges and solutions

As a rather innovative and unconventional initiative, Dublinked sits outside the traditional 'comfort zone' of the City Administration. Projects with such characteristics tend to fail in traditional cost-benefit and risk assessment analysis; moreover, they hardly fit the established regulatory environment. There were significant resistance and barriers before launching the initiative (a 'you can't do it' feeling). So, what were those barriers and how were they overcome?

First, Dublinked raised significant legal and regulatory challenges. Before the start-up stage, the design of the partnership and the relationship with IBM were particularly problematic. IBM was strongly involved since the early beginning and helped to shape the vision, but directly teaming up with IBM would constitute, in the eyes of EU law, a form of privileged treatment and unfair advantage (e.g. for future procurement tenders). In this case, an acceptable solution was to establish the core partnership between DCC, NUIM and three county councils, while signing a complementary agreement with IBM for the implementation of parts of the platform, under many legal clauses that could counter the regulations' straightjacket. The role of the City Manager was pivotal in the process, linking and bridging different interests and potential partnerships towards the development of a jointly viable solution.

Second, there were (and still are) legal and technical issues related to the more complex datasets (e.g. data streams). Thus, the launch of Dublinked had to be done with the already available, easier-to-collect datasets. Starting up with such data allowed the momentum to be maintained while demonstrating the business case for open data – the difficulties involved with larger data streams would have raised too many difficulties from the early beginning. However, in order to pool together a significant amount of data from the beginning, the metropolitan cooperation of the four councils in the Dublin region was very relevant in bringing the initiative to an adequate scale.

Third, many City departments (e.g. within DCC) were hesitant about releasing their own data ('it will embarrass us'; 'people will pick up holes'). An important challenge was thus to identify the right department in the City to act as 'central facilitator' for sharing the data and driving internal processes, ensuring cooperation, dialogue and liaison with other departments, within and outside the City. A first try was made by DCC's IT department but it was not the most appropriate, namely due to its orientation towards formal and structured projects and 'inside-the-box' thinking. To solve this issue, the City Manager successfully assigned the task to a newly created and more flexible 'design-thinking' unit of DCC ('The Studio'), with less standardised routines and strong communication skills.

4.2 Implementation

As referred to, Dublinked was formally launched during 2011 and it is thus (at the time of writing, December 2013) at an early implementation stage. However, there have already been already promising achievements (see Section 5), and the initiative is steadily running. So how has it been implemented?

The implementation of Dublinked has been run by its founding members, with the technical support of IBM. DCC set the partnership and mobilised the necessary networks, taking the lead in coordinating data collection and releasing efforts with other partners. All the councils (or local authorities) have worked together to provide data and actively cooperate in overseeing, defining and implementing the vision of the initiative. Dublinked's coordination, management and customer-facing portal services are in the hands of NUIM, who among other tasks is responsible for the front-end of the platform, data curation and the organisation of networking events. Finally, IBM is formally involved by providing the back-office of the platform (structuring the data, query tools, etc.).[8] Yet, beyond this partnership, what critical enablers have made it happen, and how? Which challenges still lie ahead?

4.2.1 Leadership

During the early implementation stages, namely when the initiative's design was mainly 'in the people's minds', leadership played a key role. The initiative started with a small group of persons within DCC, namely the City Manager John Tierney and his senior assistants within the Economic Development Unit. In this case, beyond the necessary *inspiration* and *passion*, leadership was key to ensuring 'connectedness', both outside and inside DCC.

First, the leadership of DCC (John Tierney) has been relevant to keep up the momentum and re-shape the partnership as difficulties emerged, namely 1) by challenging the legal status-quo and 2) by being able to quickly involve and convince high-level contacts in the other county councils, NUIM and IBM, whose relations had been established before in other settings and strategic thinking platforms (e.g. the Creative Dublin Alliance).

Second, high-level leadership was central to legitimate such an 'out-of-the-box' initiative within the organisation, for example by 'making sure' that departments would share their data for fast implementation. Since the project was backed at such a high level within DCC, it could overcome the control of more bureaucratic departments, delegating responsibilities to 'lighter' departments (The Studio – see Section 4.2.4), with less crystallised routines and interests, making it easier to speed up and smooth the implementation process internally, namely by establishing faster and more informal communication channels. By being backed at a higher level, a group of executive managers (with the right technical and communication skills) could emerge and act as central facilitators within the organisation, driving the process forward on a daily basis. At the current implementation stage, high-level City managers 'moved back' and left the executive implementation in the hands of such managers.

4.2.2 Finance

So far, Dublinked has been running on a rather small budget (under €100,000 per annum) and, therefore, finance has not been (yet) a critical issue in the implementation of the initiative. This amount goes to website and data hosting, plus the organisation of small events such as diffusion and workshop meetings, 'data camp' events and competition prizes.[9] Staff in each of the four councils collect the data, NUIM curates it and IBM does the technical management of the platform and advises on data issues. The lion's share of the budget currently comes from NUIM which pays for its staff from central research funds as an investment towards this new infrastructure – however these funds are limited. The Dublin regional authority has recently contributed to the local authority costs for this project.

Apart from the start-up costs, there will be continued costs in running the innovation events and workshops, the operation of the legal agreements for those accessing the Research Zone, and in assisting data providers to release their data. Thus Dublinked considered asking members for a small membership fee to support these activities, based on their size and ability to pay, resulting in differentiated membership fees (see Table 3.1). Access to the data will remain free.

As referred to, Dublinked does not collect money or royalties from IP development. This would be undesirable for the innovators and could result in a loss of interest in the initiative at an early implementation stage. Moreover, it would be undesirable for Dublinked as well,

Table 3.1 Membership fees per annum

Membership type	*Fee (€)*
Large company (>1,000 employees, >100M€ turnover)	5,000
Medium-sized company (>100 employees, >10M€ turnover)	2,500
Small company (trading)	250
Start-up company (<5 employees, no turnover, less than 3 years)	100
Individual	100
Data provider only	0
University or other third-level institution (with funded research)	2,500
University or other third-level institution (without funded research, and certified)	250
Charities	0

Source: Dublinked (n.d.)

since that would raise many legal issues, block 'early wins' and slow down the implementation overall. Representatives of DCC and Fingal consider that there are other more appropriate financial streams to recover the investments, such as taxes, local service improvements, etc. Moreover, according to one interviewee, 'we get for free the apps that otherwise we would have to develop ourselves'.

4.2.3 Organisation and networking

The implementation of Dublinked has been largely benefiting from the involvement of a diverse yet limited set of stakeholders. As previously mentioned, the shape of the partnership was important to overcome legal (state-aid regulation) issues. Additionally, it largely contributed to bringing complementary resources to the initiative.

First, as already referred to, the partnership between the four local authorities helped Dublinked to gain scale (more scale-relevant data available, more 'regional feeling'), which was essential bearing in mind its innovation and (regional) problem-solving objectives. Moreover, the council of Fingal brought in their previous experience with open data platforms and subsequently the possibility of showcasing concrete evidence of its potential (e.g. the apps already developed out of Fingal's data).

Second, NUIM brought their reputation and expertise in the field, assuming the technical leadership and coordination of the initiative; as a trustworthy player, it could also give a more independent feeling to the initiative, showing that it was not something that 'belonged' exclusively to the councils but was open to society and to innovators.

Third, despite the legal blockages hampering more direct involvement, IBM has been playing a key role in the implementation of Dublinked (in addition to its emergence and design, already analysed in Section 4.1). IBM is formally responsible for the technical back-office and design of Dublinked's platform; yet, their effective role goes way beyond this.

On the one hand, the data 'requests' of large players such as IBM signal the types of relevant (i.e. profitable) data – e.g. specific types of water, planning, traffic or energy flows. However, since Dublinked runs on an equity basis and free access, once collected the data becomes available for everybody in the platform and not only for the companies who could afford (and have enough power) to access it. Therefore, out of their own interest, IBM has contributed to spur the diffusion of relevant data and competition for the 'best innovations' in the region.

On the other hand, IBM (and other large companies) can contribute to the fast prototyping of solutions and are well positioned to demonstrate the business case of open data. They are also better positioned to make the connections between open data and other city applications and to create advocacy for open data benefits around the world, through their extensive business network.

Finally, a central organisational feature in the implementation of Dublinked refers to user involvement – the final beneficiaries of the initiative ('quadruple' helix). Who will use the data collected? How will it be used? How can the data collection and release be improved? To get answers to these questions, Dublinked organises different types of events, not only to speed up the implementation but also to improve the initiative.

One popular type of event is the thematic workshop e.g. on opportunities and challenges in using spatial data. These are attended by Dublinked managers, data providers and an emerging community of open data users. There are presentations from experts, users and feedback discussions about the quality of the data, and new challenges are raised. Moreover, such events constitute networking platforms where ideas on the data are exchanged and new projects and cooperation may emerge.

Other popular, more hands-on types of events and user-involvement platforms are 'hack-days' or 'data camps' and 'app development contests'. During these events, which can have different durations, teams of IT developers (eventually together with designers and marketers) are invited to use the available data provided to develop applications and solutions out of it, and are awarded prizes. Such events are popular within the IT community in Dublin. Beyond the technical dimension, these events are relevant networking moments as well as cradles for new projects and commercial partnerships. They also contribute to create open data advocacy. To organise such events, Dublinked has been actively collaborating with the National Digital Research Centre and more recently with Dublin's Science Gallery (see Box 3.3).

Box 3.3 Data camps and hack-days

Dublinked was informally prototyped during a 'data camp', jointly organised with Inventorium (an initiative to support the development of new innovative businesses, hosted by the National Digital Research Centre in Ireland). During 48 hours, 20 (yet to be published) Dublinked datasets and data from Fingal Open Data were 'given' to a community of 120 developers (e.g. from the Irish Internet Association) to see which potential uses and solutions would emerge out of that, in an 'open-innovation' fashion.

A follow-up yet more elaborate event took place in late 2012, jointly organised by Dublinked, IBM and the Science Gallery as part of the exhibition *Hack the City*. It was a specific 'idea camp' running on Dublinked data held in a vacant city-centre derelict building. It received more than 600 worldwide applications. Out of those, five teams were selected to participate in the camp. Each team was given €1,500 to participate in a ten-day idea camp (one orientation day on Dublin's problems and issues; a four-day workshop to develop the solution, during which the general public could enter the venue and see the teams working; two days to fine-tune with feedback from the users, e.g. people from the transport authority; two days refining and prototyping; one day of presentation to venture capitalists). Some broad topics were previously defined as a starting point for the idea development: dereliction, crowd sourcing in public services, safety and well-being, and open data transport services.

4.2.4 Communication

Communication skills played an important role in the implementation of Dublinked, namely to ensure that relevant data could be easily collected with the council administrations, and that proper liaison was established among external-to-Dublinked stakeholders (e.g. collecting users' feedback – see above). Internally to the public administration, there were concerns that data would reveal their flaws. Open data was simply too big a step for many bureaucratic units within and outside the City.

Under this context, the actions of a new unit within DCC called The Studio (directly reporting to the Dublin City Manager, see Box 3.4) were important to better communicate the project's aims, reducing resistance within the administration. As a new design-thinking unit, it lacked the crystallised routines and vested interests of other City departments, making it easier to establish cooperation with other units and communicate with the general public. The Studio has been so far a good fit for the project's implementation (e.g. collecting and preparing data to be released), namely because open data was new to the City and to DCC, requiring a combination of knowledge management with a customer-oriented perspective (the future users of the data). Together with The Studio's coordinator, there are managers within each department responsible for collecting and authorising the release of data. Each manager was required to release 30 datasets for the launch phase.

In the beginning, data collection and releasing was done in a flexible way, with limited concerns for integration issues. Moreover, the first datasets were relatively 'low-hanging fruit': no personal data, no privacy concerns, easy to collect. This had a communication rationale as well, since starting with too complex data would bring more barriers, making it difficult to demonstrate and showcase the value and the possibilities of open data. Moreover, Dublinked's early implementation has been benefiting from a number of apps already developed under the recent Fingal's app contest, which are the 'physical' evidence of the potential value of open data.

Box 3.4 The Studio – innovation and design-thinking unit within DCC

'The Studio' is a recent unit of DCC, directly established under the responsibility of the City Manager in order to generate and prototype new solutions and work at engaging staff and customers using design-thinking tools. It is staffed by a multi-disciplinary team of designers, planners and architects, among others. The Studio started as a way to internalise the knowledge which had been recently acquired by seven City officers in two nine-month collaborative projects called Designing Dublin: Learning to Learn and Love the City. As well as Dublinked, The Studio is involved in service design projects and partnership projects with Irish and other universities and the Science Gallery. A key role is the creation of new ideas to solve city problems.

The Studio invites DCC staff from multiple levels in the joint pursuit of new ideas that would benefit the City. Included in the staff ideas that were prototyped in 2011 were the design of new products, information and knowledge sharing and ways to save city resources. Subsequently a new round of idea prototyping took place. The studio supports people to test their ideas across the organisation. The City Manager has been backing the functioning of The Studio even in a context of strong financial pressure and budget cuts within the City.

4.3 Results

What are the direct and indirect results of Dublinked so far? At the time of writing (May 2012), Dublinked is still in an early development stage; therefore, 'hard' results cannot be fully assessed yet. Moreover, the philosophy of Dublinked and open data, by involving shifts in established mind-sets is *per se* a more long-term endeavour, which takes time to run and show its full potential. Open data is not a quick win.

However, and having said that, it is already possible to identify a number of promising first achievements. By doing so, it becomes clear how Dublinked is starting to contribute to improve Dublin's structural assets or 'capitals', towards a new stage of more sustainable competitiveness.

First, although not initially developed under the umbrella of Dublinked, Fingal's open data initiative (recently federated into Dublinked) gave rise to 36 new app ideas and, out of those, 21 apps and solutions were developed by small companies and entrepreneurs involved in the 'apps4fingal' contest. Most of them are web-based, but others run on mobile gadgets (two-thirds run in Android and one has reached the iPhone platform). One example is 'Fingal day tripper', which used the available data to provide a number of tourism routes and places to spend the day, linked with existing amenities. Another is the (already popular) mypp.ie, an application using planning data for the whole Dublin region that identifies on a map where particular types of developments (roads, buildings, infrastructure, etc.) are being planned. So far, namely because of the data already released (more large flow-based datasets are still to come), most of the developed apps are focused on modest improvements in local service provision;[10] yet they already contribute to paving the way and communicating the potential of open data and Dublinked.

Second, another indication of the early success of Dublinked is the number of visitors and data downloads. Dublinked reported that in the first six months of implementation, the platform and datasets had more than 6,600 views, from 46 countries and 250 cities, mainly in Europe and the US but also in Asia and Australia. However, perhaps a better indicator of the interest and effective demand is the number of downloads per dataset, which already reached the thousands during the first month of implementation. Figure 3.2 illustrates the 12 most downloaded datasets, with water-related, traffic and planning data at the top. A dataset on the journey times across Dublin City has also generated much interest.

Third, the contribution of Dublinked to the creation of new communities of innovators around open data in Dublin and Ireland can also be accessed by the number of participants in:

- the 'data camps' (120 participants during the launch of the initiative; 600 tenders for the 'hack-the-city' day, from all around the world);
- the Dublinked technical workshops. Qualitative information from our fieldwork shows emerging evidence of new project networks and innovation communities, with the support of Dublinked and other partners (e.g. Inventorium, Science Gallery) as brokers and boundary spanners, breaking silos, creating 'buzz' and helping entrepreneurs and smaller partners to plug into new innovation networks. From October to December 2013, Dublinked organised 13 events; in excess of 670 people attended these events and the Dublinked mailing lists contain over 700 names.

Fourth, by providing access to relevant data and communities of innovators, Dublinked is contributing to reinforce the position of the local subsidiaries of leading firms *vis-à-vis*

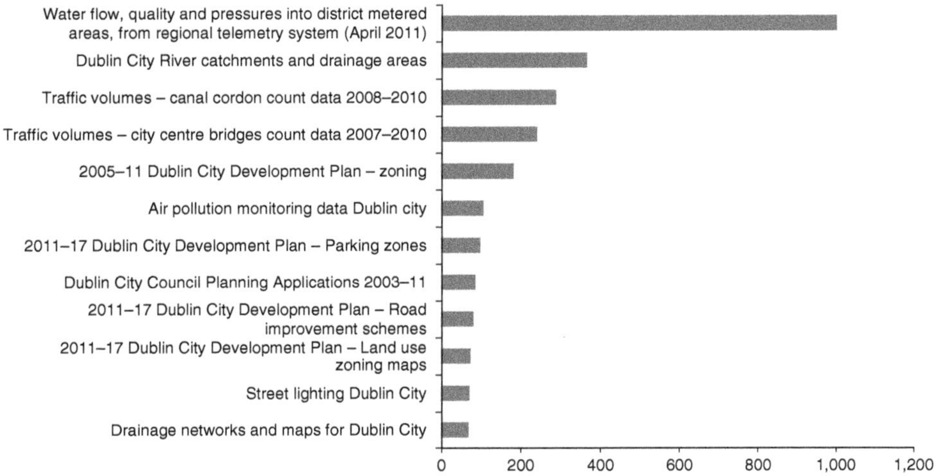

Figure 3.2 Dublinked, downloads per dataset (top 12), 28 October 2011.

Source: Adapted from Dublinked/IBM (2011).

other global R&D departments and headquarters. Dublinked and the committed support of DCC in releasing relevant data are important resources in the global competition for resources and intra-corporate R&D investments. The announcement in March 2010 of Dublin as the location for IBM's global 'smart-city' lab (200 R&D jobs) is a good indication of the value for the city of open data and the development of supportive networks of innovators.

Fifth, Dublinked is contributing to better governance between public authorities and to raising awareness about the value of open data. During the first months of implementation, the number of datasets released by the four city councils doubled. Associated with this are better data standardisation procedures, and importantly, better inter-department and inter-institutional communication. Since the inception of Dublinked (October 2011) up to the time of this writing (December 2013), the range and number of datasets published on the Dublinked data store has increased significantly and the data publishers, or contributors, have expanded beyond the local authorities to include national bodies, such as the National Transport Authority and the Railway Procurement Agency, among others.

Sixth, Dublinked has contributed to form new advocacy for the development of an open data movement at the national level. Since the launch of Dublinked, open data has been increasingly seen as a major driver for economic growth, improved transparency, accountability and performance of public services through more informed policy development for the Irish Public Service. Government policy statements, such as the eGovernment strategy of April 2012, have shown national-level commitment to open data. The Irish Department of Public Expenditure and Reform established the Office of the Government Chief Information Officer in 2013. For example, Ireland is expected to officially join the Open Government Partnership (OGP) in April 2014 and will host the OGP European regional meeting in May 2014. As part of this, the Minister for Public Expenditure and Reform has announced a number of measures to accelerate open data in Ireland.

5 Conclusions

5.1 Broad assessment

Dublinked is an ambitious, state-of-the-art open data initiative, developed in partnership between the four city councils of greater Dublin region and the National University of Ireland Maynooth (NUIM). The development of Dublinked can be seen as a response to the recent challenges confronting Dublin in the face of the global economic downturn and financial meltdown, tapping into the city's key structural assets and capitals: 1) strong IT innovation ecosystem; 2) excellent knowledge base; and 3) increasing capacity to organise economic development initiatives at the local level.

Dublinked's ambitions changed over time, but they always found a good fit within the City's general vision and Economic Development Action Plan. This has catalysed the initiative and proved important when gathering political support to implement it. Although Dublinked aims at improving the efficiency and quality of public service provision in Dublin, its main aims have always been local economic development, innovation and growth, namely nurturing new technologies and solutions grounded in the use of publicly released data. This makes Dublinked distinct *vis-à-vis* the more conventional initiatives emerging worldwide, which are primarily focused on transparency, accountability and citizen participation.

The first years of implementation of Dublinked reveals promising achievements, such as the development of new communities of innovators, cultural change (e.g. releasing data), the national growth of the open data movement and the expansion of corporate R&D labs to explore and prototype new solutions out of urban-related data. Such results do contribute to improvements in Dublin's sustainable competitiveness. However, these achievements should be seen in the context of a more long-term initiative, not a quick fix. There is a substantial learning dimension involved. Larger and more transformational impacts on the city's assets and 'capitals' will take time to materialise, but the potential pay-offs are worth the challenge.

5.2 What have been critical factors?

Despite Dublinked's early age, a few critical success factors can be already identified.

First, the push of IBM and other leading firms interested in open data was important to ignite the initiative, showing the relevance of the data and the opportunity for the city. The requirements of such companies helped to set priorities on the most 'valuable' types of data to collect and release. The involvement of leading firms and their knowledge was also relevant in shaping the design of Dublinked. Yet, Dublinked's proponents also recognised that too much private involvement could result in rent seeking and unfair advantages *vis-à-vis* other smaller players – thus making the data accessible on a neutral basis, combining the best of two worlds: private push and equitable access.

Second, the kick-off and early implementation of Dublinked largely benefited from the leadership of a high-level city official (the City Manager) and his senior assistants. We found this leadership capacity associated with four key dimensions:

1. *formal position* (e.g. important to legitimise such an 'out-of-the-box' initiative within the bureaucratic structures of DCC, and to empower a recently created City Unit to coordinate it);
2. *connectedness* (e.g. to form a high-level partnership to steer the project, bridging interests between high-level representatives of other city councils and the university);

3. *capacity to dare* (e.g. to go against the status-quo and cultural blockages while challenging legal interpretations; investing in uncertain projects in times of austerity); and, finally,
4. *capacity to step back* (e.g. follow at a distance but leave the implementation power in the hands of others).

Thus, leadership here was not so much about ruling and controlling, but much more about legitimising, nudging, connecting and daring.

Third, and related to the above, was the role of the university (NUIM) in acting as a neutral independent party between public and private, contributing to solving many of the issues and challenges faced during the start-up process. Namely when finance was an issue, the university stepped up making an important contribution, which was critical in terms of the overall progress.

Fourth, finding and agreeing on the right scale for the initiative (Dublin region) was a strong plus. That scale helped to collect data at the relevant scale for the users, and also to start up with a large and more consistent amount of data. Moreover, it facilitated learning from the more advanced open data initiative of Fingal.

Fifth, the project largely benefited from The Studio as the anchor unit within DCC and as the connection with other partners and organisations outside DCC. The Studio had the right user-involvement thinking, knowledge management and communication skills necessary to convince departments within the organisation to release their data, doing this through faster communication channels than more bureaucratised units would allow. However, the best anchor unit seems to vary across organisations – for example, in Fingal the IT departments excelled in performing the same tasks which failed in Dublin.

Sixth, the launch and early implementation of Dublinked benefited from smart *prioritisation*. It started with the collection and release of less complex datasets, with limited privacy issues, technically easy to deal with, and almost exclusively from the public sector. This helped to generate early enthusiasm and give a good dimension to the data repository, while testing the user's feedback (paving the way for a second, more demanding stage: the Research Zone). Starting with the most complex and controversial pieces would slow down the process and create many blockages from the early beginning. Overall, it was about the capacity to select and pick the 'low-hanging fruit'.

What can be said about the transferability of Dublinked and its success factors to other urban contexts? Other cities willing to develop open data initiatives can certainly learn from Dublinked's early experience and general success factors, but should bear in mind that Dublin provided a particularly good 'soil' for the development of such initiatives. Particularly relevant here is the strong alignment between an 'advanced' local economic vision, proactive leaders and the economic structure of the city (an almost unrepeatable concentration of large and knowledge intensive IT companies). Moreover, Dublin's size (not too big, not too small) and its excellent knowledge capital provides fertile ground for the development of new ecosystems of innovators (e.g. easiness to connect, knowing 'who is who', pushing new projects and interactions, etc.).

Notes

1 See, for example, van Winden *et al.* (2012) and Cudden (2011) for recent and detailed city profiles.

2 A good business climate (e.g. tax) and (post-crisis) moderated office and property prices helped, but the key driver still seems to be the presence of a highly qualified labour pool.

3 As of December 2013, the strategic axis framing Dublinked has evolved, yet retains the key ambitions, namely, in this order, job creation and innovation, product and service development, global leadership in the field and supporting other domains such as eHealth or eGovernment. Overall, the new axis suggests a stronger emphasis on global ambitions while stressing further the economic development and innovation drive of the initiative.
4 As a relatively small university, NUI has always tended to branch out and cooperate with external stakeholders.
5 Even if there are slightly different, more cautious visions within Dublinked promoters (some would be happy if it would simply help to better organise the data within the City Council), this vision and aims are broadly supported by the proponents.
6 'Polemic' datasets are, for example, those that can raise privacy issues (e.g. allowing individuals and companies' behaviour to be identified), legal issues (e.g. ownership of valuable data) and/or are technically difficult to be made accessible (e.g. large data streams).
7 Such a fee was designed to cover the larger operational costs of running the Research Zone; moreover there are different scales according to the types of members (e.g. large companies should pay €5,000 per year, more than universities (€2,500) and individual users and entrepreneurs (€100). The philosophy is that the Zone has controlled access but it should not be limitative.
8 The efforts of these organisations have been divided into several working groups, namely a regional group (with each City Council's anchor man/woman), a larger coordination group and a technical group.
9 For example, for Fingal's 'app contest', the combined prize fund was of €11,500, shared by the Fingal County Council (FCC) and other sponsors.
10 There are many difficulties in measuring the value of newly developed apps out of open data. For example, should the impact and value be based on the cost of producing them? Or based on the value of take-off? Or based on the number of users? IBM is currently developing with Dublin and Fingal new methods to assess the effective value of such apps.

References

CDA – Creative Dublin Alliance (2009). *Economic Development Action Plan for the Dublin City Region*. Dublin: Dublin City Council.

Cudden, J. (2011). *Dublin Economic Profile*. Overview developed for the project 'Cities as Engines of Sustainable Competitiveness'. Rotterdam: Euricur, Erasmus University Rotterdam.

Dublin City Council (2011). *City Development Plan 2011–2017*. Dublin: Dublin City Council.

Dublin City Council – The Studio (2012). Dublinked, open data and the Research Zone. Dublin: Dublin City Council. Accessed at http://scribd.com/doc/81709459/PDF-Pivot-Dublin-Interview-with-Deirdre-Ni-Raghallaigh (12 May 2012).

Dublinked (2010). Dublinked Memorandum of Understanding. NUIM, Dublin City Council, South Dublin County Council, Fingal County Council and Dún Laoghaire Rathdown County Council.

Dublinked (n.d.). Dublinked – Membership fees. Accessed at http://www.dublinked.ie/?q=node/31# (12 May 2012).

Dublinked (2013). Dublinked – Vision. Accessed at http://www.dublinked.ie (12 December 2013).

Dublinked/IBM (2011). Dublinked – downloads per data set at 28 October 2011.

IDA Ireland (2012). End of the Year Statement (by CEO Barry O'Leary). January 2012. Dublin: IDA Ireland.

NYC Global Partners (2011). *Best Practice: Creative Dublin Alliance – Creating the Leadership Consensus*. 19 July. New York: NYC Global Partners. Accessed at http://www.nyc.gov/html/unccp/gprb/downloads/pdf/Dublin_CDA.pdf (19 March 2012).

Science Foundation Ireland (2010). *Annual Census 2010*. Dublin: Science Foundation Ireland. Accessed at http://www.sfi.ie/news-events/publications/orgnisational-publications/ (15 May 2012).

Van Winden, W., Carvalho, L., van Tuijl, E., van Haaren, J. and van den Berg, L. (2012). *Creating Knowledge Locations in Cities*. Abingdon: Routledge.

4 Integrated Housing Development Programme (Addis Ababa)

Alberto Gianoli and Alexander Otgaar

1 Introduction

This chapter describes and analyses the Integrated Housing Development Programme, implemented since 2005 in Addis Ababa with the objectives of drastically increasing the supply of housing, creating employment, promoting the development of micro and small enterprises, regenerating inner city slum areas, and promoting homeownership for low-income households. Since the 1970s Addis Ababa has been characterised by an acute shortage of residential housing, particularly for low- and middle-income households. This housing deficit has been growing in the recent past as a result of population growth, migration to urban areas and the dilapidation of the existing housing stock due to poor maintenance. The Integrated Housing Development Programme was developed against this background.

The Integrated Housing Development Programme, one of Africa's most ambitious, is characterised by three innovative aspects. First, the programme is based on a large-scale approach to addressing the current housing deficit, the poor quality of the existing housing stock, and future housing needs due to continued urbanisation. Second, the programme explicitly focuses on low- and middle-income households and their private ownership and aims at greatly improving their living conditions, security of tenure, and access to basic services. Third, and most importantly, the programme is based on an integrated approach directed at jointly stimulating the economy, creating employment and improving the capacity of the construction and financial sectors. The chapter, while showing that the Integrated Housing Development Programme has been effectively implemented and has proved to be a valuable tool for affordable housing delivery at a large scale, also highlights some of the key challenges that it has faced.

The chapter is structured as follows. Section 2 analyses the structure and recent dynamics of Addis Ababa across a number of dimensions or 'capitals' for sustainable competitiveness. Section 3 outlines the city's broad vision and strategy. Section 4 constitutes the core of the chapter and analyses the Integrated Housing Development Programme implemented in Addis Ababa, focusing on its content, implementation process and results. Section 5 concludes with some lessons learned and the challenges ahead.

2 City profile: capitals for sustainable competitiveness

2.1 Performance

Addis Ababa, 'flower' in Amharic, became the capital city of Ethiopia in 1889 under the reign of Emperor Menelik and later, under Emperor Haile Selassie, experienced a higher degree of development compared to its surrounding regions. Since its establishment it has undergone many changes in terms of size, population, spatial organisation and economic structure. During the socialist regime period, from 1974 to 1991, Addis Ababa experienced a phase of low urban development and economic growth. This situation has radically changed in the last two decades which have been characterised by, among others, a more prominent even though still limited role of the private sector in the development processes of the city. Very significant has been the impact of the Addis Ababa City Development Plan 2002–2010 (City Government of Addis Ababa, 2002), which has supported economic growth, particularly through private sector and infrastructure development, and has initiated a city management reform to improve its governance arrangements.

Today, with a total land area of some 540 square kilometres, Addis Ababa is the largest city in Ethiopia and, situated on a plateau at an altitude of 2,408 metres, is the third highest city in the world, after La Paz and Quito in Latin America. The population is estimated at about 4 million people, which is approximately 4 per cent of Ethiopian population and less than 30 per cent of the country's urban population. The city is inhabited by numerous nationalities with foreigners constituting less than 1 per cent of the population. The major ethnic group is the Amaras (49 per cent of the population), followed by Gurages (18 per cent) and Oromos (17 per cent). Addis Ababa is the federal capital of Ethiopia and a chartered city and is the political, commercial and cultural centre of the country (UN-Habitat 2008, 2010).

It should be noted that the rate of urbanisation in Ethiopia is still relatively low. It is only over the past decade or so that a considerable wave of migration has led to the urbanisation of poverty, with the consequence that high urban unemployment has became a key issue at the forefront of the political and policy agenda. The fact that the city economy has been unable to generate enough jobs, coupled with the continued influx of young semi-skilled and unskilled job seekers, has created a significant burden on the job market. These dynamics explain why in recent years in Ethiopia there has been a clear shift from a development approach essentially focused on rural development to the realisation of the importance of the urban dimension, and in particular of the following reform pillars: land, governance, housing, and job creation.

Over the past decade, Ethiopia has attained high economic growth, averaging 10.7 per cent per year. In 2012, Ethiopia was the twelfth fastest growing economy in the world. According to the latest *Ethiopia Economic Update* report (World Bank, 2013), if the country is able to sustain its growth performance, it could potentially reach middle-income status by 2025. This, in turn, may require an adjustment in economic policy to phase in the private sector as an additional engine of growth. To date, the growth strategy which has been followed has emphasised a strong expansion of public investment. By way of illustration, the public investment rate of Ethiopia is the third highest in the world, while the private investment rate is the sixth lowest. Unfortunately, there are no data available on the rate of growth of the urban economy of Addis Ababa. However, as the capital and largest city of the country, it is reasonable to assume that its rate of growth and growth strategy are in line with those at national level. According to the *Global Urban Competitiveness Report* (Ni *et al.*, 2013), which measures the ability of a city to attract and transform resources, to control and dominate the market, thus creating more wealth for its citizens, Addis Ababa ranked 445 in 2010 and 435 in 2011.

As shown in Table 4.1, Addis Ababa is expected to be one of the fastest growing large cities in Africa in terms of population over the period 2010–20.

Regarding infrastructure provision as measured by the percentage of households with access to water, sanitation and electricity, Addis Ababa has achieved a reasonable performance when compared with other large African cities (Table 4.2).

Table 4.1 Absolute growth projections in selected African cities

City	*Absolute Growth (2010–20 projections)*
Kinshasa	4,034,000
Lagos	3,584,000
Luanda	2,308,000
Dar Es Salaam	1,754,000
Nairobi	1,669,000
Ouagadougou	1,548,000
Cairo	1,539,000
Abidjan	1,375,000
Kano	1,100,000
Addis Ababa	**1,051,000**

Source: UN-Habitat, 2010

Table 4.2 Infrastructure provision

City	*Improved water*	*Improved sanitation*	*Electricity*
Dar Es Salaam (2004)	81.1	55.6	59.8
Kampala (2006)	92.6	100.0	59.0
Lagos (2003)	88.2	72.8	99.8
Maputo (2003)	82.8	48.8	28.8
Nairobi (2003)	93.3	82.9	71.4
Addis Ababa (2005)	99.9	71.8	96.9

Source: UN-Habitat, 2010

2.2 Economic structure and knowledge

Addis Ababa is the hub of major services, industries and commercial activities of the country and attracts the lion's share of the national investment capital. In terms of real estate development, Addis Ababa has a share of over 60 per cent of the overall real estate investment. It provides the major transportation links between the different parts of the country. Nevertheless, the city still faces a host of problems, mainly related to inadequate infrastructure, public services and housing. The public sector accounts for the bulk of urban formal employment and much of the private expansion is in microenterprises. A large part of the local economy is informal and there is a high level of poverty. The urban labour market is segmented into a relatively high wage public sector, a small private formal sector, and a large informal economy (City Government of Addis Ababa, 2005; World Bank, 2011).

According to the limited data available, while up to about 2004 urban economic activity overall grew slowly, in more recent years there has been an acceleration of urban economic growth, particularly in the banking and insurance, transport and communication, and construction sectors. According to World Bank (2013) data, over the period 1999 to 2005, growth in value added in industry of almost 3 per cent per annum was eroded by productivity falling by 0.6 per cent per annum, whereas in transport and communications a significant increase in productivity has occurred, with output increasing despite little growth in employment.

These trends are consistent with the gradual emergence of Addis Ababa as a services centre. However, even with the acceleration in activity, especially in services, urban job growth seems to have slowed overall, with a consequent rise in urban unemployment. The unemployment rate, which is exacerbated by rural–urban migration dynamics, was around 24 per cent in 2005, well above the national average of 13.5 per cent and that of the next largest city Dire Dawa at 18 per cent (Haregewoin, 2007; World Bank, 2011). The 2011 Urban Employment Unemployment Survey (CSA, 2011) shows that the highest level of unemployment in Ethiopia at 25 per cent is recorded in Addis Ababa.

2.3 Social conditions

Addis Ababa is one of the fastest growing cities on the continent (see also Section 2.1 above). The number of inhabitants has nearly doubled every decade: in 1984 the population was just below 1.5 million, in 1994 it was 2.1 million and it is currently estimated to be 4 million. According to UN-Habitat, this number will continue to rise, reaching 12 million in 2024. It should be noted that Addis Ababa's growth rate has been moderate in comparison to smaller towns as is reflected in its declining share of the total population in urban areas from 40 per cent to 30 per cent over the past 15 years. The geographic location of the city, combined with its political and socio-economic status, has created a strong pull factor for urban–urban and rural–urban migrants in search of employment opportunities. This has exacerbated problems related to the high rate of unemployment, concentration of slum dwellings, poor housing, and lack of infrastructure and sanitary facilities (Haregewoin, 2007; UN-Habitat, 2008; City Government of Addis Ababa, 2010).

Some basic data illustrate the extent of the issues the city is facing. More than 25 per cent of the houses have no toilet facility (without counting informal dwellings), 33 per cent of households share toilet facilities with more than six families, 29 per cent have no separate room for cooking, 34 per cent of the residents depend on water from frequently interrupted public taps. A high volume of wastage due to faulty piping also contributes to water shortages. The existing sewerage system is inadequate. On a more positive note, health service provision in Addis Ababa has improved in the recent past. The current health coverage in the capital has increased to 85 per cent, the distance travelled to find a health facility is less than 2 kilometres, and vaccination coverage is 95 per cent. These positive results in the health sector are a direct consequence of the involvement of the private sector which has been playing a substantial role in improving the physical coverage, even though it should be pointed out that this has raised issues of affordability (Haregewoin 2007, UN-Habitat 2008, City Government of Addis Ababa 2010).

2.4 Environment

Addis Ababa faces major problems related to the quality of its urban environment, particularly due to the magnitude of informal settlements. The water, soil and air qualities in these areas are greatly depreciated. This challenge is considered as the core problem that affects the city's livelihood and quality of the urban environment. Informal settlements, high urban growth rate as well as instability and the limited technical and financial capacity of the city administration have prevented the implementation of sustainable solutions to the issue. An estimated 30–40 per cent of the potential water supply does not reach the end consumer and is wasted due to leaking pipes and aging distribution infrastructure. Few parts of the central city are served by the 110 kilometre sewer network, which constitutes only 4 per cent of the total network needed by the city. The infrastructure supporting public transport in Addis Ababa is also highly inadequate. Currently, taxis, city buses and private cars altogether account for 30 per cent of the modal share of which 26 per cent is constituted by bus services, 70 per cent by taxis and 4 per cent by private cars. Seventy per cent of urban travels are made on foot. Large segments of the city area have evolved without basic urban planning and the overall condition of the roads is poor. The relative rise in automobile ownership together with the poor condition of the roads and the poorly functioning traffic management system have resulted in a high level of congestion particularly at peak hours. To partly address these issues plans have been discussed to introduce an urban mass-transit system involving electric trams and buses and a light railway is currently under construction (Dubbale *et al.*, 2010).

2.5 Image

The picture depicted so far of a fast-growing city that is overwhelmed by problems, including extensive urban poverty, joblessness, inadequate housing and infrastructure, is in stark contrast with the image of Addis Ababa as the so-called administrative or diplomatic capital of Africa, seat for many headquarters and branch offices of international organisations and non-governmental organisations, including the African Union and the Economic Commission for Africa (KPMG, 2012). The two images are coexistent and simply reflect a complex reality in which the attempts to reconcile the needs and demands of the growing population have not always been successful. It should be noted at this stage that one of the objectives of the Integrated Housing Development Programme was indeed to enhance the city image by improving slum conditions and providing better housing, thereby also contributing to making Addis Ababa attractive to international investors.

2.6 Institutional conditions

Since 2003, the Addis Ababa city administration has undergone a radical reorganisation which has introduced a more decentralised structure. The Mayor who led the reform process, Arkebe Oqubay, eventually received the respected ‘African Mayor of Year 2005’ award for his effort to bring fundamental change to the city. According to the revised Charter of the City (Proclamation No. 361/2003), currently Addis Ababa has three administrative levels: the city government of Addis Ababa, ten sub-cities, and 99 localities, which are the smallest administrative units. A sub-city encompasses an average of 300,000 inhabitants while a locality has less than 50,000 inhabitants. Furthermore, as part of the decentralisation process, one-stop shops were introduced to provide local services to citizens and a community-based neighbourhood development programme was launched with the intention to give citizens a greater say over their own political and developmental priorities. The introduction of decentralisation was also accompanied by an administrative modernisation involving the rotation of personnel, stricter recruitment criteria, flatter organisations, customer orientation, shorter processing times and outsourcing of some previously government functions like public parking management and garbage collection (Haregewoin, 2007; UN-Habitat, 2008; Knebel and Kolhatkar, 2009).

All three administrative levels have elected councils. The city government has the elected city council which is its main organ of power. In addition to the city council, other city organs of power are the mayor and the city cabinet. The city government has the power to make laws and to exercise judicial powers specifically conferred on it by the city charter as well as executive powers and functions over matters that have not specifically been included in the details of the powers and functions of executive organs of the federal government. The city council is elected for a term of five years and constitutes the legislative organ of the city. It approves the budget of the city government and adopts policies on matters over which it has jurisdiction, including for example the master plan of the city. It also levies taxes and duties as well as sets service charges upon financial matters falling under the power of the city government in accordance with the city charter. The mayor is elected for five years by the city council and is accountable to the city council but also to the federal government. The city government undertakes its functions through eight bureaus which are accountable to the mayor and 11 agencies, institutes and offices organised under the city manager office.

Sub-cities comprise a sub-council, a sub-city administration and an executive body. Their tasks among others include the approval of economic, social development and municipal service plans, the election of the chief executive and deputy chief executive of the sub-city, and the allocation of the budget set aside to it by the city council. However, the role is mainly to implement policies decided at the city government level. Localities comprise a council and a chief executive. In addition they have their own social courts. They are centres for development and participation and for the delivery of basic services. Given their very limited resources, autonomy and power, their function is mainly to implement city and sub-city decisions (Haregewoin, 2007; UN-Habitat, 2008; Knebel and Kolhatkar, 2009).

Sub-cities and in particular localities face a number of challenges. First, they have insufficient capacity as responsibility and authority are delegated at lower levels but without the corresponding needed means in terms of working procedures, organisational arrangements and manpower. Second, the pay scales in place do not allow the sub-levels to attract and retain qualified professionals, and are not commensurate with the extent of responsibility of the employees. Third, there is a significant mismatch between what has to be carried out and the funds available to do so. The ability to use the collected revenues to finance local expenditures is very limited. Fourth, both reporting and accountability procedures are not clearly established (UN-Habitat, 2008).

It should also be emphasised that even though the legal framework shows that power is decentralised, in practice inter-governmental relations between the federal and city government are characterised by a top-down approach of control and supervision.

2.7 Conclusions: main issues

Addis Ababa is a fast-growing city which is characterised by a number of imbalances in its economic growth trajectory. Most notable is the predominant role played by the public sector in relation to the overall level of investments as well its role in employment creation. Social and environmental conditions in the urban area need substantial improvements, also to take into account the influx of migrants in search of employment opportunities. The dual image of the city epitomises the tensions that affect Addis Ababa in its attempt to present itself as the administrative and diplomatic capital of Africa while dealing with the challenges of providing adequate infrastructure services to its inhabitants.

Against this background, the rest of this chapter focuses on the Integrated Housing Development Programme, a government-led initiative which had the multiple objectives of addressing the housing deficit for low- and middle-income households and at the same time generating employment opportunities by strengthening the capacity of the private sector.

3 City-wide vision and strategy

3.1 Vision

The Addis Ababa City Development Plan 2002–2010 (City Government of Addis Ababa, 2002) provides a long-term vision of the city's future development. To improve its effectiveness and impact, a hybrid approach was adopted in the form of a structural plan focusing on the big picture and comprising a general framework for infrastructure development and for investment, and a more action-oriented development plan. According to the broad vision, as Ethiopia aspires to become a middle-income country and boost its competitiveness in global urban markets, urbanisation is seen as an important factor towards the achievement of this goal. Addis Ababa and the surrounding Oromiya region are expected to play a decisive role in the process. Addis Ababa already has a strong position in terms of attracting large-scale investment projects, particularly foreign direct investments. The medium- to long-term objective of becoming Africa's diplomatic capital and an internationally competitive city is complemented by the need to provide adequate housing, employment opportunities, and social and infrastructure services to a growing population.

3.2 Strategies

The realisation of the vision is based on a strategy consisting of the following pillars (City Government of Addis Ababa, 2002):

- Urban economy: The urban region comprising Addis Ababa has to establish itself as a centre for competitive and innovative businesses to facilitate the transformation of Ethiopia into a middle-income country. The long-term prosperity of Addis Ababa and its residents depends on the capacity to take advantage of the opportunities for sustained and rapid economic growth. Therefore, in addition to poverty reduction interventions, the city needs a pro-growth economic strategy. This should promote accelerated and sustained labour-intensive industrial development, for example by increasing private sector investment in the manufacturing sector, and also developing the service sectors relevant to international markets. Innovative projects linking employment creation with housing and infrastructure development need to be further promoted and implemented.
- Environment: The waste management and pollution control practices should ensure that water and air quality as well as noise levels are within acceptable standards. Green structures should be planned based on ecosystem services and the minimum green areas per capita should meet internationally accepted requirements, so that adequate, accessible, networked and functional green spaces can be provided to the urban population. Settlements in areas subject to high risk of natural disaster (e.g. floods and landslides) should be prevented to reduce exposure, and regulations pertaining to developments on disaster-prone areas should be developed and enforced.
- Centrality: Addis Ababa should develop as a polycentric city with a metropolitan centre complemented by hierarchically organised sub-centres. The multi-nucleated city structure should be supported by ongoing key projects such as the light rail transit system. The metropolitan centre should be characterised by a distinguishable high-density central business district with business corridors along the main mass transit routes.
- Transport: Land use and urban transport should be better integrated to reduce traffic and congestion in Addis Ababa. The mass transit system to be established in the city is

intended to provide high accessibility to the main centre, sub-centres and corridors that connect high residential, industrial and social services areas.
- Housing: Addis Ababa aims at guaranteeing affordable and decent housing for its inhabitants through different delivery options and to achieve compact development to ensure the economic use of land and infrastructure. For decades the city has suffered from a severe shortage of affordable housing that has particularly affected low-income households, which has resulted in a growing rate of informalisation and a poor quality living environment.

The Integrated Housing Development Programme which was implemented in Addis Ababa as from 2005 is clearly related to the city strategy focusing on the housing sector. As explained in the next section of the chapter, the Integrated Housing Development Programme was explicitly developed in order to address the acute shortage of residential housing, particularly for low- and middle-income households. With its emphasis on employment creation and on the provision of support for small businesses involved in the construction sector, the Integrated Housing Development Programme was directed as maintaining the livelihood of the poor in Addis Ababa and at substantially improving their living environment and standards.

4 The project

4.1 Vision and content

Since the 1970s Addis Ababa has been characterised by an acute shortage of residential housing which has resulted in severe congestion and overcrowding. This housing deficit has been growing in the recent past as a result of population growth, migration to urban areas and the dilapidation of the existing housing stock due to poor maintenance. The supply of housing has been constrained by restrictive land policies, the marginal role played by the private sector in housing development and by severe shortages of key construction inputs (Rijkers, 2007). Other drivers of the demand for housing include a lack of alternative investment opportunities and speculation (World Bank, 2009). In 2004 the housing shortage in Addis Ababa was estimated at 300,000 housing units, and the shortage was increasing by some 40,000 units each year. Furthermore, about a third of the total housing stock of 640,000 units was of poor quality. By way of illustration, approximately 120,000 housing units administered by the city administration were built with mud and straw and were older than 30 years, their estimated lifetime (Haregewoin, 2007; World Bank, 2009; UN-Habitat, 2011).

The Integrated Housing Development Programme (IHDP) was implemented in light of this massive shortage of low- and middle-income housing in Addis Ababa. The programme, one of Africa's largest and launched in 2005 by Arkebe Oqubay, built on the experience of the Low Cost Housing Project, a pilot project implemented by the German Organisation for Technical Cooperation (GTZ) and the Housing Development Programme Office (HDPO) in the Bole Gerji area of Addis Ababa. The aim of the IHDP was to construct 192,500 new houses, generate 80,000 jobs, support 1,300 existing Micro and Small Enterprises (MSEs) and create 1,000 new ones. In addition, the programme was directed at substantially reducing slum areas, developing 1,200 hectares of land, strengthening the construction sector, promoting low-cost construction technologies and minimum construction standards, as well as developing the required institutional capacity. It can be stated that the multiple objectives of the programme included the redevelopment and privatisation of public housing, maintaining the livelihood of the poor in the city centre, promoting effective use of scarce land, encouraging the use of low-cost construction technology, facilitating housing finance through simple and affordable payments modalities, creating job opportunities, promoting MSEs, and promoting the collaboration between the public and private sector. Land being a scarce resource in Addis Ababa, its economic and efficient use had to be a key pillar of the programme. At the time of its implementation the city had already utilised more than 75 per cent of its potential urban area. This explains why the programme aimed at achieving densification and construction of vertical units, which were also meant to open the possibility of creating green areas, playgrounds and other communal facilities, in line with the principles of sustainable urbanisation (Haregewoin, 2007; Knebel and Kolhatkar, 2009; World Bank, 2009).

Besides its impressive targets, this ambitious government-led programme had the following key characteristics (UN-Habitat 2011):

- Large scale: the programme was based on a large-scale approach to addressing the housing shortage, the low quality of the existing housing stock, as well as the future housing needs due to the continued urbanisation process.
- Pro-poor: the programme explicitly focused on low- and middle-income households and private ownership. Through the construction of fully serviced housing units meeting minimum quality standards, the programme aimed at improving the living conditions of the less well-off, granting them security of tenure and access to basic services. The programme was also directed at facilitating access to credit for low-income households.
- Integrated approach: the programme had the objective of stimulating the economy, creating employment, and strengthening the construction and financial sectors. The adoption of cost-effective construction techniques (e.g. pre-cast concrete elements) substantially reduced construction costs compared with conventional systems, improved the speed of construction, and fostered the development of MSEs involved in the production of construction elements. New construction management systems were introduced to further reduce construction costs and material wastage. The fixed price system forced firms to sell their outputs below the market price in exchange for the support they received.

It can therefore be concluded that the aim of the programme was not just to close the existing gap between demand and supply of housing units, but also to reduce urban poverty and boost the city economy using the construction sector as an engine. Regarding employment creation, the programme was characterised by a number of key elements. First, the programme stimulated the creation of MSEs in the construction sector by selecting qualified workers and training them on how to establish small business enterprises. Second, the programme contracted the newly established MSEs to work on the housing projects using innovative low-cost technologies in order to build affordable housing. Third, the programme actively supported the MSEs by facilitating access to land and credit, providing machinery leases at favourable conditions, and offering skills training (Haregewoin, 2007; World Bank, 2009).

4.2 Implementation

4.2.1 Leadership

The Mayor of Addis Ababa Arkebe Oqubay was the driving force behind the programme and its goal to build low-cost housing in Addis Ababa. He decided to involve the German Organisation for Technical Corporation (GTZ), which set up an office in Addis Ababa and was involved in the design and implementation of the pilot condominium housing project in the area of Bole Gerji. The pilot project consisted of 750 residential units in addition to office and commercial units and was efficiently implemented in terms of cost and time. As a result of this, Addis Ababa city government decide to scale up the pilot project of Bole Gerji and to build at least 40,000 houses per year. Following the decision of GTZ not to be directly involved with project design and implementation but to have an advisory role, in 2005 the city government established a new office specifically for housing development, the Addis Ababa Housing Development Project Office (HDPO). In the early days of the programme, a steering committee was set up comprising the Bureau Heads of the Addis Ababa city administration and representatives from GTZ and an Ethiopian design and engineering firm. Subsequently, the HDPO assumed full management of the programme under the control of a Board of Governors containing all members of the former steering committee with the exception of the GTZ representative (Haregewoin, 2007; UN-Habitat, 2011).

4.2.2 Organisation

The key role played by Arkebe Oqubay in the promotion and implementation of the programme should be emphasised. His commitment, charisma, capacity to mobilise political support at the level of federal government, attract and retain the most qualified staff to key positions and deal effectively with the inevitable obstacles were all instrumental in the successful execution of the IHDP.

A crucial role in the implementation process was also played by the Addis Ababa city administration and by the Commercial Bank of Ethiopia. The Addis Ababa city administration through the HDPO was the managing and coordinating agency for the programme. The office was set up to ensure the successful implementation in three core areas and phases of the programme: design, construction, and housing transfer and administration. In particular, it was responsible for the selection of new sites, preparing land and ensuring the supply of infrastructure facilities, executing design works, procurement and distribution of major construction inputs, MSEs support, securing funding from the city's budget to finance construction, the acquisition of bonds from the Commercial Bank of Ethiopia to pay for all other factors including the infrastructure costs and design-team costs, the compensation of all households displaced by inner-city renewal, and coordination and supervision of the activities of the various institutions involved in project work. Some of these tasks were delegated to sub-city project offices, specifically set up to coordinate the programme at sub-city level, including the administration of construction sites, collection and delivery of construction materials, supervision, selection of MSEs, and land preparation. The HDPO also worked closely with other government offices such as the Department of Trade and Industry, the Ministry of Urban Works and Development, the Ministry of Finance, and the city administration MSEs department, in charge of MSEs support to all the priority sectors, including construction and municipal services. The complex and multi-phase process, from infrastructure provision to transfer of housing units, and the number and diversity of the actors involved inevitably called for a strong coordination role by the HDPO (World Bank, 2009; UN-Habitat, 2011).

4.2.3 Finance

The Commercial Bank of Ethiopia, a government-owned bank with the mandate to provide finance for commercial purposes, played a very prominent role in the programme. Initially the city of Addis Ababa used its own budget, almost 50 per cent of the total capital expenditure over the period 2004–05, to finance the IHDP. As no additional donor funding was available, the city government started selling bonds to the Commercial Bank of Ethiopia to finance the programme. An agreement between the city administration and the Commercial Bank of Ethiopia ensured that the HDPO could directly receive the bonds for the programme implementation, and as a result the HDPO already borrowed USD 230 million in three separate phases, at an annual interest rate of 8.5 per cent (Knebel and Kolhatkar, 2009; UN-Habitat, 2011). It is important to note that the Commercial Bank of Ethiopia also agreed with the HDPO to provide loans to all condominium beneficiaries based on the following arrangement: the Commercial Bank of Ethiopia paid 80 per cent of the unit price on behalf of the beneficiary at the handover of the property (the beneficiary having to cover the remaining 20 per cent down payment) and the beneficiary entered into a loan agreement with the Commercial Bank of Ethiopia to pay back this amount plus interest. With only a small percentage (5–10 per cent) of condominium beneficiaries able to pay the full 100 per cent of the cost, a substantial amount of buyers entered into a loan agreement with the Commercial Bank of Ethiopia (UN-Habitat, 2011).

4.2.4 Technical aspects

A key characteristic of the programme was the type of housing to be developed, namely condominium housing: multi-storeyed housing units for several households where communal areas are jointly owned and managed. Even if one of the main objectives was to carry out slum upgrading, in the initial phases the programme implementation mainly affected open spaces or land reserved for purposes other than housing. The major reason for this was to avoid the social and political costs of slum clearance, as well as the delays in programme implementation that this may have caused. At the same time, in line with the principles of compact urban development and to minimise the costs related to the provision of additional infrastructure services, the Ethiopian government restricted all new developments under the programme to within the existing city boundary (Haregewoin, 2007).

Building contractors were hired to build the structure of each block. Secondary tasks such as fitting out the buildings, including the plumbing and electricity works, were carried out by sub-contractors consisting of MSEs. This was done in an effort to enable the capacity and development of MSEs, to keep a close control on costs, and increase competition with the aim of both improving construction quality and reduce costs. External consultants were also hired to oversee construction and maintain quality standards in the whole process. The construction cost of a condominium housing unit on the private market was estimated to be USD 193 per square metre. The target cost of the pilot project in the Bole Gerji area was USD 61 per square metre and the actual figure achieved was USD 77 per square metre. It has been estimated that approximately 10 per cent of this reduction in construction costs was due to the type of technology used, and 40 per cent could be ascribed to the improved management practices. Besides being a direct consequence of a production process that deviated from the one conventionally used, the low-costs arose from the fact that the type of housing built was homogeneous, novel construction technologies were used, cheaper inputs were produced by firms selected to participate in the programme, and fixed-price contracts and a standardised production procedure, permitting greater specialisation, were widely utilised (World Bank, 2009; UN-Habitat, 2011).

4.3 Results

It is reasonable to argue that the IHDP has been successful in many respects, not least by introducing the idea of housing as a process rather than as a product. Although the large-scale programme has not met all of its original targets, it has built 171,000 housing units, a remarkable achievement considering the context within which it was implemented and the previously limited capacity of the Ethiopian housing sector. The programme has considerably increased the number of homeowners and, in addition, has benefited the housing market by increasing the supply of owner-occupied housing and rental units. The programme has also built the technical capacity of the construction sector and manufacturing industry, particularly MSEs, and has been a significant generator of employment opportunities with some estimates indicating that up to 200,000 jobs were directly created. Even though as a result of the programme only half of the target number of housing units have been built, according to estimates by the Ethiopian government the programme has contributed a significant 11.5 per cent to GDP growth. These are of course very important results for a country like Ethiopia, particularly in light of the previously underdeveloped construction and financing sectors and resulting low capacity to address the housing shortage (Haregewoin, 2007; World Bank, 2009; UN-Habitat, 2011).

Considering the housing market in Addis Ababa, as a result of the programme the market rent level has decreased due to the higher number of rental properties on the market, and the housing shortage has been reduced. According to an assessment of the programme conducted by the World Bank (2009), there is evidence that the IHDP has generated jobs and that the support provided to MSEs seems to have enabled firms to grow more quickly than non-programme firms of the same size. Haregewoin (2007) argues that the programme promoted the creation of more than 2,000 MSEs and directly contributed to the setting up of over 100 real estate companies. Businesses that previously operated in the informal sector were able to receive tailored training and to access specialised credit facilities that helped them move into the formal economy. It is reasonable to argue that through its effects on job creation, the programme contributed to maintaining social stability, which could have been potentially threatened by the persisting high level of unemployment.

From the perspective of the residents, those that moved into condominium housing benefited from an improved physical environment, access to basic services and infrastructure, as well as security of tenure. Most importantly, as emphasised above, the programme has achieved its goal of delivering housing at low cost, with units costing around USD 77 per square metre against a market price of USD 193 per square metre. It should also be noted that condominium ownership has become a means to generate significant income through renting as owners are allowed to move into dwellings with cheaper rent and use the difference for productive activities. In this way, condominium ownership, at least for those who can access the programme as explained below, has turned into a poverty reduction strategy. Even though exact figures are not available, some estimates point out that up to 70 per cent of owners rent out their property to increase their income (Haregewoin, 2007; UN-Habitat, 2011).

The programme has also contributed to changing the saving culture of Addis Ababa inhabitants by providing, at least to those able to afford it, a concrete opportunity to become property owners. In this regard, the Ethiopian Government started to implement a new housing finance scheme in 2013 in Addis Ababa supported by the Commercial Bank of Ethiopia. The scheme aims to encourage low- and middle-income households to deposit money for a given period of time with the goal of reaching sufficient savings to allow for the purchase of a housing unit. When the savings reach a predefined percentage of the estimated housing cost, the saver is entitled to a house. The scheme is based on three main options available to the saver: the '10/90', the '20/80' and the '40/60' options. The 10/90 scheme targets low-income households where the potential house owners are able to save up to 10 per cent of the estimated house price and after three years the government long-term loan covers the remaining 90 per cent. The 20/80 scheme targets lower-middle-income households where savers can deposit a predefined amount for five years until the amount reaches 20 per cent of the estimated total housing cost with the government long-term loan covering the remaining 80 per cent. The 40/60 scheme targets upper-middle-income households with savers expected to save for five years before the long-term loan is provided for 60 per cent of the housing cost. In the 40/60 scheme individuals may also pay upfront a 40 per cent down payment and a loan is arranged to cover the remaining balance (World Bank, 2013).

A number of unanticipated challenges have however characterised the programme. The most relevant one concerns the issue of affordability of the units particularly for low-income households. Cost increases in the price of condominium houses make them now unaffordable for many low-income households. Particularly the poorest citizens of Addis Ababa have been unable to benefit from the IHDP due to their inability to cover the initial down payment and debt service payments. The inability to pay the monthly mortgage and service payments forces low-income households to move out of their unit and rent it out rather than risk losing the property because of bank foreclosure. This may also explain the high percentage of owners renting out their property, which clearly conflicts with the interpretation that this should be viewed as an effective poverty reduction strategy. The fact that the low-cost housing programme, which required a substantial investment on the part of the city government, has been unable to fully address the financial constraints of the lowest income groups, the majority of households in Addis Ababa, constitutes a difficult and pressing political issue (Haregewoin, 2007; UN-Habitat, 2011). The new financing scheme mentioned above that introduced more flexibility in terms of options ('10/90', '20/80' and '40/60') was introduced precisely to address the issue of affordability.

Another challenge concerns the location of most condominium projects in the peripheral areas of Addis Ababa, particularly because of the lack of employment opportunities and the costs related to infrastructure provision. The lack of available land in the central areas of the city combined with the initial objective to avoid undertaking construction projects in slum areas because of resettlement issues implied the development of peripheral areas, problematic in terms of livelihoods and infrastructure provision. Further away from the city centre of Addis Ababa there are insufficient employment opportunities, and transportation to access the city centre is costly and time consuming, thereby placing further financial strain on the inhabitants of condominium projects. Also, by exercising a strong pull factor and attracting labour from rural areas to be employed in the construction sector, the programme may have actually exacerbated the overcrowding of existing poor areas due to the lack of housing opportunities available to these migrant workers. Furthermore, the cost of infrastructure provision is greater in peripheral areas, which contributes to increasing the overall project costs. The quality and design of condominium blocks and their efficient management are also critical factors that must be addressed to improve the long-term sustainability of the programme. Other problematic issues raised regard the quality of construction, not always in line with minimum acceptable standards, the cultural difficulties that some of the new residents have faced in adapting to living conditions in multi-storeyed apartment blocks, and the sustainability of the MSEs created beyond the duration of the programme (Knebel and Kolhatkar, 2009).

Concerns have also been expressed that the financing approach characterising the programme is unsustainable. Due to the lack of donor involvement, initially the programme relied heavily on funding from the city government, subsequently complemented by bond funding through the Commercial Bank of Ethiopia. However, the high level of capital investments required has raised the issue of what city government activities have been reduced or discontinued in order to finance condominium projects which require significant amounts of financial resources. In addition, programme financing may run the risk of owners being unable to meet mortgage repayments as well as the commercial units being unsuccessful (UN-Habitat, 2011). An alternative would be to establish a mortgage bank dedicated to housing and to stimulate and support the potential role that commercial banks could play in the housing finance market. Furthermore, microfinance institutions in Addis Ababa have started to issue loans to low-income households to address the still partly unresolved issue of affordability of the programme.

Regarding the unquestionable success of the IHDP in terms of job creation, the World Bank (2009) assessment mentioned above expresses some reservations about the sustainability of the jobs created as, in spite of signs of a construction boom, there appears to be underutilised capacity in the sector and the MSEs created by the programme might have difficulties adjusting to novel demand conditions. It also stresses that from the point of view of the employment creation effects of the programme, elements such as the introduction of new technologies in the construction sector and building skills which might make firms sustainable in the future need to be strengthened.

5 Conclusions

The IHDP has proved to be a highly effective strategy for affordable housing delivery at large scale. The use of condominium development for urban renewal and slum upgrading has the potential to be replicated, under certain conditions, in other areas of Addis Ababa characterised by poor housing conditions or a lack of housing availability. There are many advantages to scaling up small housing projects to city-level programmes to reach more beneficiaries, relying on a coordinated effort that also contributes to the development of the construction sector. As the experience of the IHDP clearly shows, this should be based as a key precondition on a strong political commitment at all levels of the public administration.

Nevertheless, some key challenges remain (Haregewoin, 2007; UN-Habitat, 2011). First, even though the private sector in Ethiopia has failed to deliver housing for the poor, and real estate developers have often bought land merely for speculative purposes, this may be due to the fact that the right conditions and frameworks were not in place and that a highly politicised environment characterised by lack of trust has prevented the formation of strategic alliances and partnerships between the public and private sectors. There is a clear rationale for housing delivery not to be monopolised by the public sector, which could focus on the provision of housing to those who cannot afford it, and therefore for promoting other supply-side instruments and arrangements and for supporting actors such as housing cooperatives. At the same time, more participatory processes should be adopted that take into account the views and preferences expressed by citizens as beneficiaries of the housing programme. Opportunities for beneficiaries' involvement in planning and design should be created and actively promoted in an effort to create greater ownership and make the built environment more responsive to local needs and cultures.

Second, programme implementation can also be improved by optimising some of the key processes on which it is based, for example by finding cheaper ways of building and reducing the need for expensive inputs, improving quality control through better reporting and monitoring and by making more extensive use of information and communication technologies, and carrying out slum clearance and resettlement only when necessary and based on due processes to avoid forced evictions.

Third, regarding the potential of the IHDP to be replicated elsewhere, following the experience in Addis Ababa the programme was implemented in 58 towns in Ethiopia across its nine regions, as well as in the other chartered city and second largest urban area of Dire Dawa. The towns were jointly selected by the federal and regional governments. However, the programme was ended by the federal government in 2009 due to problems related to affordability, low demand, lack of capacity and human resources, and poor programme implementation. This shows that such a complex programme heavily relies on specific contextual conditions and local circumstances and that its centralistic orientation may be difficult to reconcile with replicability and sustainability in other urban areas.

References

CSA – Central Statistical Agency (2011). *Analytical Report on the 2011 Urban Employment Unemployment Survey*. Addis Ababa: Central Statistical Agency.

City Government of Addis Ababa (2002). *Addis Ababa City Development Plan 2002–2010*. Addis Ababa: Office for the Revision of Addis Ababa Master Plan.

City Government of Addis Ababa (2005). *Socio-Economic Profile of Addis Ababa for the Year 2004*. Addis Ababa: Finance and Economic Development Bureau.

City Government of Addis Ababa (2010). *Addis Ababa Atlas of Key Demographic and Socio Economic Indicators 2010*. Addis Ababa: Finance and Economic Development Bureau.

Dubbale D. A., Tsutsumi J. and Bendewald M. J. (2010). Urban Environmental Challenges in Developing Cities: The Case of Ethiopian Capital Addis Ababa. *World Academy of Science, Engineering and Technology*, 42, 397–402.

Haregewoin Y. M. (2007). Integrated Housing Development Programs for Urban Poverty Alleviation and Sustainable Urbanization – The Case of Addis Ababa. Paper presented at the International Conference on Sustainable Urban Areas, Rotterdam.

Knebel, N. and Kolhatkar, M. (2009). The Dilemma of Urban Governance in Africa between Western and Eastern Development Models: A Case Study from Addis Ababa. Paper presented at the International Conference City Futures, Madrid.

KPMG (2012). *The Role of Cities in Africa's Rise*. KPMG Africa.

Ni, P., Kresl, P. K. and Liu, W. (2013) *The Global Urban Competitiveness Report*. Cheltenham: Edward Elgar Publishing.

Rijkers, B. (2007). The Employment Creation Impact of the Addis Ababa Integrated Housing Program. Unpublished paper.

UN-Habitat (2008). *Ethiopia: Addis Ababa Urban Profile*. Nairobi: United Nations Human Settlements Programme.

UN-Habitat (2010). *The State of African Cities 2010*. Nairobi: United Nations Human Settlements Programme.

UN-Habitat (2011). *Condominium Housing in Ethiopia: The Integrated Housing Development Programme*. Nairobi: United Nations Human Settlements Programme.

World Bank (2009). *Ethiopia – The Employment Creation Effects of the Addis Ababa Integrated Housing Program*. Washington, DC: World Bank.

World Bank (2011). *Towards a Globally Competitive Ethiopian Economy: The Role of Services and Urbanization*. Global Development Solutions, LLC. Washington, DC: World Bank.

World Bank (2013). Ethiopia Economic Update – Laying the Foundation for Achieving Middle Income Status. Press release. Washington, DC: World Bank.

5 Kasongan pottery cluster (Yogyakarta)

Jan Fransen and Erwin van Tuijl

1 Introduction

The Kasongan pottery industry is one of the most competitive in Indonesia. It has been able to sustain its competitiveness despite a range of shocks: the Bali bombing in 2002, a devastating earthquake in 2006 and the financial crisis since 2008. We have studied an innovative cluster project that aimed to support sustainable competitiveness of Kasongan pottery firms in the period 2009–11. The project combined a bottom-up cluster programme of local government, with a top-down programme that is nationally funded, called 'One Village, One Product'. Indonesia has a rich tradition of cluster projects, but many fail to create sustainable competitiveness (Tambunan, 2006a, 2006b; Mawardi *et al.*, 2011).

The project allows the fleshing out of a number of dimensions in cluster development. First of all, Yogyakarta provides an exceptionally enabling and decentralised environment, linked with its regional autonomy and the role played by the Department of Industry, Trade and Cooperation, bridging actors and combining roles and projects of different ministries – most cluster projects in Indonesia are only top-down and remain disconnected from local visions, strategies and practices (Tambunan and Supratikno, 2004). Second, the project combines formal leadership of local government with informal leadership of the entrepreneur chairing the business association of the cluster. Formal leadership has enabled integration of support from various government layers and departments, while informal leadership has ensured that the voice of the private sector has steered the project. It is uncommon for informal leadership of the private sector to play such a vital role in a government-led project. Third, it builds on trust and reciprocity created during reconstruction after the earthquake; trust between the public and private sector is also uncommon in Indonesia.

This chapter is structured as follows. Section 2 analyses the profile and recent dynamics of Yogyakarta across a number of dimensions or 'capitals' for sustainable competitiveness. Section 3 sketches major city-wide economic strategies and the relation with the cluster supportive project. Section 4 is the core of the chapter and explores the Kasongan cluster in depth, namely its start-up history, vision and content (4.1) and the key features and enablers that 'made it happen' (4.2). Section 5 looks into the prospects for sustainable competitiveness of the Kasongan cluster in relation to the project(s) under analysis. Section 6 concludes by raising lessons and challenges ahead.

2 City profile: capitals for sustainable competitiveness

2.1 Performance of the city and the pottery cluster

2.1.1 Yogyakarta

Yogyakarta, with its about 3 million residents, benefits from the stable and high economic growth rate of Indonesia (Shima *et al.*, 2006). Located in central Java, Yogyakarta has excellent access to a large market with a quickly growing middle class. It is generally assumed that Jakarta is the only powerhouse of the country, but this is a misconception. In fact, smaller towns and cities are growing at a faster pace than Jakarta. Contrary to many other Asian countries which depend on manufacturing exports, economic development in Indonesia is driven by domestic consumption, creating a measure of resilience to the global financial crisis. A growing middle class and rapid urbanisation drive economic development in cities, while higher labour productivity drives growth of agriculture in rural areas. A rapid but temporary increase in working-age population creates a window of opportunity for continued economic growth (Oberman *et al.*, 2012).

Yogyakarta takes a special place in Indonesia. It was founded as a kingdom in 1755 and was the centre of the independence struggle and national capital from 1945–49. The sultan was rewarded for his role in the struggle with relatively extensive powers within the 'special province of Yogyakarta'. The sultan rules by benign degree, focusing on inclusive growth, education, culture and stability. Yogyakarta has the highest life expectancy and lowest infant mortality nationwide. It has a low cost of living, a clean and safe living environment and affordable tariffs of basic infrastructure services (Special Province of Yogyakarta, 2005; World Bank, 2012). Despite these achievements, there are considerable geographical differences within Yogyakarta special province: residents of Yogyakarta city are over three times as wealthy as residents of Kuon Progo, the poorest rural part of the province (BPS, 2012). The richer urban agglomeration comprises Yogyakarta city, Bantul and Sleman. Rural–urban migration has created the highest density outside Jakarta (over 1,500 per square kilometre in the province).

Yogyakarta is renowned as a centre of classical Javanese art, including batik, ballet, drama, music, poetry and puppet shows. Art feeds an exceptionally large and dynamic creative industry, comprising among others wooden batik, filigree silverware, bamboo, designed furniture, lamps, handbags, clothing and garments. In 2003, about 40 per cent of the enterprises in the industry sector (30.587 units) were involved in the creative economy, representing 12 per cent of the total national export value. The sector is an important generator of jobs and income, especially through small- and medium-sized enterprises (SMEs) (Special Province of Yogyakarta, 2005). Artists and creative industries are clustered in a number of neighbourhoods and villages. Until about two decades ago, these clusters were rather static. Their competitiveness was based on local materials, traditional skills of local craftsmen working in micro, home-based enterprises and strong social ties (Ismalina, 2011). In 1990, Out of Asia entered the scene. Out of Asia is a large exporting company, owned by an Australian designer. This new development triggered innovations in clusters, as Out of Asia worked with small firms to improve their design and quality. Many other exporters followed. As most exporters combine products from different clusters, they have created networks among clusters. Today, clusters are no longer isolated, but strongly embedded in global value chains and networks of industries comprising Yogyakarta city and the surrounding regions Sleman and Bantul (Tambunan and Supratikno, 2004; Fransen, 2013).

2.1.2 Kasongan

Kasongan is located in the Bantul area of Yogyakarta Special Province and is part of Yogyakarta metropolitan area. In Bantul, the creative industry is dominant, with about 1,800 SMEs located in 73 clusters (interview data). Most of these clusters are not very dynamic, but some have potential as tourist villages. Five clusters are specifically recognised and supported as innovative clusters by the Department of Trade, Industry and Cooperation. These are the Kasognan pottery cluster (the focus of this chapter), Jipangan bamboo cluster, Krebet wooden batik cluster, Lemahdadi stone statue cluster and the Gendeng dolls cluster.

Kasongan has about 450 households and an almost equal number of SMEs, most of which are home-based. Ninety-five per cent of the labour force comprises potters and other craftsmen. The other 5 per cent are primarily civil servants, some of whom are crafting as a second job. The history of Kasongan dates back to 1825, when government reports Kasongan products sold in Bavaria (Jakarta); but the shapes were basic and the quality poor. In 1975, two people were instrumental in further developing pottery in Kasongan: Mr Septo Mukojo, a gallery owner, and Dr Salianto Sulaman, a lecturer. They organised monthly exhibitions, invited international buyers and trained the present, somewhat elderly generation of crafters. Kasongan enjoyed a golden age from about 1995 to 2006. The most important agent of change was a new exporter, Dr Kimboel, who introduced new contemporary designs, new finishing techniques and combined ceramics with other materials, such as glass, wood and paint. With government support, open burning was slowly replaced by closed stoves and gas burners, and suppliers moved from treating clay by hand to slap machines. Over time, the new designs, techniques and technologies became commonplace.

During the past decade, however, Kasongan has been hit by three subsequent crises. On 12 October 2002 the 'Bali Bombing' about halved the number of international tourists visiting Yogyakarta. This impacted on Kasongan by reducing local sales, but was swiftly overcome by increasing exports. On 27 May 2006 a devastating earthquake hit Kasongan, and about 90 per cent of houses and working places and 80 per cent of all stock was destroyed. As a consequence, pottery production declined by as much as 50 per cent. In order to reconstruct Kasongan and other areas in Bantul, the 'Two Years, Move On' project was initiated. The project successfully reconstructed Kasongan and created trust among partners.[1] By 2009, Kasongan was back on its feet with more potters than before the earthquake (Dewangga, 2012). But just when business prospered once again, the global financial crisis hit Kasongan. The effect was disastrous, as Kasongan was highly dependent on exports of one product. Again, Kasongan had to show its ability to adjust, this time by replacing exports by local sales. This required once again different product designs, different marketing channels and a slow but significant diversification to tourism and diverse tourist products. Timboel, the largest exporter, was the first to respond, refurbishing his shop in order to attract local sales. Other firms followed suit. This has transformed Kasongan from an industrial centre to an upcoming shopping and tourist destination.

So far, the Kasongan pottery cluster has proven to be highly resilient to crises. One of the main reasons is that firms in Kasongan are structured to operate in highly volatile global markets with short product cycles. This includes working in small firms that work closely with global buyers, using local materials and highly skilled, low-cost labour, keeping investment costs low, and subcontracting production to even smaller firms. As will be discussed in the remainder of this chapter, the project under analysis has contributed to enhance further the sustainable competitiveness of the cluster.

2.2 Economic structure and knowledge

The main economic sectors of Yogyakarta are education, tourism and light industries. These are strongly interrelated: education provides highly educated staff for tourism and industries as well as a demand for products; tourism creates demand for light industries; and light industries create employment and offer products for local and global markets. These dynamics drive local competitiveness (Tambunan and Supratikno, 2004). Yogyakarta has about 270,000 students in over 100 universities and is the second most important tourist destination in Indonesia, with over 1 million local tourists annually and 100,000 international tourists.

Tourism banks on the rich cultural heritage, reflected in ancient monuments and historic sites. Industrial sectors are based on knowledge and skills related to designed furniture, interior design, metal, leather, textile and clothing. Yogyakarta is known for its creativity and design. Yogyakarta has a large artist community, many industrial designers and a working culture where new designs are easily absorbed (Indarti, 2010; Kusumaningrum, 2012). Non-tangible assets are political stability, the highest ease of doing business in Indonesia (World Bank and IFC, 2012), relative equality and its peaceful, tolerant and healthy living environment (Special Province of Yogyakarta, 2005).

2.3 Image

Yogyakarta brands itself as a prime centre of education, culture and creativity (Shima *et al.*, 2006). This image is strongly engrained in all actors, as expressed by our respondents. The rich cultural heritage and high level of decentralisation has enabled Yogyakarta to build up its image over a sustained period of time. Yet, although the brand may be strong within Indonesia, internationally the image is less visible. As a medium-sized city, Yogyakarta has as yet been unable to firmly position itself as the centre for creatively designed pottery and other crafts. As we will discuss later, one of the reasons is the policy focus on weaker creative industries instead of supporting and showcasing the most innovative firms and designs.

2.4 Institutional conditions

The organising capacity of Yogyakarta is characterised by a high level of decentralisation and participation. However, like all local governments in Indonesia, it is also plagued by corruption, nepotism and collusion. These problems appear to be somewhat less in Yogyakarta province (Lewitschnik, 2012; Shima *et al.*, 2006), but firms still don't fully trust the government (Fransen and Helmsing, forthcoming). The decentralisation laws of 1999, revised in 2004, have created three tiers of government: national, provincial and city. Below the level of the city, Yogyakarta has established the level of a neighbourhood. Decision-making power is increasingly decentralised to the provincial and city levels. Decentralisation laws stipulate participation of people in public meetings (Shima *et al.*, 2006). After Indonesia's decentralisation, Yogyakarta local government had been especially active in promoting collaboration between government, private sector and civil society, including industrial associations, NGOs, universities and research institutions (Shima *et al.*, 2006). Political leaders at the neighbourhood level coordinate annual planning meetings with residents, while a participatory forum organises monthly meetings to discuss plans at a lower level. These formal participatory processes functioned relatively well during reconstruction, but appear dormant in the absence of special reconstruction activities.

Social capital is strong among firms in Indonesian clusters, but tends to be weak between the public and private sectors (Brata, 2009, 2011; Ismalina, 2011; Shima *et al.*, 2006; Mawardi *et al.*, 2011). Traditional clusters in Yogyakarta, including Kasongan, are characterised by strong social ties, trust and reciprocity among residents. Those with firms in Kasongan, but living outside, are not embedded in these social ties, which impacts on their business networks and social networks (Ismalina, 2011). Social ties among residents were further strengthened by the reconstruction of Kasongan after the earthquake, as collaboration created trust.

2.5 Conclusions: main issues

Competitiveness in Yogyakarta is first of all strongly influenced by its location. Yogyakarta benefits from high and stable economic growth rates in Indonesia and from the large market of the growing middle class on Java. In addition, Yogyakarta is endowed with an abundance of high-quality natural resources. Understanding the sustainability of Yogyakarta's competitiveness, however, requires us to look beyond factor endowments towards knowledge and innovation. The economy of Yogyakarta banks on education and creative industries. The creative industry is labour intensive, but highly dynamic and innovative. It has transformed from a traditional industry to a specialised cluster in global value chains. It thrives on social ties, based on trust and reciprocity (Ismalina, 2011; Fransen and Helmsing, forthcoming). Reconstruction after the earthquake has created strong social ties between the public and private sector. As we shall see, this has positively influenced the development of the project under analysis.

3 City-wide vision and strategy

Yogyakarta's city-wide vision 2020 is to become a prime centre of education, culture and tourism. Sustainable economic growth is rooted in 'the unique cultural heritage and natural environment, its higher education systems and skilled human resource base, its healthy living conditions and political stability and its strategic location and public infrastructure assets' (Special Province of Yogyakarta, 2005, p.4). These days, such strategies are commonplace due to the hype of creative cities, but in Yogyakarta these policies have been in place for the past four to five decades. This long-term, highly consistent vision has provided a stable and reliable environment for educational institutes and industries to prosper.

Yogyakarta adopts a pro-poor, endogenous growth strategy (Shima *et al.*, 2006). In Bantul, where Kasongan is located, the core of the strategy is to upgrade four clusters of creative industries and develop them as tourist destinations. One of these touristic areas is the axis of Kasongan, a craft market (Benusan) and a sugar factory. These are all a few hundred metres apart. The local Planning Department (Bappeda) is working towards a plan to ease transport between those areas and creating a more attractive tourist destination. This contributes to social, physical, economic and environmental objectives of Yogyakarta's vision (Special Province of Yogyakarta, 2005):

- *Social*: to reduce regional poverty based on linkages between economic growth sectors and labour-intensive sectors, linkages between formal and informal sectors, strengthening of small- and medium-sized enterprises, and creation of social safety nets for the poor. Employment creation and social inclusion in labour intensive, traditional industries.

- *Physical*: urban – rural linkages, as local natural resources are used and creative industries are often produced in rural and peri-urban areas, while knowledge intense activities are located in the city centre.
- *Economic*: to effectively and efficiently use local resources in global and local markets for inclusive growth.
- *Environmental*: a clean and green city.

4 Kasongan pottery cluster project

4.1 Vision and content

The project's vision is to use pottery as a 'spinning wheel' for sustainable economic development. Reconstruction had illustrated that much could be achieved by collaborating. It was realised that additional support was needed in order to create sustainable competitiveness during crises. The vision has two interrelated components: economic and social empowerment of potters as the main industrial sector; and economic diversification towards tourism and related creative industries. The political leader of the neighbourhood notes: 'Economic development of Kasongan is the key. Pottery and tourism go hand-in-hand, but this demands much better marketing and branding'.

The programme comprises a bottom-up cluster programme and the top-down 'One Village, One Product' programme, which are closely interwoven. Their main characteristics are summarised in Table 5.1.

4.1.1 Cluster programme (2009–10)

Indonesia has a long history of cluster support, dating back to the 1950s. Support has mainly been channelled through business associations, resulting in the proliferation of associations. In total 64 different policy instruments have been applied, ranging from training to supporting business associations and micro credit. But support is often centralised and not very effective (Tambunan and Supratikno, 2004; Ismalina, 2011). By contrast, not many local governments have deliberative cluster programmes (Mawardi *et al.*, 2011).

The cluster programme is managed by the Department of Trade, Industry and Cooperation in Yogyakarta, supported by the Department of Culture and Tourism. It aims to empower

Table 5.1 Comparison of the two programmes

	Cluster programme	*One Village, One Product*
Funding agency	Local government	National government
Implementation agency	Department of Industry, Trade and Cooperation	Department of Industry, Trade and Cooperation
Target group	All firms in Kasongan	Innovative potters in Kasongan
Main activities	Support information centre Training Exhibitions and trade fairs Infrastructural works Guarantee fund	Technology transfer Training

Source: Own elaboration

both potters and related industries in Kasognan, with a strong focus on innovation. The project empowers individuals, firms and networks of firms. Project activities comprise:

- Support of the information centre in Kasongan (UPT), staffed by four people who coordinate activities of local government departments in Yogyakarta. A key activity is to promote Kasongan and run a lab with advanced technology. It is one of the three information centres in Bantul region.
- Technology and quality control from the National Ceramic Board in Bandung.
- Training courses.
- Financial support for joining exhibitions and trade fairs. Bantul annually provides Rp 1.6 billion for exhibitions.
- Infrastructural works: beautification of the entrance and conservation of the riverbank as a key attractive environment for tourists.
- A guarantee fund was established at the business association for a micro credit and saving scheme (Rp 150 million). The scheme is being reconsidered at the moment, as new regulation stipulates that only a saving cooperative can provide these services.
- Annual festival and fishing competition, aiming to attract buyers and tourists to Kasongan.

4.1.2 One Village, One Product (OVOP; 2010–11)

OVOP is a regional development programme that originates from Japan. The programme supports one product with a high value added per region. The programme has been implemented for about 300 products in Japan and is widely used in countries across the globe, including China and the US. In Indonesia, OVOP is a national programme of the Ministry of Industry, with support by the Ministry of Cooperation. The programme comprises about 70 projects across Indonesia, operating as a national fund for which regions can apply. In Bantul, the programme supports the Krebet wooden batik cluster and Kasongan cluster. The programme is locally implemented by the Department of Industry, Trade and Cooperation.

The aim of OVOP is to strengthen core industries, in this case pottery. Focus of support is to improve industrial design, product quality and innovation. This is in line with both the vision of Yogyakarta and the present need of volatile global markets. In reality, however, the programme specifically targets weak potters, mainly subcontractors, as the Department feels that larger exporters do not need funded support. Support thus targets the weakest link in the global value chain. An exception is that a few higher productivity potters were supported with closed burning systems, in order to reduce CO_2 emissions.

The programme has two main activities:

- *Technology transfer*. The largest part of the programme is the transfer of advanced and basic technology. This comprises three distinct elements: 1) Advanced technology was transferred to the information office in Kasongan. This includes radiology technology, recommended by the University of Gadja Mada, as an advanced tool to control quality. The information centre constructed a lab, which is also used as a training centre. 2) The information centre also received computers for their offices. 3) Three groups of potters submitted proposals for equipment, of which one was approved. A group of potters received basic equipment, such as a compressor and a closed burning system for pots.

- *Training*. The information centre ran a training course for one week for 20 potters. The course dealt with industrial design, market trend, regulations, health issues, production technologies and e-marketing. Participants were selected and approved by the Department, in consultation with the information centre and Dr Timboel, leader of the business association. Training was conducted by Dr. Timboel, the National Ceramic Board, University of Gadja Mada, Institute of Art (ISI) and others.

OVOP is nationally coordinated and locally implemented. It adheres to a bottom-up process, whereby local proposals are nationally approved. However, the role of the Department of Industry, Trade and Cooperation is overarching. The process functions as follows. First, potters are informed by the information centre of a funding opportunity. They are also informed informally during visits of the Departnment and/or by an Annual Development Plan meeting organised by the political leader of the neighbourhood. The information centre and a host of training courses inform potters of the format of proposals, which is straightforward and simple to apply. Second, proposals for activities are written by the information centre, groups of potters and/or the business association. Joint proposal writing and implementation indirectly stimulates networking. Third, the project team discusses the proposals within the Department of Industry, Trade and Cooperation and consults informally with Dr Timboel as an expert. The Department subsequently selects which activities it wants to forward for funding. And fourth, the project team annually reports on their activities and proposes new activities at the National Coordination Committee. The National Committee subsequently decides.

4.2 Implementation

4.2.1 Finance

Appreciating financial arrangements is crucial in understanding the selection of activities, targeting of firms and project successes and failures. Three issues are of particular importance: 1) differences between national and local funding sources; 2) financial targeting; and 3) the (dis)connection between the public project and market. Financial support ended in 2011, due to a positive project evaluation and lack of funding.

NATIONAL VERSUS LOCAL FUNDING

While the cluster programme is financed and controlled by local government, OVOP is fully funded by national government based on tight criteria and processes. Bantul won two of the 70 OVOP programmes, while 300 regions are eligible for funding. This indicates Bantul's strength of clusters and organising capacity. National government keeps tight control of finance: funding for project activities is approved by a national committee, based on proposals of the Department of Industry, Trade and Cooperation. Also, the programme does not disburse funds, but disburses equipment and training. OVOP activities and processes are thus firmly restrained by financial arrangements.

It is uncommon that a nationally funded programme, such as OVOP, is embedded within a local project. National programmes in support of clusters tend to have no impact due to tight financial arrangement and weak local capacities. The relative success of this case study is due to the coordinating hand of the Department of Industry, Trade and Cooperation. The Department is large, with about 30 staff, and is relatively capacitated and networked. As the

Department has written the proposal for OVOP funding and has a coordinating role in implementation, OVOP has been integrated into its own cluster programme.

Nevertheless, two distinct differences arise between the cluster and OVOP programme. First, whereas the cluster programme targets firms with weak capacity within the cluster, the OVOP programme is primarily directed at *innovative potters* that can potentially benefit from advanced technologies. In selecting potters and technologies, a clash arises between national and local priorities. The project coordinator noted: 'Support is mainly for those who need it, not for the big guys …'. Contrary to OVOP objectives, therefore, innovative firms were excluded from participating in OVOP training and proposals for new technologies. Second, OVOP funding specifically aims for *advanced technologies*, whereas the cluster programme adopts a bottom-up approach. Based on expert advice of the University of Gadja Mada (UGM), the latest radiological technology was bought. Unfortunately the technology is not (yet) used by potters, as they were not trained on how to use it.

FINANCIAL TARGETING

The project specifically targets support to firms with weak capacity. They often operate at the bottom of global value chains. Box 5.1 illustrates that firms with weaker capacity often cannot absorb knowledge and technologies received, which severely reduces the impact of the project. This is strengthened by absence of training on the new, advanced technologies.

Box 5.1 Targeting errors: the case of Putra Onsan Keramik

Putra Onsan Keramik is a small-scale pottery firm, which mainly produces for the local market, including tourists. The firm was exporting based on innovative product design until a few years ago, but has stopped exporting due to the financial crisis and now focuses on standardised, low-cost products. The entrepreneur is a highly respected and networked member of the community, who is active in the business association and mosque.

Over the years, the firm has received a lot of support. This includes subsidies to attend six international exhibitions in Jakarta and over ten training courses, including OVOP training. During the courses, he learned among other things about contemporary product design, advanced finishing techniques and e-marketing. During the OVOP course, he was asked to submit a proposal for new technologies under OVOP funding with four other small firms, which he did. The proposal was approved and he received a compressor for himself and a kiln for all four firms. He wrote and won many other proposals as well. As a member of the business association, he receives micro credit, benefits from relatively low-cost raw materials and has been given a spot to sell his products at the shop of the association. He also gets credit from a bank (BRI).

The result of all this support is difficult to see. The website and e-marketing which were developed are not working. The firm does not produce new contemporary designs, as the firm decided to focus on a more conservative market segment. And exhibitions have not resulted in more international sales, on the contrary. At best, support has prevented bankruptcy.

PUBLIC–PRIVATE (DIS)CONNECTION

Whereas publicly funded support targets firms with weaker capacity, private funding and knowledge is channelled through exporters with high levels of absorptive capacity. The reason is that global buyers demand sophisticated and trendy products from their exporters. Exporters have access to the latest knowledge on markets and contemporary products, due to their contacts with global buyers such as IKEA. Exporters subcontract production and actively support their subcontractors' innovation processes (Fransen and Helmsing, forthcoming). Exporters, therefore, are a strong agent of change. Government support is complementary to the role of exporters, as government supports the weakest links in global value chains. As government selects subcontractors to be supported jointly with Dr Timboel, the selection may be appropriate. However, government support may also target those firms that in the end will disappear in processes of creative destruction.

4.2.2 Leadership

Formal leadership lay firmly with the project leaders at the Department of Industry, Trade and Cooperation. They approve the cluster programme's local proposals and select which proposals to forward for national OVOP funding. They also direct the information centre in Kasongan. However, Dr Timboel is the undisputed informal leader of the project, as he drives processes from within. He is the largest exporter, responsible for about 25 per cent of all exports. Many subcontractors and suppliers work for him. He is also considered 'the most creative' and has introduced many new designs and technologies. He is held in high regard, in part due to his title, his lecturing position at the Institute of Art (ISI) and the books he has written. As a result he is asked as lecturer for training around Indonesia and as consulting expert of the Department of Industry, Trade and Cooperation. As chairperson of the business association he chairs meetings and is responsible for the association's shop and the saving and credit programme. His network extends far beyond the cluster and far beyond that of the Department. An example of his informal leadership is his ideas on the role of the information centre. Dr Timboel said:

> UPT (the information centre) has much equipment, which is hardly used. So I asked myself: can I make better use of the equipment? I try to link the business association to UPT and spend money of the association to make better use of the equipment. It is slowly working.

4.2.3 Organisational arrangements and networks

Key organisations of the project are the Department of Industry, Trade and Cooperation, the information centre (UPT) and the business association. The business association ties local firms to the project. Yet, a host of other organisations are also involved (Figure 5.1).

The Department of Industry, Trade and Cooperation, located outside Kasongan, is the linking pin between the public and private sector and between the national and local levels. Its coordinating role is crucial. Merely a decade ago, a multitude of departments supported Kasongan, each with their own programme. Now support is coordinated and largely implemented by one department. UPT is the eyes and ears of the Department in Kasongan. It was set up as a bridge between potters and government in 1995. These days it employs four staff and focuses on training on marketing and design, using the technology available in their offices. There are no regular meetings with potters, only when needed. Its role in other activities tends to be limited.

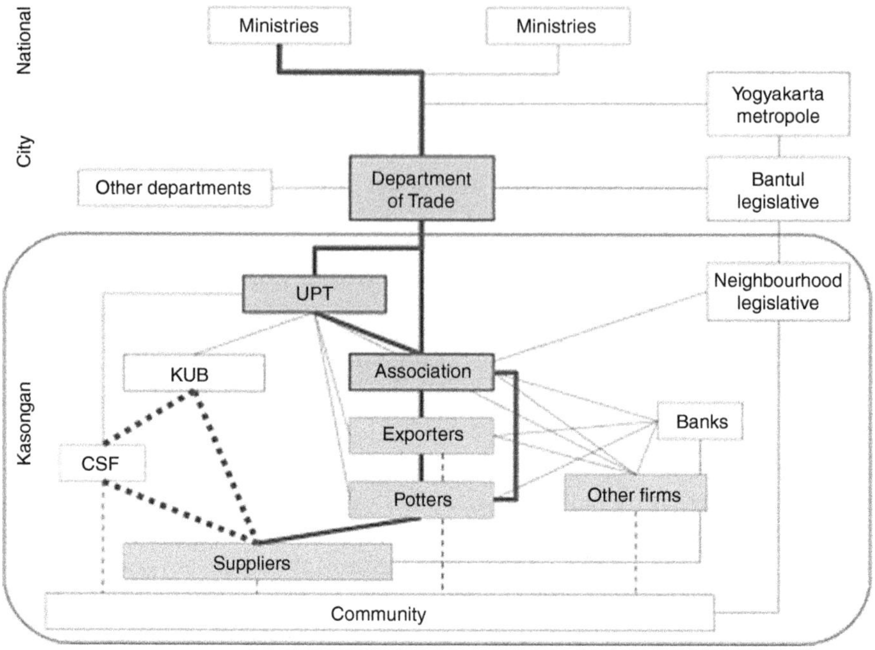

Figure 5.1 Organisations involved and their relation to the project

Source: Own elaboration

Within Kasongan, the business association is influential. It has 582 members and aims to solve local problems, such as those related to copying to design and streamlining of government support. It fulfils a wide range of activities, ranging from micro credit and savings schemes, to a shop and display of products, supply of clay and wood to potters and advising the Department of Trade, Industry and Cooperation who to send to exhibitions and training courses. The association meets every two months and whenever problems arise.

The project is manageable due to its straightforward institutional arrangements, yet creates institutional thickness due to a large number of associated organisations:

- *Department of Tourism and Culture.* This Department proactively supports the Department of Industry, Trade and Cooperation in marketing and tourism promotion. The Department supports the annual festival and fishing competition in Kasongan by inviting mass media, famous artists and international students. The festival is marketed on the internet as well. As a result, the festival expanded between 2010 and 2013. Tourism is promoted by advertisements at airports and on websites, as well as by inviting travel agencies to visit Kasongan. This works to a limited extent, as potters cannot afford commissions comparable to other attractions in Yogyakarta (they can afford 5 per cent and firms in the silver cluster 50 per cent). The Department of Tourism and Culture also collaborates with the Planning Department (Bappeda) and others in preparing a plan to improve the touristic potential of Kasongan and the related market and sugar factory.
- *Department of Planning (Bappeda).* The Department of Planning has lost much of its coordinating role over the past decade, but it still develops physical plans, integrated

with social, economic and environmental objectives. It especially plays a leading role in infrastructural works.

- *Legislative*. There is a parallel political structure ranging from Yogyakarta metropole to Bantul and the neighbourhood. The legislative plays a role in approving neighbourhood plans and in communication channels among actors. The Annual Development Plan meetings provide an opportunity for residents to be informed and voice their opinions.
- *Other associations: KUB and CSF*. There are a number of associations in Kasongan. The main one is Sabyo Bawara, which we refer to as the business association. Another small but influential association is KUB, an association of suppliers of clay and bamboo producers with its own exporting cooperative (CSF). KUB and CSF represent a separate and equally successful example of cluster support. They are also supported by donor agencies. Other local associations include Seinn Kliwan, a small association of potters, and AKRAB, a youth association.

One organisation is hardly represented in the organigram: *universities*. Formally, they play a role in training and education, but not as a partner in research activities. However, Dr Timboel is a lecturer at the Institute of Arts (ISI) and supports a change in curriculum of ISI towards entrepreneurship and applied, commercial art. As such, Dr Timboel acts as a bridge between academia and the pottery sector and can feed the cluster with new concepts and ideas.

NETWORKING

Networking among all actors is best practice in the Indonesian context. Not only are different levels of government jointly contributing towards one vision, but firms are actively involved and informal leadership of an entrepreneur is fully accepted by the public sector. The programme contributes to sustainable networks among the various organisations. For instance, in the cluster programme, there was a focus discussion group with various local government departments, universities and associations; funding proposals were prepared in formal and informal networks of firms and non-firm actors; the Department of Industry, Trade and Cooperation networks formally and informally with many actors; the information centre in Kasongan has been strengthened as a linking pin between public and private initiatives.

A precondition for networking is providing adequate information. This is mainly done by the information centre, but this knowledge is highly skewed towards technology, training and national regulations. Meetings of local political leaders with support of the Department of Industry, Trade and Cooperation also inform firms and residents on local policies, strategies and funding opportunities. However, market information is missing from the equation.

4.3 Results

To what extent has the project been successful in sustaining the competitiveness of Kasongan pottery cluster? The Department of Industry, Trade and Cooperation has evaluated the project as very successful. They note that skills improved, industry diversified and value added in the pottery sector improved due to better finishing. As a result, they have reclassified Kasongan as a mature, developed cluster, which no longer needs support. Instead, government support moved on to weaker, embryonic clusters. Due to its perceived success, the project structure, objectives and activities will be replicated.

We partially confirm the positive project results. In particular, the organising capacity of the cluster has been improved, resulting in strong institutional arrangements. Public cluster support has been set up, well integrated with business association(s), the main business association runs a micro credit and saving scheme, and networking among organisations has been, and still is, exemplary. The main organisational weaknesses are the limited role of universities and research institutes in innovation, an isolated and marginal role of the information centre in Kasongan, absence of a business development service from a market perspective and over dependence on one person, Dr Kimboel.

From an economic perspective, pottery firms in Kasongan have been able to sustain competitiveness, despite a compilation of crises. Over time, they have been able to adjust to new market demands and shocks: skills and procedures related to quality and finishing have improved, firms have been able to replace diminishing exports with increased local sales and Kasongan has continued its slow diversification towards related industries and tourism. The prime role of the project was that it has stimulated skill development and networking among local firms, which increases innovation and creates economies of scale and scope. Also, attendance to trade fairs is financed, which significantly improves access to knowledge and export markets. On the other hand, the effect of new technologies on economic competitiveness is marginal, due to mistargeting. A major weakness is that support focuses on smaller firms with limited capacity instead of support for new innovative ideas that will boost competitiveness.

From an institutional perspective, despite being part of volatile global markets and a larger local innovation system in Yogyakarta area, Kasongan still functions as a cluster based on strong social ties. The focus of the cluster project on firms with weak capacity, which limits economic sustainability, has at the same time created a safety net for firms. Support for the business associations and networking between business associations, local government and national government has built up trust and reciprocity. However, interviewees were of the opinion that the high level of trust and reciprocity during reconstruction has been reduced ever since. Governance has slowly returned to business as usual. Also, employment in pottery is slowly reducing and youngsters, often with higher levels of education, are looking for other opportunities. An open question is how the changing social and economic fabric will impact on sustainable competitiveness in the long run.

5 Conclusions

5.1 Broad assessment

The Kasongan pottery cluster is one of the most competitive clusters of Indonesia and has been able to sustain its competitiveness despite three subsequent crises: the Bali bombing, an earthquake and a financial crisis. We have studied a combination of initiatives that supported the sustainable competitiveness of the cluster in the period 2009–11. The project combined a bottom-up cluster programme of local government with a top-down programme that is nationally funded. The two programmes were integrated by the Department of Industry, Trade and Cooperation, as implementing agency. This was no easy feat, due to national regulations and different target groups. Also, Indonesia has limited experience with decentralised projects. Nevertheless, the project managed to integrate the two programmes and benefit from their complementarity.

The project has contributed to sustainable competitiveness in a number of ways. Most importantly, it has strengthened institutional arrangements in support of the cluster. Capacity and networks of the business association, information centre, local government, central government and other actors have improved considerably. Second, the project has supported firms with weaker capacity.

While the project has significantly contributed to the remarkable resilience of the cluster, it is not the main cause. The main cause is that the industry is structured from top to bottom for flexibility and resilience. Industrial structure is characterised by small firm sizes, subcontracting, low capital investments and a high capacity for change. Changes in product quantity and quality can be implemented quickly with limited risk and investment. Exporters are the gatekeepers, always on the watch for new contemporary markets, designs and technologies. They subsequently train and support subcontractors in improving designs, processes and markets.

5.2 Critical factors and challenges ahead

We identify three reasons for the success of the project, which all point towards an exceptionally high level of organising capacity, especially related to decentralisation, participation and public–private partnership.

The first reason is the steadfast vision of Yogyakarta as a city of education, culture and tourism. For over four decades, the vision has been engrained in all actors, carefully building up the city's identity. Cultural industries benefit from the stable policy environment and city branding. Decentralisation in the late 1990s and early 2000s have been a crucial next step, as it enabled Yogyakarta to put its vision into practice.

Second, the Department of Industry, Trade and Cooperation has been mandated and capacitated to develop and manage local projects in line with the vision and strategies of Yogyakarta. The Department firmly integrates and manages the local cluster programme and national OVOP programme. This integrative role is possible due to the integration of services under one roof. Moreover, in addition to formal leadership of the Department of Industry, Trade and Cooperation, all actors accept and value the informal leadership of Dr Timboel. As the largest exporter and leader of the business association, he bridges the public and private sector.

The third critical factor is trust between the public and private sector. The collaborative reconstruction of Kasongan and its firms after the earthquake has created trust, in a country where public–private partnership is characterised by corruption, collusion and nepotism.

Sustaining competitiveness is a never-ending process, as global markets are highly volatile. Any project falls short in such a never-ending process, let alone a project of just two years. For that reason, the institutional arrangements that have been developed by the project are of pivotal importance. The main challenge is to sustain these institutional arrangements. This is particularly difficult in developing economies, as funding is scarce, priorities are many and bureaucracy reduces responsiveness. This project illustrates these risks, as priorities shifted to other clusters and support reduced afterwards.

A second challenge is the dependency of clusters on one product, in this case pottery. Such a high level of specialisation may be useful in creating an innovative and competitive cluster, but it is highly vulnerable to market volatility. The programme 'One Village, One Product', which is implemented in many countries, specifically supports this particular industrial structure. It is a challenge for cluster projects to balance the wish for short-term value added and long-term sustainability.

A third challenge is that the economic reality in cities in developing countries changes very rapidly. In Kasongan, changes have included the multiple crises, entrance of exporters, new technologies and rapidly changing contemporary designs. Kasongan is no longer standing on its own, but is part of global value chains and much wider economic networks in

Yogyakarta. The challenge of cluster projects is to adapt to these rapid changes. This is particularly difficult in bureaucratic environments.

What lessons are learned and can they be applied to other cluster projects that combine different sources of funding? Lessons learned are the importance of a steadfast vision, decentralised project coordination under one roof, active participation of the private sector and trust between the public and private sectors. These lessons are not new, but are scarce in the elusive context of developing economies. It demands building an organising capacity over a longer period of time.

Note

1 In Kasongan, Rp 953,600,000 was available to support 315 firms. They received Rp 15 million each for slap machines, temporary buildings, houses and malls (prefab forms). In addition, public infrastructure was reconstructed, houses were rebuilt and firms continued exporting.

References

BPS (2012). *Trends of Selected Socio-economic Indicators in Indonesia*. Publication Number: 03230.1201. Jakarta: BPS.

Brata, A. (2009). *Innovation and Social Capital in the Small-medium Enterprises: A Case of Bamboo Handicraft in Indonesia*. Munich: MPRA, 15696.

Brata, A. (2011). *Social Networks and Innovation (Handicraft Industry in Bantul, Yogyakarta)*. Munich: MPRA, 28032.

Dewangga, K. (2012). Kasongan pottery surviving the major earthquake. Accessed at http://www.tnol.asia/arts-culture/14325-kasongan-pottery-surviving-the-major-earthquake.html (23 September 2013).

Fransen, J. (2013). *Innovation of Small Firms in Indonesia. The Case of Handicraft Firms in Yogyakarta*. Rotterdam: Institute for Urban Housing and Development Studies, Erasmus University Rotterdam.

Fransen, J. and Helmsing, A. H. J. (forthcoming). *Absorptive Capacity as a Mediator: Innovation of Handicraft Exporters in Yogyakarta, Indonesia*.

Indarti, N. (2010). *The Effect of Knowledge Stickiness and Interaction on Absorptive Capacity: Evidence from Furniture and Software- Small- and Medium-sized Enterprises in Indonesia*. Groningen: University Library Groningen.

Ismalina, P. (2011). *An Integrated Analysis of Socioeconomic Structures and Actors in Indonesian Industrial Clusters*. Groningen: University Library Groningen.

Kusumaningrum, F. A. (2012). *Exploring Innovation of Handicrafts: The Case of Silver Craft in Kotagede, Yogyakarta*. Rotterdam: Institute for Urban Housing and Development Studies, Erasmus University Rotterdam.

Lewitschnik, L. (2012). Happy Valley Yogyakarta. *Monocle*. June.

Mawardi, M. K., Choi, T. and Perera, N. (2011). The factors of SME cluster development in a developing country: the case of Indonesian clusters. *2011 ICSB World Conference*. Stockholm: International Council for Small Business.

Oberman, R., Dobbs, R., Budiman, A., Thomson, F., Tosse, M. (2012). *The Archipalego Economy: Unleashing Indonesia's Potential*. McKinsey & Company. Accessed at http://www.mckinsey.com/global-themes/asia-pacific/the-archipelago-economy (23 September 2013).

Shima, K., Onishi, T. and Leksono, P. (2006). Challenges and scopes toward the decentralized regional development framework in southeast Asian countries: Several lessons from Yogyakarta special province, Indonesia. Conference paper, 42nd ISOCARP (International Society of City and Regional Planners) Congress. Accessed at http//www.isocarp.net/Data/case_studies/768.pdf (23 September 2013).

Special Province of Yogyakarta (2005). *Regional Agenda*. Yogyakarta: Special Province of Yogyakarta.

Tambunan, T. (2006a). *Development of Small- and Medium-scale Industry Clusters in Indonesia.* Jakarta: Kadin Indonesia. Accessed at http://www.kadin-indonesia.or.id/enm/images/dokumen/KADIN-98-1572-02032007.pdf (23 September 2013).

Tambunan, T. (2006b). *The Role of Government in Supporting Transfer of Technology to and Technology Diffusion among Indonesian Non-farm Small and Medium Enterprises. Review of empirical studies and case study of Tegal.* Jakarta: Centre for Industry and SME Studies, Faculty of Economics, University of Trisakti, Working Paper Series No. 4.

Tambunan, T. and Supratikno, H. (2004). The development of small and medium enterprises clusters in Indonesia. *Gadja Mada International Journal of Business*, 6, 29–44.

World Bank (2012). *Indonesia. The Rise of Metropolitan Regions: Towards Inclusive and Sustainable Regional Development.* Washington, DC: The World Bank.

World Bank and IFC – International Finance Corporation (2012). *Doing Business Subnational. Doing Business in Indonesia.* Washington, DC: International Bank for Reconstruction and Development/The World Bank.

6 Mobile World Capital (Barcelona)

Alexander Otgaar and Luís Carvalho

1 Introduction

In this case study we analyse Barcelona's strategy to use its status as Mobile World Capital as a catalyst for the development of economic activities related to mobile technology and applications. Barcelona received the right of being Mobile World Capital (MWC) for seven consecutive years (2012–18) from GSMA: the association that represents mobile operators worldwide. As we will explain in more detail later, MWC is a comprehensive initiative that not only entails an annual conference, but also an annual festival, a permanent visitor centre and – most importantly – an on-going economic development and innovation programme.

Barcelona's ambition to use MWC for changing the city's economy fits into a long tradition of using events for strategic ends, with the Olympic Games 1992 and the Universal Forum of Cultures 2004 as recent examples. While these events mainly enabled fundamental changes in the urban fabric – opening the city towards the sea and improving internal accessibility – MWC is expected to further upgrade the city's economy, for example by stimulating the development of innovation platforms and networks (new forms of cooperation). As such, MWC provides an interesting addition to the growing body of knowledge on how to use (global) events for sustainable urban development.

An important feature of MWC is the strategic involvement of businesses in the development of a strong innovation ecosystem. Private companies are expected to have a prominent role in the Foundation that runs the initiative. MWC thus also responds to the trend of private-sector involvement and corporate leadership in local economic development. The initiative adds a new chapter to the famous 'Barcelona model' (Blanco, 2009), which has been described by several scholars as a way to use major events for urban regeneration, but also as the adoption of a participatory style of governance involving several layers of government, as well as businesses and citizens.

MWC can be described as a permanent (seven year) industrial event that promotes the city and its industries, targeting various audiences: professionals, citizens and visitors. The activities are expected to further improve the reputation of Barcelona as a business location and place to live for knowledge and creative workers. The initiative also clearly fits into the trend of using urban areas as living laboratories for the development and testing of new technological applications in various sectors (e.g. health, travel and finance), promoting the involvement of users in the design of these applications.

The structure of this chapter is as follows. In Section 2 we first take a closer look at the local context in which the MWC initiative has emerged: relevant trends in the regional economy, demography, urban planning and governance, and changes in 'capitals for sustainable competitiveness'. Section 3 introduces the vision and strategy of the city discussing the plans and actions that are related to MWC. In Section 4 we present the key findings of the empirical study, dealing with the set-up of the project (4.1), the implementation of the project and the factors that enabled progress (4.2), and the (expected) results of MWC (4.3). Finally, Section 5 presents lessons and challenges ahead.

2 City profile: capitals for sustainable competitiveness

2.1 Performance

With a population of around 1.6 million, Barcelona is the second-largest city of Spain. The Barcelona metropolitan area counts 4.8 million inhabitants, making it one of the ten largest metropolitan areas in Europe. The city is the capital of the autonomous region of Catalonia, which is one of most prosperous regions in Spain and an important engine of the Spanish export industry. While the region accommodates 16 per cent of the Spanish population, its share in GDP adds up to 20 per cent and its share in exports to 26 per cent. With 1.65 per cent of GDP being spent on R&D, the region has a higher performance in innovation than the Spanish average (1.39 per cent) but a lower one than the EU average (2 per cent).

In comparison with the EU average Catalonia has experienced a more than average growth since 1986 when Spain entered the Union. Between 1986 and 2009, the average per capita income (based on purchasing power parity) went from 86 per cent of the average (EU-15 = 100) to one that surpassed the EU average by 20 percentage points (EU-27 = 100) and surpassed the eurozone average by 12 percentage points (eurozone = 100) in 2009 (Nel. lo, 2014).

2.2 Economic structure and knowledge

Like many other regions in Europe, Catalonia has experienced a process of economic restructuring. Between 1977 and 2010 the share of industrial jobs in total employment decreased from 40.3 per cent to 19.2 per cent (8.9 per cent in Barcelona) while the shared service sector jobs grew from 41.6 per cent to 69.7 per cent (86.8 per cent in Barcelona). Table 6.1 provides a more detailed analysis of the city's economic structure. It shows that Barcelona is relatively strong in business services, health and social services, education and ICT (among others).

The City of Barcelona plays an important role in the transition of Spain towards a knowledge economy (City of Barcelona, 2011). According to the Catalan Department of Enteprise and Labour, 30.1 per cent of the Barcelona-based companies is 'knowledge intensive', employing 48.4 per cent of the working population. The share of workers in Catalonia with university studies is lower than the Spanish average: 36.9 per cent versus 37.8 per cent (EU average: 30.7 per cent). In total there are 12 Catalan universities of which eight are located in the Barcelona metropolitan area. The region is home to 21.6 per cent of all innovative companies in Spain while 17.1 per cent of all patents are registered in Catalonia, with Barcelona as the obvious centre of innovation.

For this case study it is particularly interesting to look at the relevance of the ICT sector, although – as we will explain later – the event also aims to have an impact on other industries such as health, transportation, finance and hospitality. In Catalonia the ICT sector employs about 77,500 people: 2.7 per cent of the working population. Together they contribute 2.1 per cent to the regional GDP. In Barcelona this percentage is significantly higher: 3.5 per cent. One of the leading firms in this sector is Telefónica, one of the five largest mobile operators worldwide and the second-largest firm in Spain. In 2011, the company inaugurated

Table 6.1 Contribution of economic sectors to GDP, 2012

Sector	*Barcelona*	*Catalonia*	*Spain*
Manufacturing industry and mining	3.5	6.4	6.1
Energy and water	0.4	0.4	0.7
Construction	9.8	14.4	14.5
Trade and repair	19.9	22.5	24.2
Transport and storage	5.2	7.1	6.6
Hospitality	6.3	7.6	8.9
ICT	3.5	2.1	1.7
Finance and insurance	2.3	2.1	2.2
Real estate	6.8	5.6	4.0
Professional, scientific and technical services	18.5	11.9	11.8
Administrative and auxiliary services	6.9	5.4	5.0
Education	2.8	2.3	2.5
Health and social services	7.2	5.3	4.4
Arts, leisure and entertainment	2.7	2.2	2.3
Other services	4.4	4.7	5.0
TOTAL	100.0	100.0	100.0

Source: Department of Statistics, City of Barcelona

the new global headquarters of Telefónica Digital, the firm's R&D division, in Barcelona (Madrid being the home of its corporate headquarters).

Telefónica Digital is located in the 22@ district, like many other firms in the ICT sector. This district, also known as Poble Nou, has been the focal point of Barcelona's transition to a knowledge economy for more than a decade.[1] The City of Barcelona describes 22@ as 'a new model for a compact city, where companies on the cutting-edge of innovation co-exist with universities, research, training and technology-transfer centres, as well as homes, infrastructures and green areas' (City of Barcelona, 2008, p.2). Poble Nou has been redeveloped into a high-density, mixed-use district with economic activities in a range of clusters: media, ICT, medical technology, energy and design (City of Barcelona, 2011).

Barcelona has been developing into an important test-bed for new technologies: adopting and adapting technologies (for the local market) instead of developing new ones. The city now has the ambition to 'enter the next level' by promoting the development of new technologies. In the first stage of Poble Nou, the aim was on bringing the assets of the region together; now the focus is on linking companies to the outside world. Barcelona wants to be a living lab for local and foreign companies, attracting external knowledge to the region. As we will explain later, MWC fits in with this ambition.

2.3 *Social conditions*

Nel.lo (2014) gives an overview of some important social and demographic changes in the Barcelona region. First, it becomes clear that the City of Barcelona has gained population since the mid 1990s (growing from 1.51 to 1.62 million), after a period of population decline between 1975 and 1996 (from 1.75 to 1.51 million inhabitants). Statistics show that not only the City of Barcelona but also other municipalities in Catalonia – particularly smaller towns with between 20,000 and 50,000 residents – have seen their population grow. A closer look at demographic data provides us with an insight into why Barcelona's population has increased: between 1996 and 2010 the number of foreigners grew from less than 30,000 to more than 280,000 (17.4 per cent of the population). Like other cities in Spain, Barcelona experienced a massive inflow of job seekers in a period of economic growth.

Of course, the current crisis has a major impact on the socio-economic conditions in Barcelona and Catalonia. In Catalonia the unemployment rate increased from 9 per cent in 2008 to 22.6 per cent in 2012, and a similar trend can be observed for Barcelona (from 8.7 per cent to 22.6 per cent). Barcelona and Catalonia are performing slightly better than Spain as a whole, with an unemployment rate of 25 per cent.

2.4 Environment and accessibility

Barcelona is well connected to global transportation networks. The El Prat airport is Europe's tenth largest, with around 30 million passengers a year (City of Barcelona, 2011). Excellent accessibility, an attractive urban environment and a favourable climate probably explain why Barcelona is one of the most popular destinations for meetings and events. In 2010, Barcelona hosted 148 international meetings, making it the second city for organising such meetings worldwide (City of Barcelona, 2011). In the same year, the Barcelona Tourist Board and the City of Barcelona counted in total 616,833 delegates attending conferences, symposia, courses, conventions, etc. Barcelona is one of the top European cities in terms of quality of life, which not only makes the city attractive as a meeting place, but also as a place to live and work. For several years now, the city has been leading on this indicator in the well-known European Cities Monitor (e.g. Cushman and Wakefield, 2011).

2.5 Institutional conditions

Catalonia is one of the autonomous communities in Spain. Its elected government (the Generalitat) has formal planning competences, also supervising plans developed on the level of municipalities. The community authority takes care of infrastructure, housing, education and health and has the right to collect income tax, adding to its financial independence from Madrid. Catalonia consists of four provinces that mainly function as departments of the state. On the lowest level, municipalities – such as the City of Barcelona – enjoy a relatively high level of autonomy.

Administratively, the Metropolitan Area of Barcelona comprises Barcelona and 34 surrounding municipalities in the so-called 'first ring'. Within this area municipalities cooperate on issues that cross municipal borders such as transport, housing, service delivery, water supply and waste management. This area has a population of around 3.2 million. In its regional plan, the Catalan government recognises a metropolitan 'ambit' housing more than 5 million residents and including also a second and third ring of municipalities around Barcelona.

Several discussion partners refer to the political stability as an important context factor in the development of Barcelona through the last decades. For a long period (1979–2011) the city was governed by left-wing socialists. Only recently, in 2011, a Catalan nationalist party (Convergence and Union) in favour of (more) independence succeeded to beat the Socialists' Party of Catalonia. The left-wing orientation of Barcelona clearly differs from the relatively stable right-wing majority of the Catalan government. Partly due to the stable political situation the financial process has been organised with extreme rigour. In comparison with other Spanish cities the financial position of Barcelona is relatively strong. Previous experiences with large debts and austerity measures (in the early 1980s and the early 1990s) now prove to be valuable in the current crisis.

Barcelona has often been praised for its organising capacity. The city is known for the 'Barcelona model' (see Arbaci and Tapada-Berteli, 2012; Blanco, 2009), which comprehends a style of urban management with strong public–private interaction, leadership of the City in the design and management of urban transformation projects and active promotion of Barcelona as a global city. Possibly it is the limited availability of urban space that explains why Barcelona has always been very careful with transforming the city, attaching much value to preservation, urban vibrancy and diversity (mixing functions).

Another element of the Barcelona model is the use of events as catalysts of urban regeneration. Barcelona has a history of using events to transform the urban fabric and the economic structure of the city. The severe crisis of 1977–84 gave incentives to develop a common agenda for opening the city towards the sea and improving infrastructure and the quality of housing and the living environment significantly. The Olympic Games in 1992 stimulated stakeholders on various levels (including the national government) to implement the transformation in time. After the mega-event the city became more popular as a tourist destination, but also a place to live and to locate a business. A second crisis in the early 1990s made actors aware of the need to not only change the urban fabric but also the economic structure, turning Barcelona into a city of knowledge and creativity. This time the city invented a new event – the Universal Forum of Cultures – to align different stakeholders. Although the event has been criticised for its strong business orientation (García, 2008 cited by Martí-Costa and Miguel, 2012) and the fact that it attracted less visitors than expected, it has played an important role in the transformation of Poble Nou.

2.6 Conclusions: main issues

In this section we analysed Barcelona's capitals for sustainable urban competitiveness. Our analysis provides some insight into the main issues that explain the development of the project to be discussed in Section 4. One of the main issues is the fact that the city – despite its strong performance in the Spanish knowledge economy and attractive living environment – has not been able to attract many knowledge workers from abroad. The share of workers with university studies is relatively low. Another observation is that the city is relatively strong in adopting and adapting technologies (incremental innovations), but not so much in creating new technologies (radical innovations). After the successful redevelopment of Poble Nou, emphasis has shifted from physical restructuring to socio-economic transformation. The main challenge is to enhance Barcelona's position in the global knowledge economy, notably by attracting talents and innovative firms. It will be discussed in the following sections whether a business event such as MWC can be instrumental to that end.

3 City-wide vision and strategy

3.1 Vision

In 2008, the Metropolitan Area of Barcelona (AMB) presented a document that formulates a vision for the development of the region until 2020. More than 650 people have participated in various working groups to discuss the (economic) future of the metropolitan area, which resulted in the following vision:

> By 2020, as a global metropolis and capital of Catalonia, the AMB will have reinforced relationships with the emerging cities of the world and the capitality of the Mediterranean.

> It will exercise economic and social leadership in line with its values and competitiveness factors; Sustainability is one of these values.
>
> (Pla Estratègic Metropolità de Barcelona, 2010, p.8)

In view of the present case study, two aspects of the vision are particularly relevant: one is the ambition to develop 'new growth-driving sectors'; another is the strong emphasis on sustainable development and competitiveness. As we will explain in the next section, the MWC initiative fits well in this vision, notably because it aims to stimulate the development of new sectors but also because mobile solutions could play a role in enhancing the sustainable competitiveness of the Metropolitan Area of Barcelona.

Also the Barcelona City Council sees the MWC as an important instrument to create jobs and to stimulate economic growth, and more specifically to transform Barcelona into a city of business. The project has been identified as one of the 'main priorities' in the economic strategy, which has also been translated into the allocation of staff and budget.

3.2 Strategy

Vision 2020 has been translated into six strategic lines representing different challenges for the region:

1. sustainability and climate change;
2. position the AMB as a benchmark in the new global framework: greater presence in countries which lead the world and capitality of the Mediterranean;
3. global leader in some growth-driving sectors of knowledge;
4. getting beyond the 'bio' companies. Updating and strengthening traditional industry and sectors;
5. being among the most attractive European regions for innovative talent;
6. being an interesting and socially balanced city: a social response to the crisis.

For each strategic line, a commission takes care of initiating concrete projects. In the field of sustainability and climate exchange, for example, there is a project that deals with improving the energy efficiency of buildings. The idea is to make existing buildings more energy-efficient (retrofit), thus meeting EU targets and creating jobs. Fifteen different institutions signed an agreement to participate in this project. Another project under this line involves the introduction of indicators to measure the sustainability performance of the city.

The commission that works on the third strategic line (becoming global leader in growth-driving sectors of knowledge) aims to develop the metropolitan area into an international centre for training and innovation, attract talents from abroad and produce new products that can be exported to the world. Vision 2020 highlights the importance of public–private collaboration. It states that governments should move the traditional 'subsidies model' to a model in which private actors get incentives and are encouraged to take initiatives themselves and become more creative. The document specifically identifies the creative industries, design and architecture, health and sport as examples of emergent sectors in the Barcelona area. The third strategic line is obviously strongly related with the fifth strategic line, which formulates the ambition to attract more talents to the Barcelona region and to develop more entrepreneurial spirit. As we will further explain in the next section, MWC fits in the third and the fifth strategic lines of Vision 2020.

In 2011, another initiative was launched which is highly relevant for reaching the ambitions of MWC: Barcelona Growth. This organisation, located in the same building as the Mobile World Capital Barcelona Foundation (MediaTIC), wants to promote economic development by further enhancing the Barcelona brand and supporting entrepreneurs. Barcelona Growth is a platform for cooperation involving partners such as the City of Barcelona, the Chamber of Commerce, Fira de Barcelona and Turisme de Barcelona (the tourism organisation). During a roundtable in July 2011 participants agreed on five measures to boost economic growth. One of these measures is to convert the MWC into an economic motor for different sectors.

4 The project: MWC

4.1 Introduction

In the present chapter we discuss the 'project' Mobile World Capital. We explain how it evolved from a four-day conference to the status of being capital for seven consecutive years, with four pillars instead of only one. We discuss the role of the City of Barcelona, but also look at the role of Telefónica and other private actors, reviewing how the initiative is organised and financed.

4.1.1 Brief story of the initiative

The Mobile World Conference is an initiative of GSMA: a global association of mobile operators and related companies. The event consists of an exhibition for the mobile industry and a conference with speakers who represent various stakeholders in the mobile industry, such as operators, manufacturers and content owners. The Mobile World Conference is the largest event in the industry worldwide.

Originally, the Mobile World Conference was hosted by the French city of Cannes under the name 3GSM World. However, because this city lacked sufficient capacity to host the conference, it was decided to look for another city from the year 2006 onwards. GSMA accepted the proposal of Fira de Barcelona to host the conference for four years in their venue at Montjuïc. It allowed the conference to receive a record number of 50,000 participants in 2006.

Four years later, in 2010, GSMA announced the launch of MWC, enabling the conference to grow into something even bigger. This concept would not only include the annual business-to-business congress, but also provide permanent research and business development opportunities, as well as events and meeting places for business-to-consumer relations in a city. According to John Hoffman, the CEO of GSMA at that time, the MWC city would become 'the "home" of the mobile industry, synonymous with cutting-edge mobile technology that benefits not only citizens of the city, but also the world'.

Barcelona was one of the cities that prepared a bid to become the first capital. There was strong competition from other European cities, including Amsterdam, Cologne, Milan, Munich and Paris. The Fira prepared a bid book that was supported by the City, being one of their shareholders. But also the province and the state supported the bid, with the advantage of having the same political party (socialists) leading the three levels of government. Recently this all changed with the nationalist party and conservatives coming into power, but the commitment has remained.

Already at the launch of the initiative, GSMA had some ideas of what the MWC would include. Apart from the congress, the association thought about:

- The Mobile World Centre, to be located at a strategic position in the city as a permanent flagship of MWC. This Centre could accommodate an exhibition, a museum, retail facilities, offices, etc.
- The Mobile World Festival, which targets consumers rather than businesses. This festival had to take place separately from the congress and include a range of activities such as concerts, musical and movie festivals, competitions and awards, etc.

In 2010, GSMA expected an economic impact of more than €1.2 billion and 6,000 part-time jobs for the city to become the capital for five consecutive years. Moreover MWC would also produce intangible benefits for the hosting city, in terms of an improved image and ability to attract and retain firms and talent. In the summer of 2011 GSMA announced that Barcelona was selected to become the first MWC, for an even longer period than originally envisaged: 2012–18. This longer duration was part of the Barcelona bid and was accepted by GSMA.

4.1.2 The concept

While the Mobile World Congress is 'only' a four-day event, Mobile World Capital is a more or less permanent status, at least for seven years (2012–18). In line with the original ideas of GSMA, Barcelona has developed a concept that is much broader and more comprehensive, with considerable impact expected, also after 2018 (legacy). New elements have been added to spread mobile technologies in other economic sectors and society at large. The concept is based on the idea that innovations in mobile technologies and applications change the way people and organisations behave and operate. Apart from the Congress, the Festival and the Centre, Barcelona also developed the Mobile Word Hub. Together they constitute the 'four pillars' of MWC Barcelona (Figure 6.1).

The Mobile World Congress is a four-day business-to-business event that takes place in Fira Gran Via. From 2013 onwards, the congress has taken place in this new venue, which is double the size of the venue on the Montjuïc mountain (240,000 square metres including 90,000 square metres of exhibition space) and located closer to the city centre.

In line with the vision of GSMA, the Mobile World Festival is a series of business-to-consumer events to spread the use of mobile technology and its benefits. An example of such an event is the Phonetastic Film Festival, which promotes the use of mobile devices as a tool for the production and distribution of films. Other examples are Sonar (an international festival for advanced music and new media art) and the Mobile Summer Week (a music and leisure event). The ambition is to communicate Barcelona's status as MWC 365 days a year, e.g. by incorporating 'mobile elements' in other cultural and sports events such as the Formula 1 championship, football matches of FC Barcelona and tennis tournaments. This will be done through strong cooperation with private event organisers and the City's cultural promotion department.

The Mobile World Centre is a physical place in the city that provides a business-to-consumer platform to spread and demonstrate new mobile technologies and solutions to citizens. The building that accommodates this centre is a store of Telefónica, one of the private-sector partners in MWC, located at Plaça de Catalunya near one of the busiest pedestrian crossings of Europe. The building comprehends a showcase and retail store of

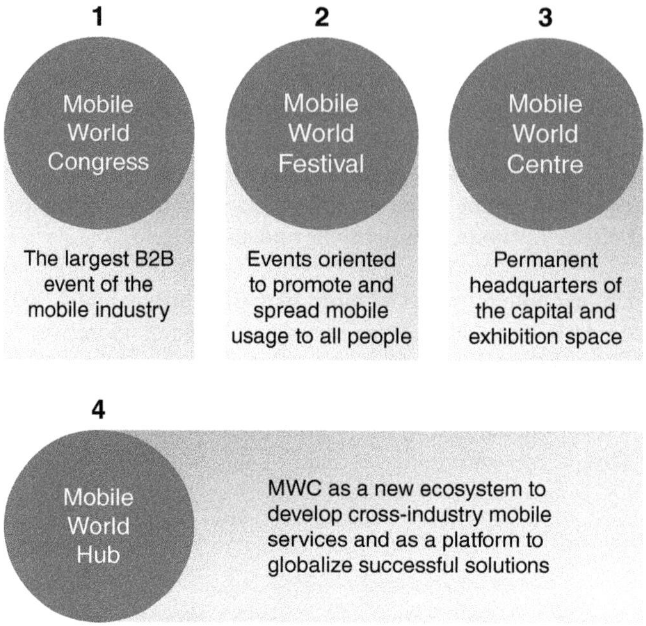

Figure 6.1 The four pillars of MWC Barcelona

Source: Innovation Strategy Model (Government of Catalonia, 2013), adapted by the authors

Telefónica (ground floor), a permanent exhibition of MWC (first floor) and space for mobile events to engage the public (second floor, 440 square metres). The Centre is a place where consumers can experiment with new mobile technologies and services and where producers can launch new products or ideas. It is open to visitors, free of charge, six days a week.

The Mobile World Hub is a permanent meeting place for businesses, allowing companies from the ICT sector to get in touch with companies from other sectors. The Hub is located in the MediaTIC building in the 22@ district. Its mission is 'to develop a new industrial sector around mobility and transform Barcelona into the world's leading mobile city' (MWCB, 2012b). The idea is to launch various competence centres (also referred to as 'solutions') around areas or sectors where the initiators expect significant impact of mobile solutions. One of these centres is mHealth, which promotes the use of mobile technology and applications in healthcare, for example related to the vaccination of children, clinical tests or a system to follow patients.

Table 6.2 provides an overview of the competence centres to be developed in cooperation with relevant public and private stakeholders.

Apart from the competence centres or solutions, the Hub also provides value-added services to accomplish its mission. These capabilities are listed in Table 6.3.

Another initiative of the Hub is the mSchools programme (also referred to as mEducation). This programme has been developed in close cooperation with GSMA and the Catalan government. Its aims are 'to lower dropout rates, improve student attainment in schools across Catalonia and throughout Spain and, ultimately, to better prepare students as they pursue further education and employment in today's digital world' (MWC, 2013). GSMA will invest €2.5 million in this programme.

Table 6.2 Competence centres of MWC

Competence centre	*Mission*
mWallet	To promote the use of smartphones as virtual wallets with related 'smart services'
mSmart City	To integrate urban services by way of mobile technologies
mHealth	To promote the use of mobile telecommunications in health care services
mContent	To improve customers' experience with enriched multimedia mobile services
mTravel	To enhance travelling experiences and safety
m4All	To facilitate technologies for everybody, despite poverty or disabilities

Source: http://mobileworldcapital.com

Table 6.3 Capabilities of MWC

Capability	*Mission*
mProgram	To stimulate the development of mobile projects with the participation of all relevant agents from the public and the private sectors
mLaw	To enhance the know-how of regulators and standardisation bodies for the use and development of mobile solutions
mNews	To spread news on mobility knowledge and innovative mobile initiatives
mTalent	To create a knowledge base linked to the territory to be capitalised in talent attraction and generation
mEntrepreneurship	To promote start-up creation and relocation in Barcelona with the required business, technology and financial support
mCluster	To promote the co-creation of mobile solutions while at the same time reinforcing innovation in clusters related to the mobile field

Source: http://mobileworldcapital.com

The idea of the Mobile World Hub is to use the city as a living laboratory for creating products that can be exported to the world. According to one of the discussion partners, the challenge is to connect global firms such as Microsoft and Google with new companies in the Barcelona region, in order to develop applications in health, education, travel, etc.

4.2 Implementation

In this section we discuss the implementation of the MWC concept. We will explain how various actors and organisations play a role in this process.

4.2.1 Organisation and finance: the MWCB Foundation

In January 2013, the Mobile World Capital Barcelona Foundation (MWCB) was officially founded with the mission to develop through the Mobile World Capital Barcelona a new industrial sector around mobility and transform Barcelona into the world's leading mobile city with high value-added services and technology clusters (MWCB, 2012b). MWCB's founders are the Ministry of Industry, Energy and Tourism (State), the Catalan government, the City of Barcelona, Fira de Barcelona and GSMA. These organisations are represented on the Foundation's board.

Three levels of government support the foundation financially with an annual contribution of €15 million in total: €5 million each. In addition there is a comprehensive programme for private-sector sponsorship aiming at an annual contribution from companies of around €15 million as well. One of the donors is telephone company Telefónica (see Box 6.1). In 2012 they signed an agreement to make a contribution of €7.8 million for various projects to be realised between 2012 and 2014. This agreement included the realisation of the Mobile World Centre in the Telefónica store at Plaça de Catalunya. The MWC Foundation confirms that Telefónica will contribute an amount that is slightly lower (€6 million) until 2015. At the end of 2013, Gruppo Damm, a Catalan brewery, was the only other private sponsor with a contribution of €1 million (for February 2013 until March 2014). So far, the goal to match the public funds with private funds has not been fully achieved: the Foundation is still looking for additional funds.

The Foundation directly employs around 30 people. Their main task is to communicate and cooperate with other institutions and individuals who can help to achieve the mission of the initiative. MWCB (2013) considers 'joint and coordinated action' as one of the two ways to 'make it happen' (the Foundation uses these expressions in its external communication). Interaction is also needed in view of the second way to make it happen: the development of integral solutions.

4.2.2 Platforms for cooperation

Within the Foundation, the competence centres function as platforms for cooperation. MWCB offers various tools as listed in Box 6.2.

Companies are invited to join the initiative, by participating in the competence centres and using the MWC logo in advertisements. Apart from some tax benefits, the Foundation emphasises the strategic advantages of corporate involvement in MWC. It is presented as a

Box 6.1 Telefónica

Telefónica is one of the world's largest operators in the telecommunications sector, with presence in 24 countries in Europe and Latin America. The company wants to be an 'engine for economic sustainable development and innovation and to help overcome social divides' (Telefónica, 2012, p.11). They do this, among others, by a concept called Wayra. In various cities, including Barcelona, the company provides a platform where researchers and entrepreneurs can meet each other. Telefónica not only provides financial assistance to start-ups (through venture and seed capital), but also technical, legal, administrative and management advice/support and even office space in its buildings.

In the last decade Telefónica has become more active in the Barcelona region, partly because the Catalan government urged large firms to take their responsibility. The company has been advised by a board of external advisors (with representatives of the Chamber of Commerce, universities and governments) on how to implement its corporate social responsibility strategy. This process resulted in three important achievements: 1) Telefónica set up an R&D centre in Catalunya with more than 300 researchers from all over the world; 2) they opened a corporate university in the region with courses for managers of the firm; and 3) the company moved to a new building, a new symbol for the city.

Box 6.2 Tools offered by MWCB

MWCB offers participating companies differential tools to take part in industrial projects throughout the value chain and to launch cross-industry mobile solutions:

1. MWCB facilitates the mobile innovation process: a one-stop showcase for mobile innovation and a meeting point for establishing relationships that help to achieve common business objectives.
2. MWCB aggregates partners for strategic mobility agendas: companies from different points of the mobile value chain can jointly address common mobile challenges and cooperate to develop innovative projects and solutions.
3. MWCB facilitates the efficient sharing of (complementary) competencies between partners: they all add value to the mobile initiative development, from shaping business models to launching solutions, from market-oriented to technology-oriented assets, from intangible know-how to physical infrastructures and from human resources to funds.
4. MWCB shapes the ecosystem in order to facilitate the development of mobile initiatives, by adapting the regulatory framework, making available and promoting new technological infrastructures, using Barcelona as a real-life mobile testing laboratory, giving incentives for company involvement and investment, promoting mobile virtues and benefits and raising people's awareness, and creating a knowledge base in the region.
5. MWCB transforms the value chain by the development of cross-industry projects (resulting in solutions with significant business, social and economic impact), involving all relevant stakeholders from the different industries and helping them to work together, taking advantage of the Capital environment and capabilities.
6. MWCB helps to launch and market mobile solutions developed within its framework, promoting them worldwide.

Source: Forward Mobile (MWCB, 2012a)

platform for innovation that allows companies to bring mobile solutions to the market in a shorter period of time, with less risk and greater ease, with lower costs and more revenues (a higher impact) (MWCB, 2012a).

Apart from businesses, universities and business schools are also invited to join the programme. Various commissions have been set up to bring public and private actors together. One of the examples is the Commission for the mSmart City in which the City Council and Catalunya's government are leading the process. They succeeded in involving various other private and public partners, such as Indra, a Spanish global technology solutions provider, and ATM, the Metropolitan Transport Authority. Another example is mWallet in which the bank La Caixa cooperates with Telefónica and Indra to develop a NFC (Near Field Communication) contactless payment plan. The City of Barcelona is involved in this plan by promoting the use of smartphones for payments and setting up the technological infrastructure needed in all the City Council institutions and equipment around the city. La Caixa, one of the most important banks in Spain, contributes by promoting the contactless payment system between its clients.

4.3 Results

Barcelona has been successful in winning the bid for becoming the host city of the Mobile World Conference for several years. In 2013, more than 72,000 people participated in the conference, which is a new record for the event. Obviously, congress participants spend more than the average tourist, resulting in an economic impact of around €320 million for the region.

Together with the founding organisations, the city has managed to develop the four-day conference into only one of the pillars of a much broader 'event': Mobile World Capital. An important output is the realisation of a permanent visitor centre, which attracts around 7,000 visitors a week and mainly facilitates interaction between business and consumers. Another result is the development of the Mobile World Hub, which acts as a platform for businesses and other stakeholders in the development of new mobile applications (e.g. healthcare, travel, etc.). At the end of 2013, three competence centres and one capability were operational: mEducation, mHealth, mEntrepreneurship and m4all. Our research does not provide insight into why the other centres and capabilities are not operational yet.

The appointment of Barcelona as MWC has also strengthened the (perceived) attractiveness of the city as a place to locate, with the MediaTic building in the 22@ district as an important hub for both local and international firms. Several companies have established in Barcelona, partly because of the favourable environment offered by the city's status as MWC.

One of these companies is NXP, a Dutch company that developed NFC technology. Another example is Toro, a Taiwanese semiconductor company. In this case, MWCB accompanied the firm's landing by helping to recruit engineers and other highly qualified staff, and providing strategic and communications advice. While NXP is still located in the MediaTic building, Toro moved to its own offices after an initial stay in the Hub.

Other companies that settled in Barcelona in 2013 are Sopra Group (ICT, France), TBSCg (software, UK), King.com (mobile gaming applications, US) and Transax (mobile payment, Canada). Although these companies have no direct link with MWCB, all said that one of the reasons for establishing in Barcelona was the favourable environment offered by the fact of the Capital and the future potential of Barcelona in this area.

5 Conclusions

5.1 Broad assessment

MWC is a multi-year event that has been developed by the City of Barcelona in cooperation with the Spanish and Catalan governments, the trade fair (Fira de Barcelona) and the organising association GSMA. The event has evolved from a four-day conference with substantial but still limited economic impact to a much more compressive all year long programme with a much more sustainable influence on the urban economy (legacy).

The development of MWC responds to the city's structural assets and capitals: while the economic structure and export orientation is relatively strong, Barcelona has found it difficult to attract and retain knowledge workers despite the favourable living conditions. After several rounds of physical upgrading (with events as catalysts), the remaining challenge is to further upgrade the economy, creating an ecosystem for radical innovations instead of only adapting existing products and technologies (e.g. for the local market). A global event such as MWC can be used to develop clusters that build on local buzz (frequent interactions, exchanging tacit information) and global pipelines (less frequent interactions, needed to get access to technology, knowledge and funds).

Enhancing global connections is also an important objective for the Metropolitan Area of Barcelona (AMB). Although the region's Vision 2020 does not explicitly refer to MWC, the event does clearly fit into the vision that has been formulated and the six strategic lines for future development. On a local level, however, several partners (the City, the Chamber of Commerce, the tourism organisation and the trade fair) are involved in a platform that does recognise MWC as an important tool for economic development and branding.

After Barcelona was appointed as MWC in 2012, the actors involved have started to implement the concept. A visitor centre was opened to raise awareness among end-consumers and to facilitate interaction between businesses and consumers, and the conference was relocated to a larger facility, allowing the number of visitors and economic impact to grow. Most importantly, however, they succeeded in setting up a programme to stimulate interaction between businesses, governments, knowledge institutions and end-users. The MediaTIC building in the 22@ district acts as a physical platform for the development of innovative partnerships between actors who want to develop new mobile applications in various fields. Particularly in the field of education and health, the results are promising. In other domains, more time is needed.

5.2 What have been critical factors?

Although the MWC event (2012–18) cannot be fully evaluated yet, with several more years to come, we have been able to identify some critical success factors in the first years of implementation.

First, the broad and stable political support has strengthened the bid for organising the event and made it easier to implement the comprehensive four-pillar strategy. In the bid preparation stage, the three levels of government were of the same political colour, smoothing inter-governmental cooperation. Elections have taken away this advantage, but the three levels of government (local, regional, national) are still highly committed and support the event financially. Interviewees also refer to the strong support of the local tourism industry via Turismo de Barcelona as a factor of success.

Second, the role of Telefónica should not be underestimated. Although difficult to prove, it is very likely that the company (and its CEO) was pivotal in the selection of Barcelona as hosts of the event. Moreover, the firm has demonstrated strong commitment to the initiative by investing a substantial amount (€6 million) in the MWCB Foundation.

Third, discussion partners refer to the strong infrastructural capital: the technical facilities needed to host the Mobile World Conference. Barcelona first managed to offer a bigger venue for the event than Cannes and a few years later, the City succeeded in finding an even larger venue (Gran Via). Barcelona's good technical offer obviously played a role in beating competing candidate cities such as Munich and Milan. Although the MWC event has grown into something more comprehensive than the Mobile World Conference, the conference is still the main event, attracting more than 70,000 people in less than one week.

The relevance of some other factors is less clear. Barcelona has been able to become the first MWC without a strong concentration of companies in 'mobile industries', apart from Telefónica. While Helsinki has been successful in becoming world design capital because of Finland's design industry, this argument seems to be less valid for the case of MWC. A possible explanation is that the 'mobile industry' is less visible (for one thing in statistics) and even more dispersed than the design industry. One could also argue that the mobile industry belongs to the ICT industry, in which Barcelona is relatively strong, with 22@ an important hotspot. The decision to locate the Mobile World Hub in 22@ fits in with this view, and 'mobile activities' are now gradually emerging, partly as a result of the global event and its permanent platform for B2B interaction.

The idea to use an industry event such as the Mobile World Congress for economic restructuring is potentially interesting also for other cities. With good meeting facilities and broad support of government and leading firms, other cities could be equally successful in attracting or developing similar events. Intuitively one would advise cities to organise events for industries in which they have a competitive advantage, but the experience of MWC Barcelona suggests that this is not a very strict condition.

Similar to other innovative urban development strategies, the MWC Barcelona project still needs some time to demonstrate its full impact on the sustainable competitiveness of the city. At the time of our visit (end of 2012), discussion partners seemed to agree on the main challenge for MWC, which they referred to as 'keeping momentum'. Some interviewees indicated that the project's progress (speed of implementation) was below expectations. Both public and private investors want to see the revenues of their investments, notably regarding the ambitious Mobile World Hub programme. More than one year later (early 2014), we must conclude that the number of competence centres and capabilities of MWC has not really increased. Apparently, it is still difficult to turn the vision and strategy into actions.

Note

1 The district is named after 22a, which used to be a code for land zoning, meaning industrial land that is suitable for residential purposes as well (in contrast with 22b and 22c). Obviously the @ is a symbol for the focus on ICT.

References

Arbaci, S. and Tapada-Berteli, T. (2012). Social inequality and urban regeneration in Barcelona city centre: reconsidering success. *European Urban and Regional Studies*, 19, 287–311.

Blanco, I. (2009). Does a 'Barcelona Model' really exist? Periods, territories and actors in the process of urban transformation. *Local Government Studies*, 35, 355–69.

City of Barcelona (2008). *22@Barcelona: The Innovation District*. Barcelona: City of Barcelona.

City of Barcelona (2011). Barcelona Datasheet 2011. Accessed at http://barcelonacatalonia.cat/b/wp-content/uploads/2012/02/datasheet-eng-2011.pdf (4 February 2013).

Cushman and Wakefield (2011). European Cities Monitor. Accessed at http://www.cushmanwakefield.co.uk (4 February 2013).

García, M. (2008). Barcelona: ciudadanos y visitantes. In M. Degen and M. García (eds), *La Metaciudad: Barcelona. Transformación de una Metrópolis*. Barcelona: Anthropos.

Government of Catalonia (2013). *Innovation Strategy Model*. Barcelona: Government of Catalonia.

Martí-Costa, M. and Pradel i Miquel, M. (2012). The knowledge city against urban creativity? Artists' workshops and urban regeneration in Barcelona. *European Urban and Regional Studies*, 19, 92–108.

MWC – Mobile World Capital (2013). Teachers and students, allies through mobile applications. Press release, 21 June. Accessed at http://mobileworldcapital.com/73/ (23 September 2013).

MWCB – Mobile World Capital Barcelona (2012a). *Forward Mobile*. Barcelona: Mobile World Capital Barcelona.

MWCB – Mobile World Capital Barcelona (2012b). Mobile World Capital Barcelona: setting up a new generation mobile smart city. Keynote at Mobile Asia Expo. Accessed at http://www.gsma.com/connectedliving/wp-content/uploads/2012/05/8-Juanma-Barrionuevo-MWCB-Setting-up-a-new-generation-mobile-smart-city.pdf (4 February 2013).

Nel.lo, O. (2014). Barcelona. In L. van den Berg, J. van der Meer and L. Carvalho (eds), *Cities as Engines of Sustainable Competitiveness*. Aldershot: Ashgate.
Pla Estratègic Metropolità de Barcelona (2010). *Vision 2020*. Barcelona: City of Barcelona.
Telefónica (2012). Telefónica's Profile – June 2011 (no longer available online).

7 Pacifying Police Units (Rio de Janeiro)

Hazem Galal

1 Introduction

Rio de Janeiro is Brazil's most famous city and the second largest Brazilian metropolis after São Paulo. Historically, Rio de Janeiro was Brazil's political, cultural and financial capital and was lovingly called by its inhabitants '*a cidade marvilhosa*', or the marvellous city, for its natural beauty. After 1960, when Brasilia became Brazil's political capital, Rio de Janeiro entered a period of decline which continued until a few years ago. The city faced the challenges of substantial reduction of financial resources, poor public management and continuous loss of talent, businesses and investments.

During this period, among others, lack of investment in infrastructure and housing led to the growth of the city's '*favela*' communities. These shanty town/slum areas housed the city's growing immigrant population arriving from the northeast of Brazil. High levels of unemployment, coupled with corruption and loose policing soon resulted in the take-over of many of these favela communities by drug-traffickers and gangs. A dual economy developed with residents of favelas living under the daily control of drug dealers, with limited access to public services, economic opportunities and reduced freedom of passage. The violence of the favelas spilled out into other areas of the city, leading to alarming rates of crime and homicide across Rio de Janeiro.

The year 2008 was a turning point for the city's policies, politics and development strategy, linked with two major events. The first was full political alignment on all three levels of government (the municipal, state and federal) of the city for the first time in many years, following the election of a new mayor. The second was the city's successful bid to host the 2016 summer Olympics. In convincing the world that Rio de Janeiro was ready to be a host for the Olympics, the main issue that the city had to deal with was how to make sure the event would be hosted securely.

The proposed answer to Rio de Janeiro's security problem came in the form of the Pacifying Police Unit programme (Unidades de Policia Pacificadora), abbreviated UPP. Introduced in 2009, the UPP has been hailed as transformational in reducing violence, cutting crime and increasing security in the favelas. The UPP, delivered by the state government of Rio de Janeiro, aims at reclaiming territories controlled by gangs of drug dealers and private militias. UPP Social, a follow-up programme delivered by the municipal government, seeks to re-integrate residents of pacified favelas back into the city through coordination of and improved access to public and private services.

To date, the UPP programme has generally been regarded as successful in reducing crime in the favelas, but also due to the spill-over socio-economic development effects. The UPP programme has attracted millions of dollars in donations from companies and there are plans to expand the programme beyond the city of Rio de Janeiro to other parts of Brazil. The recent financial crisis facing Brazil has slowed down these plans. Against the growing movement to demand a global ban of 'slum clearance programmes' and other forms of mass evictions, Rio de Janeiro's UPP scheme is a welcome innovation in the global struggle against slum violence. Its focus on re-gaining control of previously 'given up on' areas and the reintegration of slum residents into the economic, social and cultural life of the city has captured the attention of policymakers the world over.

In order to explore these issues and critically assess the early development of the UPP programme in Rio, the chapter is structured as follows. Section 0 assesses the profile and recent dynamics of Rio de Janeiro across a number of dimensions and capitals behind sustainable competitiveness. Section 3 outlines Rio de Janeiro's city-wide vision and strategy, clarifying how the UPP programme contributes to the city's overall strategic objectives and priorities. Section 4 explores the UPP programme in-depth including its objectives, organising capacities involved in its implementation and early results. Section 5 concludes the chapter with some lessons learned for other cities.

2 City profile

2.1 Performance

Rio de Janeiro is the second largest city of Brazil, the third largest in Latin America and the sixth largest in the world, boasting approximately 6.3 million people in the city proper and 11.7 million people in the metropolitan area. Rio de Janeiro city is part of the State of Rio de Janeiro, one of 27 states making up Brazil. Rio de Janeiro's population has grown quickly since the beginning of the century, nearly tripling in size between 1950 and the present day. Between 2000 and 2010 the population grew by 8 per cent (IBGE, n.d.). GDP in Rio de Janeiro was estimated at approximately R$175 billion (approximately USD 88 billion) in 2009 and grew by roughly 50 per cent in real terms over the period 2005–9 (IBGE, n.d.).

For many years Rio de Janeiro's economic structure was based around its position as the largest industrial hub of Brazil, with shipbuilding industries, steel, metallurgy, petrochemical, gas, oil refineries, chemicals, textiles and others. Over the last decade Rio de Janeiro has experienced a sharp transformation in its economic profile and is becoming a major national hub of services, including media, telecommunications, arts, education, tourism, among many others. Despite recent improvements, Brazil is still one of the most economically unequal countries in the world and Rio de Janeiro is no exception. The Brazil 2010 census (IBGE, 2011) showed that 25 per cent of the population lives on an average monthly income per capita of up to US$106 and half the population up to US$211, compared to a minimum wage of US$288 in 2010.

2.2 Emergence of favelas, crime and social conditions

A favela is the generally used term for a shanty town or slum in Brazil. Favelas are characterised by urban decay, high rates of poverty and lack of formal land ownership. However, on the contrary, community ties and sense of belonging are typically very strong (Silva, 2009). Nowadays there are approximately 900 favelas in the State of Rio de Janeiro and, according to the 2010 census data, 1.3 million of Rio de Janeiro's population live in a favela (22 per cent of the city population) (IEST, 2010). This compares with 6 per cent of Brazil's population as a whole.

Most favelas started to appear in Rio de Janeiro at the beginning of the 1970s, when people migrated away from rural areas of Brazil and moved to the cities. Without finding a place to live, many people ended up setting up homes in areas where residential development was prohibited, for example on protected lands.[1] Rio de Janeiro is somewhat unusual in that many of its favela communities are located right in the centre of the city, spreading across the hillsides, often alongside some of Rio's most valuable real estate.

The prevalence of drug-trafficking and associated violence in the favelas of Rio de Janeiro dates back to Brazil's military dictatorship (1964–85). During this period the country's jails started filling up with political prisoners, who started mingling with common criminals from the favelas. There, extreme leftist inmates taught prisoners from the favela how to organise themselves, reasoning that this would give the revolutionary movement a firm foothold in the favelas (Campos and Santos, 2004). On the return to the favelas large gangs started to form, quickly mutating from politically driven groups to ruthless moneymaking mafia. During the 1980s and 1990s Rio de Janeiro's presence as a key distribution and consumer point and its presence in the international drugs market was becoming well established. During the same period, economic stagnation, hyperinflation and high unemployment were among the factors that contributed to what is known as the 'lost decade' in Rio. Lack of economic opportunities meant that it was only a matter of time before Rio de Janeiro's favelas started to fall under the control of gangs dealing drugs. By the end of the 1990s, large amounts of drugs and gun traffic controlled the favelas of Rio.[2]

As a result of this activity Rio de Janeiro's rates of crime and violence reached alarming levels. In 2005, the city and municipality of Rio de Janeiro had a homicide rate of 49.4 per 100,000 inhabitants compared with a rate of 4.1 in Buenos Aires, 7.7 in Mexico City and 25.8 for Brazil as a whole. In 2007, the police allegedly killed 1,330 people in the State of Rio de Janeiro – compared with 347 people killed by the police in the whole of the United States during 2006 (Waisefisz, 2010).

Economic opportunities in the favela are limited and many residents employ themselves in the informal economy. This can include street vending, domestic work, and prostitution. Lack of economic opportunity means that favelas are commonly seen as 'breeding grounds' for social problems such as crime, drug addiction, alcoholism, high rates of mental illness, and suicide.

Within the favelas, drug traffickers and militia gangs often control daily aspects of residents' lives. Traffickers ensure that individual residents can guarantee their own safety but only through the allegiance of their political connections to them or payment of arbitrary 'taxes', seeing their political and social freedoms severely restricted. Within this environment, the favela communities have largely been ignored by the state (Silva, 2009). The lack of a registered land tenure system prevented the government from formally providing water, basic sanitation, energy, transport and other services to such communities. Consistent neglect led to over-crowding, unsanitary conditions, poor nutrition, pollution, disease and high infant mortality rates.

2.3 Institutional conditions

Three levels of government have an influence on the city of Rio de Janeiro – federal, state and municipal. The State of Rio de Janeiro, like all other states in Brazil, has autonomous administrations, collects its own taxes and receives a share of taxes collected by the federal government. It has a governor and a legislative body elected directly by voters and independent courts of law. The municipality of Rio de Janeiro (which encompasses the city of Rio de

Janeiro) also has autonomous administration, collects its own taxes and receives a share of taxes collected by the federal and state government. The municipality has an elected mayor and legislative body but no separate court of law.

In the area of security, government policy is coordinated at the state level of government and implemented through two police bodies. The Military Police of Rio de Janeiro State is the police force responsible for preventative police work and maintaining public order. The Civil Police of Rio de Janeiro State is responsible for the investigation of crimes such as detective work and criminal forensics and investigation. The separation between military and civil policing in Brazil is fairly unusual amongst democratic and more developed countries today and has a history which dates back to the military ruling of Brazil during the period 1964–85.

Rio de Janeiro, and Brazil as a whole, has a long history of corruption in public affairs and policy-making, a factor which has led to low levels of political participation from citizens and general distrust. High levels of bureaucracy and an overly complicated government administration also lead to slow rates of decision-making and change. In recent years there have been some positive changes with moves to increase transparency and accountability. An example is the freedom of information law that was passed nationally in Brazil in May 2012 and which will add fresh challenge to corruption, inefficiency and wastefulness in the public realm. The scope of the new law is broad, covering all levels of government (federal, state and municipality level). However, there is still no set of regulations detailing how citizens can ask for information, and what municipal, state and federal officials must do to comply.

In late 2008, the election of the new Mayor of Rio de Janeiro created alignment between the state and municipality levels of government for the first time in many years due to the mayor and governor being from the same political party (the Brazilian Democratic Movement Party – PMDB). This has created greater consensus for change, making new legislations and policies easier to enact.

3 City-wide vision and strategy

3.1 Vision

In recent years, an events-driven agenda – including the FIFA World Cup in 2014 and the 2016 Olympics – has been largely shaping Rio's urban development agendas and strategies. Large public and private investments being made in preparation for the 2016 Olympics and Paralympics Games are already having a big impact on the city. The hosting of the Games requires sports arenas and support structures to be built, investment in transport infrastructure, and improvements to and expansion of the hotel network. Property in some residential districts of the city has already obtained the highest value per square metre in the country with downtown commercial properties attaining values that can exceed those of Manhattan, New York (FIRJAN, 2011).

Capitalising on the increased attention and investment that Rio de Janeiro has been experiencing, both the state and municipal governments have been working on ambitious new visions for Rio de Janeiro. In 2009, the municipal government published for the first time a strategic plan for the city which outlined detailed objectives, budgets and investments. In 2012, the strategic plan was updated for the period 2013–16. The municipal government has presented a budget of R$38.6 million to meet the objectives laid out. The budget for the execution of the initiatives is distributed amongst the different regions of the city according

Table 7.1 Political, social, economic and sustainability objectives for Rio de Janeiro, 2013–16

Economic	*Social*	*Sustainability*	*Political*
• World leader in provision of services and technology for the energy industry • Largest tourist hub in the Southern Hemisphere • National leader for the creative industry with a focus on design, fashion, performing arts and the audiovisual industry • National leader for excellence in business environment with outstanding leadership in attracting and maintaining investments	• Leading city in Brazil for public education • Leading city in Brazil for primary healthcare, both in terms of speed of attendance and efficiency of delivery • National benchmark for delivery of improvements in housing • A capital without extreme poverty and extreme inequalities and which is more integrated culturally and socially	• World reference point for sustainable transport, eco-efficient economic activity and low emission consumer habits • Seen as world benchmark for a city which successfully combines urban development with surrounding natural landscape (rivers, lakes, bays and ocean coasts) • Universal and efficient system of collection and treatment of sewage and garbage	• Recognised by the international community for its hosting of major events • National benchmark for quality of public administration and management • Recognised internationally for quality and influence of cultural sector • Highlighted for respect for human diversity

Source: Prefeitura do Rio de Janeiro (2012)

to their need with poorer regions in the northern and western area receiving greater levels of investment (Prefeitura do Rio de Janeiro, 2012). More specific political, social, economic and sustainability objectives for the Municipal Strategic Plan are illustrated in Table 7.1.

3.2 Strategies and policies impacting Rio's favelas

As previously discussed, security policy for Rio de Janeiro is controlled at the state level of government. The Pacification Police Units are the backbone to Rio de Janeiro's state strategy on security and are the focus of this case study. Beyond pacification, a range of other planned urban interventions have an important impact on favelas (particularly policies related to housing, infrastructure and sanitation). Both the state-level and municipal government play a role here. An overview of the strategies applied in both these areas is briefly discussed below.

3.2.1 Security policy

In its Plan for the State of Rio de Janeiro 2011–2014 (Prefeitura do Rio de Janeiro, 2012), the state government outlines a series of interventions to tackle the security challenges facing the State of Rio. Due to the importance of the favelas and their associated drug gangs in driving violence and crime in the State of Rio de Janeiro, the UPP programme is seen as the key policy in tackling security. Nevertheless, the state government has outlined a range of other initiatives to support the UPP.

Improving the efficiency of the police is an example of this. Concrete measures have been laid out to achieve this in the Strategic Plan: they include introducing a system of goals and performance measurement systems as well as a new plan for recruitment and training of police (Carneiro, 2012). Low valorisation of the police has long been a problem across Brazil and the new plan seeks to address this through increased investment in training and the introduction of performance-related pay. Other policies include measures to reduce corruption, modernise the police and better integrate the civil and military police.

3.2.2 Other urban policies impacting favelas

In the areas of urban development, housing, sanitation and infrastructure, both the state and municipal governments have outlined a series of policies seeking to improve favela communities. It is important to note that the implementation of many of these policies is only made possible following pacification of the favela – hence pacification is seem by its proponents as the gateway to the realisation of a range of other benefits for these communities.

At the state level, the government outlines a series of improvements in infrastructure and housing in its Plan. These include the extension of the existing PAC policy (National Growth Acceleration Programme), which has already invested R$1.68 billion in the improvement of houses, health, education, culture, sports and leisure facilities in poor communities. Also outlined are a number of specific initiatives, which will benefit favela communities related to improved roads and transport networks, water systems and energy supply. At the municipal level, policies to tackle housing and the urban environment receive by far the largest proportion of the total budget made available for enabling the Municipal Strategic Plan (31 per cent of the total). Figure 7.1 presents the distribution of investments across different policy areas as presented in the Municipal Strategic Plan.

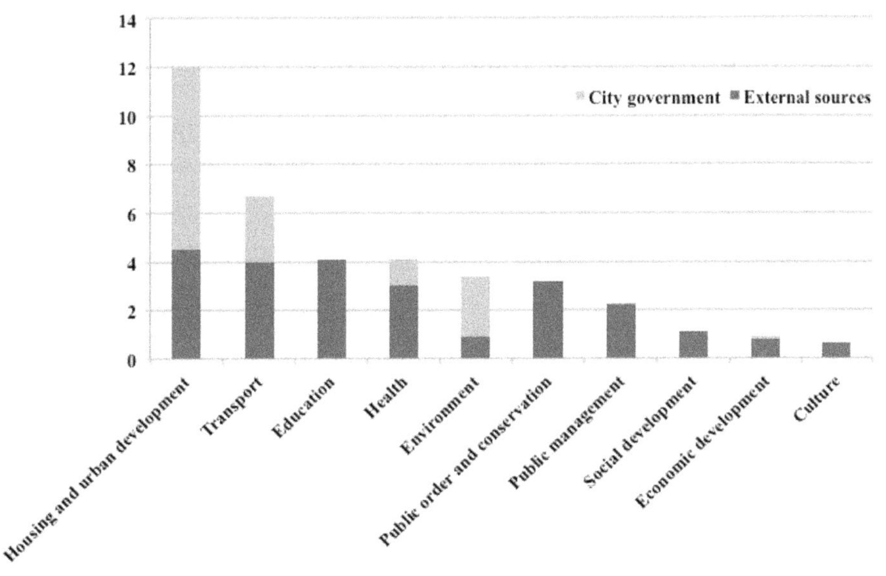

Figure 7.1 Distribution of investment by policy area, Strategic Plan for Rio de Janeiro (R$ million), 2013–16

Source: Prefeitura do Rio de Janeiro (2012)

In its strategic plan, the municipal government outlines a series of objectives for housing and the urban environment in its plan (translated to English), namely:

- achieve a 5 per cent reduction in the area of the city occupied by favelas before 2016;
- guarantee that before the end of 2016 there will be no families living in areas of high risk of landslides (living on steep sloping land);
- bring sewerage systems, water, drainage, street lighting, collection of litter and street paving to 156,000 households before 2016;
- promote the contraction of 100,000 new affordable units of housing, through partnership with the private sector and other spheres of government.

In order to realize these objectives, a series of planned interventions are presented. UPP Social, a follow-up intervention to the UPP which aims to integrate residents of pacified favelas with the rest of the city and provide them with better access to public services and socio-economic opportunities, is an important strand of the strategy and will be discussed in more detail in the following section. Other interventions presented by the municipal government in the area of housing and urban environment and intended to complement UPP Social include:

- *Morar Carioca – Urbanisation Project*. First implemented in 2010, this programme includes investments to improve urban infrastructure, sewerage systems, drainage, street lighting, collection of litter and street paving. To date, the programme has delivered improved services to 69,500 families but the programme aims to more than double this number by 2016.
- *Morar Carioca – Minha Casa Minha Vida (My House, My Life)*. First introduced in 2009, this programme seeks to provide access to affordable housing to some of the poorest residents of Rio de Janeiro, thereby moving people out of favela areas into stable housing areas. To date, 35,000 new housing units have been contracted, designed for families with monthly incomes of up to R$5,000. The programme aims to contract 100,000 units by 2016.
- *Bairro Maravilha*. The programme aims to reduce inequalities in the urban environment between Northern and Western areas of Rio de Janeiro and other areas of the city.

4 The project: Pacifying Police Units

4.1 Context

4.1.1 Previous and new policies targeted at favelas

The government in Brazil has made several attempts to improve favelas and the associated violence that surrounds them. During the 1970s, the military government attempted the eradication of favelas by forcibly removing residents and placing them in public housing projects or back to the rural areas that many emigrated from. Another attempt to deal with urban poverty came by way of gentrification. Here, the government sought to upgrade the favelas and integrate them into the inner city with the newly urbanised upper-middle class. As these 'upgraded' favelas became more stable, they began to attract members of the lower-middle class pushing the former favela dwellers onto the streets and suburbs further away from opportunity and economic advancement. In Rio de Janeiro, the vast majority of the homeless population is black, and part of that can be attributed to favela gentrification and displacement of those in extreme poverty.

In recent years the proportion of Rio de Janeiro's population living in favelas has started to drop. Decreases in the population of favelas can be accredited to a reduction in the traditional rural to urban migration flow, which has become less attractive given investments in rural Brazil that are improving living standards in other regions as well. More generally, other investments across the nation (in industry, infrastructure, tourism and social assistance) are helping to spread wealth and develop historically underdeveloped regions (like Northeastern Brazil) thus reducing the reasons to migrate to economic cores like São Paulo and Rio de Janeiro. These trends are moving an increasing proportion of the population of Brazil into the middle classes and reducing rates of poverty.

4.1.2 Vision for UPP

In 2008, State Governor Sérgio Cabral launched an innovative initiative to take back state control of Rio de Janeiro's favelas. Cabral travelled to Colombia in 2007 in order to observe public security improvements enacted in the country under Colombian President Álvaro Uribe since 2000. Following his return, he secured USD 1.7 billion for the express purpose of security improvement in Rio, particularly in the favelas. The vision for the UPP, was spearheaded by the then-incoming Secretary of Public Safety for Rio de Janeiro state, José Mariano Beltrame, with the strong backing of the Governor.

While the primary objective of the UPPs is to achieve 'pacification' and regain security control of the favelas, the programme has a broader and longer-term objective, which is to reintegrate favela communities into the economic, social and cultural life of the city. Ensuring the full reintegration of favela communities back into the life of the city requires increased state participation in a number of areas including urban and housing urban development, policies to improve social inclusion and strengthening of democracy and citizenship. However, it was recognised that pacification of the favela had to be the first step – only then could the state begin to act in other areas.

4.2 Strategy

The UPP programme began with the implementation of the first Pacifying Police Unit in the Santa Marta community in late 2008. By February 2014, 37 UPPs had been implemented reaching 257 territories and benefitting 1.5 million people. Delivery of the vision for UPP involves a 'whole-government' approach, employing the use of various bodies through a sequenced and coordinated series of interventions. The Pacifying Police Units themselves, delivered by the state government of Rio de Janeiro, aim at reclaiming territories controlled by gangs of drug dealers and private militias. UPP Social, a follow-up programme delivered by the municipal government, aims to re-integrate residents of pacified favelas back into the city through coordination of and improved access to public and private services. The strategy for UPP is based around three key stages, which are discussed in the following sections: 1) Pacification, 2) Implementation of UPP and 3) Implementation of UPP Social.

4.2.1 Pacification

Pacification involves the introduction of large numbers of security forces to clear the violent groups from the favela areas and hunt down and forcibly remove their members. It was decided early on during the planning for the programme that using the existing military police would not be sufficient for this operation. Instead, an elite division of the military police (the Police Special Operations Battalion – 'BOPE') is used. BOPE police are specially hired and trained. They have extensive experience in urban warfare and use equipment deemed more powerful than traditional civilian law enforcement. BOPE is independent from the rest of the police and specially funded by the government.

The process of pacifying a favela is notoriously dangerous and can take several weeks, using up immense quantities of police resource. The task is made more challenging by the fact that the police have little knowledge of the communities they are pacifying before entering them, due to their informal nature and the lack of documentation/information about them (for example, maps). Moreover, Rio de Janeiro's favelas vary greatly in terms of size and structure. Some have a greater stronghold of violence and gangs while the geographical location of others makes them more difficult to enter and pacify.[3]

4.2.2 Implementation of the UPP

Following pacification (removal of the violent groups and regain of control of the favela by the authorities) the BOPE give way to the UPPs. The UPPs are specially trained police who enter the favelas and remain stationed within them for an indefinite period of time providing day-to-day policing and control.

The Brazilian population has had a long and turbulent relationship with the police. During the days of the military dictatorship, the police were viewed as the 'enemy' rather than a protecting force. Following the dictatorship security budgets were cut suddenly preventing effective policing of challenging areas, such as the favelas. The public felt abandoned by the police and were distrusting of them. Thus, a strategic decision was therefore made that the UPPs should consist of new policemen, recruited particularly for the job in hand. It was argued that this would help to build trust amongst the community because new police recruits would be fresh and motivated for the job, absent of prejudices that may have developed as a result of working in favela communities before.

4.2.3 Implementation of UPP Social

Following implementation of the Pacification Police Units, UPP Social enters the favela. The city government, in coordination with the state and federal governments, communities, civil society and private sector, delivers this part of the process. Following the pacification of a favela, coordination can quickly become a major challenge. It means that suddenly, and for the first time in the history of that community, a range of agents from the public sector, private sector and not-for-profits wish to enter the community. Favelas can be rather heterogeneous (both between and within favelas) and the needs of residents are often not well understood. UPP Social was therefore developed to play a coordinating role, helping to ensure that the services provided meet the needs of favela residents. The ultimate aim of UPP Social is to ensure that pacification will give way to a full range of economic, social and political benefits for these communities.

The UPP Social initiative is launched in each new pacified favela through a community forum, designed to provide a space for open debate among the different stakeholders. A local management unit produces detailed information about the communities, identifying programmes/interventions, strengthening dialogue among residents, service providers, government agencies and the private sector. The unit then engages the necessary institutions to respond to the identified needs and provide support and supervision for their execution. An important role that UPP Social plays is in collecting and documenting basic information about the favela. An example is the mapping of streets, often never done before, which allows residents to gain an address for the first time. Helping to support the formalisation of public services is another area in which UPP Social can provide support. This includes efforts to help legalise electricity, gas, cable and internet services.

4.3 Implementation

Certain factors and organising capacities have been critical to the implementation of the UPP programme. This section discusses in more detail how the programme has been implemented and what the critical enablers were.

4.3.1 Leadership

Leadership has also been a key enabler in the implementation of the UPP programme, namely of the Secretary of Public Safety José Mariano Beltrame. A study from Carneiro (2012) identified a number of key actions and steps that enabled the UPP process. First, Beltrame (and his group of assistants) could identify a clear set of priorities and show the sense of urgency to tackle them. The collection and communication within the Secretary of relevant data was important to demonstrate that the pacification issue was not only a theme but also a serious problem that could be addressed by joint action. Moreover, Beltrame and his team could select and communicate an articulated and consistent set of ideas for the UPP policy, instead of plugging in the theoretical debate about what pacification is or should be about.

Second, Beltrame worked to negotiate and form a wide and solid coalition to support the visions behind the UPPs. This was essential to assure that the process would not get blocked by the many different sensitivities and interests around the issue (e.g. those of the police, human rights associations, community leaders in the slums, local government, etc.). Finding common points of interest was pivotal to get consensus, such as social inclusion and safety. Moreover, due to the scarcity of resources (available police staff, financial constrains), it was essential to demonstrate that the UPP process could also be financially attainable for the government.

Third, Beltrame could clearly communicate policy objectives within and outside the state government, and paid special attention to public opinion. He could clearly relate the means and the instruments to the objectives through simple and clear language instead of using conceptual jargon. Moreover, leadership was also critical at a more operational level. For example, UPP field leaders have been chosen with an eye to broker trust between the newly established police units and the local residents.

Beyond the strategic level, local leadership was also of utmost importance. Pricilla Azevedo was the Police Chief of the first UPP to be implemented in the Santa Marta and has become a reference point for the programme. She successfully commanded respect from the local community and demonstrated that the police can be a partner for citizens in the community. She has since received a range of awards in recognition for her leadership, including a prize which nominated her as one of the ten most courageous women in the world. Pricilla Azevedo commanded a troop of 120 men in Santa Marta. The choice of a woman to lead the first UPP was a deliberate strategy to help try to build relationships with the community.

4.3.2 Capacity building

Significant investment in capacity building of the UPPs was another key enabler for the programme. Previous police involvement in favelas tended to be in one-off shoot-outs. The police would quickly exit again, leaving the favela residents to deal with the aftermath. Contrarily, the concept behind UPPs is long-term community policing – a significantly different concept of policing for both favela residents and the police alike. Community policing required a new focus on building relationships as well as on the prevention of crime. Training focused on softer skills and team-working to encourage the behavioural change. New targets which included a reduction in the number of criminals shot and bonuses according to the productivity of police teams (rather than individuals) have been implemented.

Other statewide efforts to improve the working of the state police have also developed police capacity, therefore supporting the development of the UPP programme. Examples include an increased focus on results, increased valorisation of the police, technical modernisation of police infrastructure and the more intelligent use of information on crimes and communities to better allocate resources.

4.3.3 Private sector involvement

The involvement of the private sector, not just in helping to finance the UPPs but also in supporting their long-term impact and sustainability, has also been identified as important to the programme's success.

In financing, private sector support has been vital in supporting the fast-paced roll-out of the UPPs. Private sector companies have donated the equivalent of millions of dollars to the scheme but have also engaged in important Corporate Social Responsibility activities, investing in particular communities or development projects. Examples include Coca-Cola who teach two-month entrepreneurship courses for people who want to start businesses, and Unilever who sponsor Santa Marta's bright-red cable cars that transport 10,000 people a day.

The role of the private sector is also important in stimulating economic activity in the pacified communities, creating job opportunities for young residents and alternatives to drug dealing. Many private sector companies are starting to invest in favela communities for the first time following pacification in the search for untapped consumer markets. A local advertising agency has opened an office in the Santa Marta favela to do projects for clients, conduct research and lead immersion tours for marketers. Several have already visited Santa Marta including Coca-Cola, local hamburger chain Bob's and Takeda Pharmaceuticals.

The increased participation of financial services providers in favela communities is playing an important role by providing residents with access to bank accounts, credit and insurance facilities for the first time. Santander is an example and, although the bank has been present in some communities for many years since before pacification, the UPPs have facilitated its expansion and allowed it to provide a wider range of products to a broader range of consumers. Microfinance schemes operated by Santander and other commercial banks allow residents to access finance at reasonable rates supporting families who wish to invest in a new business or improvements to their homes, for example. The Bid Network supports entrepreneurs by facilitating access to investors and providing coaching and business planning support.

Local electricity companies (such as LIGHT) have also been involved in the scheme, with an eye to bring residents 'back to formality' when it comes to electricity consumption. The idea is to make sure that electricity is accessed on a regular basis and not through illegal connections. Yet, initiatives are not restricted to repression and control of illegal electricity connections. After an UPP is established, LIGHT can deploy a number of social programmes to 'tutor' residents how to consume energy in a more efficient way, as well as programmes to incentivise recycling and other sorts of electricity-related pacts.

4.4 Results

Overall perceptions of the UPP programme appear to be strong, although a recent rash of abuse and corruption scandals involving UPP police offers has shaken some resident confidence in the programme (Rodrigues, 2014). A 2013 study by the Instituto Data Favela (see Rodrigues, 2014) indicated that 75 per cent of favela inhabitants approve of the UPPs. A study of residents in Santa Marta, the first community to be pacified, found that 87 per cent of residents supported the actions of the police. In Cidade de Deus, the second community to be pacified, the figure was 93 per cent (IBPS, 2010).

In 2013, the murder and subsequent cover-up of a bricklayer lead to the arrests of several UPP offers in one of Rio's largest favelas, Rocinha. Other similar incidents have been reported. These cases have led to a shake-up of confidence of local residents in the UPP programme. In response, the City has instituted an additional policing force, the Pacifying Police Coordinating team, as part of efforts to rebuild the police force's credibility and curtail corruption.

4.4.1 Crime and violence

Evidence indicates that UPPs have had a major impact on the reduction of crime and violence in Rio de Janeiro. A study released in 2013 found that the number of reported murders in Rio's pacified favelas decreased by approximately 65 per cent between the years of 2008 and 2012. Conducted by the Institute of Public Security, 22 of the city's now 37 UPPs were included in the research. And in the first year of the UPP programme, murders in the Cidade de Deus favela dropped from 34 to 6 and car thefts from 69 to 11. Similarly, in Dona Marta, murders fell from three to none and car thefts from 25 to 14.3 (ISP, 2013).

Murder rates in Rio Janeiro (city and municipality as a whole) dropped sharply between 2008 and 2010, from 35.09 per 100,000 inhabitants to 26.7 per 100,000 inhabitants in the city, moving Rio de Janeiro from a national state capital ranking of 3 to a ranking of 23 in terms of homicide rates (Waisefisz, 2010). Nevertheless, Rio's decline in murder rates is part of a longer downward trend that started in the city at the end of the 1990s and so it is important not to attribute all of the improvements experienced here to the UPP programme.[4]

4.4.2 Health

A key benefit of UPP programme has been that it has allowed the government to standardise healthcare provision and improve coverage, particularly of primary healthcare. Prior to pacification, favela residents would often have access to healthcare; however services were frequently patchy and unreliable, with some areas having very little coverage at all. Favela residents would often have to travel outside the favela to find healthcare workers because many doctors and nurses were afraid to work inside the favelas. Turnover and absentee rates of health workers inside the favelas were high. In addition, as a result of a lack of locally provided primary healthcare, residents would tend to over rely on hospitals for healthcare needs, putting unnecessary strain on the city hospitals.

The percentage of Rio de Janeiro's population with access to primary healthcare increased from 3.5 per cent in 2009 to 35 per cent in 2012 (Prefeitura do Rio de Janeiro, 2012). Improvements in primary healthcare provision has taken pressure off the city's hospitals and led to an increase in the overall efficiency of service delivery. Alongside the increased provision of primary healthcare, the UPP programme has enabled a greater emphasis on prevention and education in health. Health education programmes have also been implemented in many of the city's pacified favelas and have been perceived as particularly important in reducing the spread of preventable diseases such as dengue.

4.4.3 Education

Improvements in school attendance and education outcomes as a result of the UPP programme have also been reported. A study carried out by the Secretary for Education revealed that between 2011 and 2012 there was a 28 per cent increase in the matriculation rate of 15 schools located in communities with UPPs (e.g. Almeida, 2014). In the Cidade de Deus favela, attendance at one local high school increased by 90 per cent. A further study released in 2012 found that schools located in neighbourhoods with UPPs had a larger increase in their test score results in the Brazilian Index of Development in Basic Education (IDEB). The largest difference was recorded in the second-segment of primary education. The eight schools located in pacified regions improved their score in the IDEB by 42.8 per cent, from a score of 2.8 in 2009 to a score of 4 in 2012 (e.g. Almeida, 2014).

4.4.4 Other public services

While levels of security appear to have undoubtedly improved as a result of pacification and there is additional evidence around health and education impacts, the introduction of other public services has been criticised as being slow. Adequate sanitation, mail delivery, garbage collection, road conditions and internet access are among key amenities that were still found to be missing for residents of many favela communities. Residual prejudices and reports of operators often unwilling or unable to fully deliver goods and services were also reported. The agency responsible for the UPP Social programme has now introduced a Rapid Participation Map (MRP) which grades pacified urban communities on specifics such as water supply, sewage, cleaning and mobility with an overall average grade then provided.

4.4.5 Economic and spillover effects

A study realised in February 2012 covering 17 communities that had received UPPs found that revenues from local retail businesses had increased by 26 per cent with the introduction of the UPPs. This figure increased to 36 per cent in the east zone of the city (Government of Rio de Janeiro, n.d.).

There is also evidence of positive spillover economic effects into communities surrounding pacified favelas. A study by the Housing Union (SECOVI) found that the rental value of houses located near UPP favelas in the south zone had increased by 149 per cent, while their sale value had increased by 59 per cent (see Monteiro, 2014). Finally, many utility providers are reaping major financial benefits from pacification as a result of the formalisation of utility supplies. The electricity provider LIGHT reported in 2013 that it had seen losses from 'gatos' (illegal electrical set-ups) drop from 64.1 per cent to 12.3 per cent, while timely payments rose from 9.6 per cent to 93.3 per cent in 15 of the 33 UPP posts (e.g. Chaves, 2011).

5 Conclusions

5.1 Broad assessment and success factors

At the time of writing (February 2014), 37 favelas have been pacified and a range of positive results has been reported including significant reductions in crime and violence (see previous sections). The overriding approval rating of the policy from favela residents is positive, as well as from the rest of the city's residents. The government has ambitious plans for the continued roll-out of the UPP programme, planning to first tackle the remaining favelas in the municipality of Rio de Janeiro before beginning on favelas in other parts of the state. The state government aims to implement 40 UPPs by the end of 2014 in 165 favela communities, requiring an estimated 12,500 officers and benefitting nearly 860,000 residents in all (Chaves, 2011). To 2016, the state government is investing R$15 million in the training of police offers alone and hopes to install 60,000 trained officers across the state.

Despite the relatively early stage in the roll-out of Rio de Janeiro's UPP scheme some success factors can be identified. These are useful for other cities considering their own implementation of similar programmes.

Establishing the UPP programme's credibility at an early stage and maintaining subsequent momentum were critical factors for its success. The programme was innovative and high risk, and so it was important to show success at an early stage to build support. The ordering of the favela pacification was an important strategy here. Santa Marta was chosen as the first recipient for the UPP programme largely due to its small and manageable size (approximately 6,000/7,000 habitants) and strategic location within the middle-class south zone community. The next series of pacifications all occurred around the south zone, gaining the approval of the middle class and attracting media coverage. Later, the UPP programme was extended to the major hills of the south side of the city and several favela communities of the north and west sides, including the complexes of Alemão and Rocinha (traditionally the largest and most dangerous favelas in the city). These 'big wins' served to maintain the attention of the public and the press, guaranteeing continued support. Next, the pacification of Jacarezinho and Marquinhos, two of the most notorious centres of the drug trade far away from wealthy districts, were announced, dispelling any cynical concerns that the UPPs are only there to benefit Rio's middle-class communities.

Strong leadership was another critical success factor in the programme. Beltrame, the incoming Secretary of Public Safety of Rio de Janeiro State who developed the vision for the programme, had a reputation for fairness and getting things done. With a background in law and experience working as a senior federal police officer, he had a passion for the programme, which he was able to communicate convincingly to serving police and civil servants, thereby inspiring them to deliver. Leadership at the local level was also important and Pricilla Azevedo, the first UPP Police Chief, has already been cited as an example.

Finally, the coordinated whole-government approach that was taken to design and implement the programme was important. State and municipal levels of government worked together to deliver different parts of the strategy. UPP Social played a particular role in coordinating the different agents involved. Active reach-out to the private sector was also critical in ensuring that they played a role in developing new economic opportunities in the favelas, opening new markets and providing residents with access to new services, such as credit and insurance products.

5.2 Challenges ahead

The main challenge facing Rio de Janeiro's UPPs now is how to build on their immediate wins of recovering physical control of some of the city's most strategic favelas and extending this to the rest of the city's communities which are still under the control of drug or militia gangs. In addition there exists the continued challenge of developing a solid basis for sustainable socio-economic development, closing the gap between those living in the now pacified favelas and the rest of the city. Some key issues are discussed below.

5.2.1 Sustainable pacification

Convincing the public that the pacification of the favelas is sustainable is a key challenge. There still remains some perception within the city that the UPP programme is not long term and that, beyond the World Cup and the Olympics, it will no longer exist. Many residents of favelas still fear that the drug landlords who previously controlled their lives will return. In some cases, this explains residents' reluctance to get involved in state programmes such as the UPP Social (for fear they will be punished later).

Moreover, an important argument commonly made against the sustainability of the UPP programme is that drug criminals who flee from pacified neighbourhoods simply move to new areas. In the nearby city of Niteroi, for example, residents have complained of increased levels of violence and crime following the pacification of favelas in Rio de Janeiro. If the long-term sustainability of the programme is to be guaranteed, then it will be necessary for the government to continue its pace of UPP roll-out to new favelas while maintaining high standards of policing in already pacified favelas.

5.2.2 Filling the void

To maintain public support for the UPP programme, the government will have to continue to work hard to fill the 'void' in public services and economic opportunities that have characterised favela communities for so long. Those living in favelas will eventually make comparisons between their lives under the control of drug dealers and under the control of the state. While the removal of drug gangs has undoubtedly improved many aspects of residents' lives there may be some things that some residents miss. For example, under an informal economy, residents would typically access many forms of utility service for free by simply plugging themselves into the network, whereas now they typically have to pay for such services. In addition, some services provided to residents for free by the drug lords in return for political allegiance now have to be paid for – an example could be funeral costs. It is thus critical that the authorities follow up the largely successful security operation with the schools, clinics, infrastructure and transport enjoyed by the richer parts of Rio. Otherwise there is a risk that the gangs will once again fill the void.

5.2.3 Promotion of civil liberties

Following pacification of the favelas, many new legislations, social norms and regulations need to be established. Whereas previously, traffickers and militia defined the role of court, communities will now need to establish their own rules. A specific example is formalised local licensing laws. In shaping the legal and social structures for establishing these new norms an important challenge is ensuring that full civic participation is encouraged so that the resulting rules reflect the needs of the local community. Some NGOs have expressed concerns around a lack of democratic process in some examples to date and the use of excessive policing to fill the current deficit in citizenship and civil rights in these territories. It will be important that the government does all that it can to promote civil participation so that the systems that define the day-to-day lives of these new communities are built collectively.

Notes

1 Favelas take their name from a hardy plant which survives in the arid northeast of the country, where most of the slum dwellers hail from.
2 As the profits and political power of the gangs grew, the temptation of bribery did too. The so-called 'rotten part' of the police soon became corrupted, along with local politicians and judges. There would be no attempt to curb the activities of the drug gangs as long as these groups were paid off, which resulted in up to 50 per cent of the drug lords' profits going to the police. An array of rival factions and militia gangs began to appear. The stealing of weapons, ammunition and uniforms became rife, and shoot-outs and turf wars became part of daily life.
3 One of the most challenging pacification operations was the pacification of Rocinha. Pacified in November 2011, the operation involved 3,000 men with support from the army, civil police and the marines. One hundred and twelve kilos of marijuana, 60 kilos of cocaine, alongside handmade bombs, fake uniforms and hundreds of rifle and machine guns were recovered as part of the programme. A key success of the operation was the arrest of Rocinha's traffic boss, one of the most wanted criminals in the country.
4 Other increasingly affluent cities in Brazil, particularly in the south of the country, have experienced similar trends over this period while other particular policies implemented in Rio de Janeiro, for example the passing of national gun control legislation in 2003, are also likely to have partly effected the change.

References

Almeida, K. (2014). UPPs e educação – possíveis impactos da implantação das Unidades de Polícia Pacificadora na segregação escolar. Master thesis, Universidade Federal do Rio de Janeiro.

Campos, L. M. and Santos, N. (2004). *O Crime Organizado e as prisões no Brasil.* Rio de Janeiro: CONPEDI, Ciências Penais, Universidade Federal de Goiás.

Carneiro, L. P. (2012). A politica de pacificação do Rio de Janeiro: um estudo sobre liderança e inovação. In A. Pinheiro and F. Veloso (eds), *Rio de Janeiro: um estado em transição.* Rio de Janeiro: Fundação Getúlio Vargas.

Chaves, R. (2011). A New Approach to Citizen Security in Brazil: Rio's Pacifying Police Units. Conference paper, 16 March. Wilson Center, Washington DC.

FIRJAN – Federação das Indústrias do Estado do Rio de Janeiro (2011). Investimentos Rio 2011–2013. Rio de Janeiro: Federação das Indústrias do Estado do Rio de Janeiro.

Governo do Rio de Janeiro (n.d.). UPP – o que é? Accessed at http://www.upprj.com/index.php/o_que_e_upp (15 April 2015).

IBGE – Instituto Brasileiro de Geografia e Estatistica (n.d.). Cidades: Rio de Janeiro – Dados Gerais do Município. São Paulo: Instituto Brasileiro de Geografia e Estatistica.

IBGE – Instituto Brasileiro de Geografia e Estatistica (2011). Brazil: 2010 Population Census. São Paulo: Instituto Brasileiro de Geografia e Estatistica.

IBPS – Instituto Brasileiro de Pesquisa Social (2010). *Pesquisa sobre a percepção acerca das unidades de Polícia Pacificadora. Pesquisa PR 004-10-UPP-25.01.* Research report, January. Rio de Janeiro: Instituto Brasileiro de Pesquisa Social.

IEST – Instituto de Estudos do Trabalhos e Sociedade (2010). *Pesquisa nas Favelas com UPP na cidade do Rio de Janeiro. Resultado consolidado.* Rio de Janeiro: Instituto de Estudos do Trabalhos e Sociedade.

ISP – Instituto de Segurança Pública (2013). Unidade de Polícia Pacificadora – dados oficiais consolidados das UPPs. Rio de Janeiro: Instituto de Segurança Pública. Accessed at http://www.isp.rj.gov.br/Conteudo.asp?ident=61 (15 April 2015).

Monteiro, C. (2014). Análise da valorização imobiliária na cidade do Rio de Janeiro: foco nos bairros de Copacabana, Ipanema e Leme. Rio de Janeiro: Escola Politécnica, Universidade Federal do Rio de Janeiro.

Prefeitura do Rio de Janeiro (2012). *Pós-2016 – O Rio mais integrado e competitivo: Plano estratégico do Rio de Janeiro 2013–2016*. Rio de Janeiro: Prefeitura do Rio de Janeiro.

Rodrigues, R. (2014). *Os dilemas da pacificação: Notícias de Guerra e Paz numa cidade maravilhosa*. Rio de Janeiro: Instituto Agarapé.

Silva, J. (2009). *O que é a favela final?* Rio de Janeiro: Observatório de Favelas do Rio de Janeiro.

Waisefisz, J. (2010). *Mapa da Violência 2010 – Anatomia dos Homicídios no Brasil*. São Paulo: Instituto Sangari.

8 Park Railway North (Antwerp)

Jan van der Meer and Alexander Otgaar

1 Introduction

This case study explores the conversion of a derelict railway yard (closed in 2000) into a green, open, recreational area. This railway yard, covering 24 hectares of land (comparable to 40 soccer fields) and including a few railway buildings, was for more than 100 years a parking, repair and maintenance site for railway workers of the Belgian national railway company (NMBS). Surrounded by a two-metre high wall, the yard worked as a barrier between densely populated neighbourhoods situated close to the city centre. Only by crossing a viaduct in the middle, could people from both sites meet without having to make a lengthy detour. The surrounding neighbourhoods are among Antwerp's most deprived. A special feature of these neighbourhoods was an almost complete absence of open and/or green space, and as a consequence, almost no playgrounds for children.

Since the early 1990s redevelopment proposals for the railway site have been put forward by all kinds of parties. Ultimately and due to strong resident lobbying and abundant grants made available by regional, national and European authorities, the city government decided to use the yard as a driving force for upgrading the neighbourhoods, in order to attract and retain families with children. The project vision adopted is to create a garden for the neighbourhood and a park for the city, showing that the ambition reached further than green space for the adjacent quarters only. In this sense the Park Railway North – or Park Spoor Noord (PSN) – project can be considered a soft location factor, indirectly contributing to raising the attractiveness and image of Antwerp in general and to tackle social problems in particular.

In this chapter, Section 2 focuses on the structure of the city and the city dynamics across a number of dimensions; this section results in an overview of main issues and challenges in Antwerp. Section 3 is devoted to the vision and the urban development strategy at the level of the city. Section 4 deals with the project that is central to this case study, the Park Railway North project. Successively, we will look at the vision, objectives and the content of the project, the implementation process, the actors involved and their roles and behaviour, and the results so far. Section 5 concludes and presents a number of lessons.

2 City profile: capitals for sustainable competitiveness[1]

2.1 Performance and economic structure

With 506,225 inhabitants (2012 statistics) Antwerp is the largest municipality in Belgium. The Antwerp metropolitan region (1.2 million inhabitants) ranks second after the Brussels capital region (1.8 million). Next to the maritime cluster and the petrochemical complex – both linked with the city's well-known port – the diamond sector is the third 'flagship sector'. In 2009, 247,000 people were employed in the city, which implies an employment-to-job ratio of 91 per cent and an employment rate of almost 60 per cent (see Sterkx, 2014).

According to van Hamme and Strale (2011), Antwerp's initial competitive advantages related to the quality of the infrastructure, the labour force and the port position have weakened since the 1990s, mainly through increased competition from cheap labour countries (in Central and Eastern Europe). Over this period some major structural changes (deindustrialisation, reconcentration to Brussels, shifts of foreign direct investment (FDI) from manufacturing to services) resulted in a rather poor economic performance of the Antwerp region if compared with the Brussels region (Table 8.1). The situation improved after 2001, but performance remained moderate and has been accompanied by rising unemployment and social polarisation (Verhetsel and Sel, 2009; van Hamme and Strale, 2011). The centre of Antwerp and the industrial areas have poor populations, while the wealthier people are located in the suburbs, outside the city boundaries.

Antwerp's industrialisation process was typical for a port region (van Hamme and Strale, 2011). Several industries could develop here due to the goods that were transported through the port, like the food industry (based on imported agricultural products), the chemical industry (based on imported raw materials), and after the Second World War, the assembling manufacturing industry (due to imported components) (Vandermotten *et al.* 2010). The current Antwerp economy is still largely dependent on harbour functions, with high shares of manufacturing, transport and communication, wholesale trade and business services (see Table 8.2). However, Antwerp has a strong and growing knowledge-intensive sector. One-fifth of the Flemish added value of knowledge intensive companies originates from Antwerp-based companies (Stad Antwerpen 2011).

2.2 Social conditions

In the early 1970s Antwerp experienced its highest population score with 550,000. In 2000 the population was reduced to 453,000. Since 2000, population has grown again with more than 50,000 inhabitants. This increase is mainly due to a positive migratory balance. Demographic figures show a strong increase (39 per cent between 2004 and 2011) of inhabitants with an ethnic background and a decrease (5 per cent during same period) of native Antwerpians. This decline of natives is the combined result of low birth rates, high mortality rates (both factors due to ageing of native people) and 'urban flight' (selective emigration from the city, namely of families with children). Compared with the Flemish average of 6 per cent, a relatively high share of foreigners live in the city (17 per cent). Moreover, the share of population with a

Table 8.1 Economic growth of Antwerp compared to Brussels and national average

	Antwerp centre	*Antwerp region*	*Brussels centre*	*Brussels region*	*Belgium*
Annual growth 1995–2008 (%)	1.10	1.88	2.25	2.38	2.27
Annual growth 2002–8 (%)	1.49	2.03	1.78	2.00	2.19
Total value added 2008 (millions)	25,046	45,159	57,665	94,195	307,877

Source: Van Hamme and Strale (2011), author's calculations based on National Bank of Belgium

Table 8.2 Economic structure (employment shares Antwerp metro region and Belgium)

Sector	*Antwerp 1995*	*Belgium 1995*	*Antwerp 2008*	*Belgium 2008*
Manufacturing industry	25.2	20.1	19.5	15.8
Food	2.3	2.6	1.6	2.0
Chemical	9.3	4.4	6.4	3.6
Metal	2.7	3.1	2.8	2.8
Machinery	1.3	1.2	1.6	1.2
Electric and electronic equipment	1.8	1.7	0.9	1.1
Transport equipment	2.5	2.0	1.2	1.0
Trade	13.8	11.8	14.7	12.9
Transportation and communication	11.1	7.1	10.8	8.3
Finance	4.6	6.2	3.8	5.2
Business services	18.2	19.8	23.9	24.1
Administration, education, health	15.2	20.7	16.2	21.2
Other sectors	27.1	35.1	27.3	33.6
Total	100.0	100.0	100.0	100.0

Source: Van Hamme and Strale (2011), author's calculations based on Regional accounts of the National Bank of Belgium

foreign background is 32 per cent. The top-three ethnic groups originate from Morocco (7.9 per cent of Antwerp's population), the Netherlands (3.3 per cent) and Turkey (2.7 per cent). The city features relatively high shares of young (0–9 years old) and very old (80 plus) people. These figures are considered a challenge for the liveability in the city (Sterkx, 2012).

Like most other European port and industrial cities, economic progress is accompanied by many social problems, boosting the emergence of a dual society, and among other things, raising feelings of insecurity. Antwerp scores high on poverty and social deprivation. Moreover, Antwerp suffers from poor housing conditions. Since the late 1980s several studies have been carried out into the phenomenon of deprivation. Already in 1987 the 'Atlas of deprivation' indicated 25 poor neighbourhoods in Antwerp. Other inquiries (Kesteloot *et al.*, 1996) confirmed the presence of underprivileged neighbourhoods. Based on a set of indicators of social backwardness, Kesteloot *et al.* (for the Flemish Social Impulse Fund programme) concluded that 35 per cent of all Flemish underprivileged neighbourhoods are to be found in Antwerp. Among these are the three neighbourhoods that benefit from the PSN project. These neighbourhoods suffered – independently of the 'usual' job, integration, housing and income problems – from an almost complete absence of public open or green space, with nearly no opportunities for children to have public playgrounds as a consequence.

2.3 Institutional conditions

Since the early 1970s Belgium has gone through a process of federalisation that was completed in 1995. Since then the Flemish government has been for many purposes the centre of address for the Antwerp administration. Next to the national government and the Flemish government there is the province of Antwerp with limited competences. The city is governed by ten local authorities: one central authority for the entire city and nine district authorities, each composed of a board and a council. The central city board comprises the elected mayor as chairperson and a number of city councillors.

The Antwerp administration falls into two independent bodies: the city government (of which port management, spatial planning and various municipal services are prominent elements) and the Public Centre for Social Welfare (Openbaar Centrum voor Maatschappelijk Welzijn, OCMW) which represents the social and care sectors of municipal policy. After the mayor, the chairperson of the OCMW holds a prominent public function within the local administration. The nationally imposed merger in 1983 of Antwerp with seven suburbs has not improved the cooperation with the municipalities in the region that managed to stay out of this merger (van den Berg *et al.*, 2003). As a consequence, regional cooperation is not yet a big deal.

The political 'visa crisis' of 2003 has acted as a catalyst for change of governance. The style had to become 'sober, open and incorruptible' with an ultimate aim of restoring confidence (Sterkx, 2014). Since this proclamation several steps were taken to adopt a more 'managerial' approach and transform the City into a more modern and professional organisation. This involved – among others – the appointment of a strategic coordinator (in charge of coordinating the policies of the entire City organisation by means of strategic planning and formulating operational objectives), the establishment of 'strategic cells' within the City organisation and the creation of autonomous entities (Autonome Gemeentebedrijven) in charge of specific competences. Together, the municipal cells and the autonomous municipal companies make up the 'Group Antwerp'. They cooperate and coordinate to achieve the objectives formulated in the coalition agreement. Moreover, the City wants to achieve its ambitions in cooperation with its citizens and with relevant organisations and associations. The attractiveness of the city is considered a common project (Sterkx, 2014). Antwerp has evolved towards co-production as a vital factor in the success of urban development projects. A project like Park Railway North, to be explored in the subsequent sections, is an excellent example.

2.4 Conclusions: main issues

Sterkx (2014) made an attempt to sum up the main strengths, weaknesses, opportunities and threats the City of Antwerp is currently facing. The study considers the integrated and global vision of urban development, the professionalised governance system and the improved local organising capacity as internal strengths. The continuing challenge to create and maintain more balanced communities in every neighbourhood is mentioned as a weakness. This challenge is linked with complex, related issues such as urban flight, gentrification, housing quality, migration processes and demographic change. Urban development, in the shape of comprehensive city renewal projects and improving the quality of public space, is used as prime policy to tackle deprivation and stimulate livability. Despite the low scores of some neighbourhoods on the livability index, Antwerp has succeeded in preventing these areas from becoming 'no-go zones' (Sterkx, 2014).

An opportunity mentioned is the relative economic stability due to port activities and logistical assets, like the excellent location. However, the economic performance of the city depends on many internal and especially external factors that cannot be isolated from each other. The lasting economic crisis has unmistakable effects on the Antwerp economy and labour market, but there are signs of recovery, especially in port activities. Another opportunity mentioned is the increased external orientation and cooperation with private and public partners of the city administration.

Sterkx (2014) refers to migration and congestion as external threats, factors that are beyond control of local governance, as they are decided on respectively by the federal and Flemish level. The City wants the federal government to take appropriate measures with respect to migration.

Finally, it can be argued that the policy attention to innovative, sustainable solutions is growing, but we have not been convinced of very strong progress in this matter. A positive sign is the participation of the city in the contest for Green Capital Award 2014. For this contest cities are evaluated on a dozen domains by international experts. Antwerp scored well on a few issues, but also disappointingly in some directions. The overall result can work as a catalyst for improvement.

3 City-wide vision and strategy

On the occasion of the merger of Antwerp with seven suburbs in 1983, the City Council decided to draw up a Global Spatial Structure Plan (GSA) for the entire territory (Stad Antwerpen, 2009). The GSA was approved in the early 1990s. For the first time this plan attempted to promote a development strategy for the city. Prime goals were to halt the processes of ongoing suburbanisation and degeneration. During the last two decades Antwerp has evolved towards integrated planning, with urban development obtaining a key position. In 1996 the Flanders government obliged municipal governments to draw up their own spatial structure plans. This obligation was reflected in the 2001–6 coalition agreement as a policy priority. In 2003 the procedure to arrive at a strategic plan for Antwerp started with the approval of the approach adopted. The final approval of the s-RSA (strategic Spatial Structure Plan Antwerp) took place in September 2006.

The s-RSA has a guiding role in the process of urban development (Sterkx, 2014). It contains a long-term vision for the spatial development of the city. It presents three types of policy that are complementary to each other: a generic, a specific and an active policy. The 'generic' policy is based on seven so-called 'inspiring images', which are relevant for the whole city. These images refer to the collective memory of the city. They should help to transform well-known, valuable qualities of the city into ambitions for the future:

1. Water city: reorientation towards the river Scheldt, better use of water to increase the quality of the living environment.
2. Eco-city: efficient use of open spaces and ecological infrastructure.
3. Port city: improving logistics and access to the port, rebalancing relationship between port and city.
4. Railway city: enhancing (rail) mobility and accessibility, mainly by improving the 'lower network' of local roads and public transport.
5. Porous city: optimising urban space, by giving relevant functions to unused buildings and plots.
6. Villages and metropolis: organisation of Antwerp according to the different functions of a polycentric (former suburbs) and metropolitan city (inner city parts).
7. Mega city: exploiting Antwerp's central location within the Northwest European metropolitan area.

The so-called 'specific' policy is executed within five concretely defined 'strategic spaces' and is as a consequence area-oriented. These five strategic spaces, where investments will be concentrated, play a key role in the city's renewal:

1. Hard spine: restoration and reinforcement between city and river; the town quarters situated along the Scheldt are main project areas for city renewal.
2. Soft spine: interconnection of green open spaces with waterways as linking element.
3. Lower network and urban centres: investing in local roads and public transport to structure the urban and metropolitan region.
4. *Groene Singel*: Green Ring Road Area, creating an open and green space connecting the outer city and the five parks of the soft spine to the inner city.
5. Lively Canal: improving the areas alongside the banks of the Albert Canal.

The 'active' policy focuses on sites, squares, green spaces, streets and buildings from different perspectives.

A key project in the city renewal process is the PSN project. This project embodies the various spatial, social, cultural and organisational elements that are inherent to integrated planning (Sterkx, 2014). Like some other large projects that started before the approval of the s-RSA – e.g. the Central Station area and the Schipperskwartier (Sailor's Quarter, since 1998; a red light district, where an integrated policy had to deal with criminality and public nuisance) – this project has been integrated into the strategic spatial plan.

In its coalition agreement, the local political coalition in charge during the 2006–12 period emphasised infrastructure investments and city renewal as driving forces for the attractiveness and livability of the city and specific neighbourhoods (Sterkx, 2014). Urban development has become the spearhead of all policies, assuming that the quality of the environment and public spaces has a positive impact on citizens' quality of life.

4 The project: Park Railway North

4.1 Vision and content

4.1.1 Vision and objectives

Park Railway North aims to improve the quality of living in a deprived part of Antwerp by creating a large urban landscape park on an abandoned railway yard. In 2000 the National Belgian Railway Company (NMBS) closed its yard because it had lost its function. The challenge for the city was to maximise the unique opportunities of this site that for more than a hundred years had divided the neighbourhoods of Stuivenberg, Seefhoek and Dam (together 35 hectares, around 32,000 inhabitants). These neighbourhoods are among the most densely populated parts of the city, characterised by unemployment, integration troubles and image problems.

Thanks to its size and form, the ambition to transform this location in favour of quality of living reached further than only the adjacent neighbourhoods. The vision for the project is that the park should not only stimulate upgrading the surrounding districts but also develop into a safe and attractive hotspot for Antwerp as a whole. In other words: to create 'a garden for the neighbourhood and a park for the city'. This vision is incorporated into the design, titled 'Villages and Metropolis', offering an open, transparent landscape that should provide citizens and visitors green space, a place for creativity, rest and recreation. Since 2009 the park has been open to the public.

It was hoped that by investing in a park the downward spiral of urban degeneration would be broken, that the 'urban flight' (of those who can afford to leave the area for a better living place) could be prevented and that private investments, new renewal initiatives, economic activities and new families would be attracted to this part of Antwerp. Another objective was to provide a meeting place for and a corridor between neighbourhoods that for a long time had been spatially separated, connected only by a viaduct crossing the yard in the middle. Moreover, although improving Antwerp's competitive position has never been explicitly mentioned as a project goal, the expectation was that the project should improve the quality of place and the image of the city. In this sense PSN can be considered a soft location factor, indirectly contributing to raising the attractiveness of Antwerp in general.

4.1.2 Features and design

THE PARK

PSN has an elongated shape with a length of about 1,600 metres and an average width of 150 metres. It covers 24 hectares, of which 18 hectares are of green space, including industrial buildings (two railway hangars, an office building and two water towers that have been restored and made ready for functions that fit well into the vision) and several leisure installations. To make the plan financially feasible, 6 hectares in the north-western part of the site (the 'head' of Railway North) are available for 192,000 square metres of commercial functions. The park is owned by the City of Antwerp which is responsible for its management, maintenance, etc.

A major aspect of PSN is its sustainable design. This is given shape by among other things the reuse of an old viaduct, industrial buildings and materials originating from the railway yard. The park also benefits from direct infiltration of rainwater, the prevention of light pollution and the use of local planting. PSN is 'basic': the infrastructure is clear and solid, and the layout is vandalism proof and maintenance friendly. It is only accessible to cyclists and walkers. Motorised traffic is not allowed.

PSN is not a classical park. Its design can be described as 'a city park off the beaten tracks'. The area is covered with sloping meadows and '1001' trees. For safety reasons the park is open and transparent: there are no bushes or middle–high planting. There is a winding network of footpaths and cycle paths. The old Dam viaduct connects the neighbourhoods on both sides of the park in the middle for motor traffic.

The restored railway buildings are important elements in the project. The large Parkloods contains two sports halls and a multifunctional expo and events hall, offices for park logistics and a catering facility. One sports hall is used by the Artesis Hogeschool (university college) for its kinetics courses. The other is available for small-scale sport activities for the neighbourhood (but not exclusively). Following a decision in 2012, the multifunctional expo and events hall (during the interview period it was in use for an art exhibition) will be developed into a hotspot for creative sector activities. The Platform is a covered, multipurpose open space that can also be closed. Since the opening in 2009 it has housed a 'summer bar', a popular bar/restaurant with terraces boarding a water garden and a children's playground. Den Boemel, a former training centre for railway workers, will be transformed into a child and youth campus, together with a primary and a middle school. Finally, two water towers on the eastern edge have been restored as protected monuments.

Although the strategy is that park users should enjoy themselves in the ways they like best, the municipal department Buurtsport (Neighbourhood Sport) provides opportunities to take part in all kinds of recreational activities in the park and in the adjacent neighbourhoods. Buurtsport reacts on 'market demand', like organising bicycle lessons if this is asked for or it supports clubs in the area. In summertime animation activities for children are organised in cooperation with partner organisations from Antwerp North.

THE 'HEAD' OF SPOOR NOORD

Next to the park there is the Kop van Spoor Noord (the 'Head' of Spoor Noord). In exchange for 18 hectares of public space the former owner of the site, the real estate developing company Euro Immo Star NV, a subsidiary of NMBS, got development rights for 192,000 square metres within a plot of 6 hectares. In 2005 the Hogeschool Antwerpen decided to build its new campus here. The investment in the campus (11,200 square metres) amounts to €64 million. In 2015 5,000 students were welcomed at the start of the first school year in this new location. The university college participates financially in the construction of the sports hall in the Parkloods building. The Plantijn Hogeschool decided to join the Artesis Hogeschool in this location.

In May 2012 it was decided to build a hospital on the site: construction activities for the Ziekenhuis Netwerk Antwerpen (ZNA, occupying 65,000 square metres) began in 2016 and it is expected to open its doors in 2018. The investment in the new ZNA hospital building is €150 million (without equipment). The Flemish and federal governments bear 95 per cent of the investment costs. The intention is to combine the hospital with a *zorgboulevard* (a complex of related healthcare services). Next to the university college campuses and the hospital, two apartment towers (70 metres high) including student housing, the (public) Customs Office (60 metres high), retail services, underground parking facilities and a park bridge (for walkers and cyclists only) connecting PSN with the neighbourhood 't Eilandje will complete the commercial part of the site.

4.1.3 Challenges and solutions

DIVIDED CITY COUNCIL

During the late 1990s the Antwerp City Council was not at all concerned with urban development, but with internal struggles and divisions (Winkel, 2011). Antwerp was governed by a 'rainbow coalition', a cooperation of not naturally likeminded political parties, mainly established to isolate the extreme right party Vlaams Blok, winner of the municipal elections of 2000 ('cordon sanitaire'). Owing to heavily divided political opinions and ideas, the advocates of a park within the municipality feared that the plan for a large urban project would not make it without a convincing vision, consensus among the partners, powerful support among residents and political support from higher layers (regional and federal government). The initiators of the Antwerp planning department managed to fulfil these requirements, but without the availability of substantial earmarked European and federal budgets it would have been questionable whether a park like PSN could have been realised.

FEAR OF GENTRIFICATION

Certain groups fear that the park could lead to gentrification in the surrounding neighbourhoods. They expect that gentrification will cause higher costs of living (especially higher rents) making it difficult for low-income households to stay in this part of the city. Gentrification would force them to leave the area in search of cheaper locations. The prevailing political opinion in Antwerp is that such a displacement process should be prevented. An answer to this is that the social balance has to be restored to make the area more livable, not only for the most deprived families. It is undeniable that some displacement has taken place, but the idea among the initiators is that the social tissue has been restored. In particular, the value of houses with a direct view of the park has increased substantially, but on average the process of gentrification shows a slow pace. A problem is that only limited information is available to show the trends, which makes it difficult to give convincing results. To control negative influences of gentrification, flanking measures in the neighbourhoods seem necessary, like providing more affordable social housing.

4.2 Implementation

4.2.1 Process[2]

From the early 1990s redevelopment proposals for the railway yard were put forward. Already in 1990 the GSA (Antwerp Master Plan) opted for an urban park. Other proposals raised were a park boulevard, a cluster of cultural products (business park for the creative sector) or housing (De Wever and Lamberts, 2003). NMBS, owner of the site, retained 'the head' for commercial development. From 1998 the site was described in the Gewestplan (Regional Plan) as a zone for urban development.

In its municipal policy statement of 2000 the city government declared that the yard should be used as a driving force for upgrading the quality of living of the deprived neighbourhoods. In 2000, when Antwerp North was selected as special attention zone in the framework of the federal Urban Policy fund, the City Council decided to submit a subsidy application for the redevelopment of the NMBS yard into a park. The federal government signed a covenant with the City providing a sizeable budget for the project. This budget enabled the City of Antwerp to take on the entire planning and, together with all partners, to produce a coherent idea about the redevelopment of the area. To manage the financial means that came available and to guarantee that the funds would be used only for the project the Autonomous Municipal Company Antwerp New North (Autonoom Gemeentebedrijf Antwerpen Nieuw Noord, AG ANN) was set up, enabling the City to save the funds and invest only when the planning phase was completed.

VISION MAKING AND FEASIBILITY PHASE (2001–2)

At the beginning of 2001, SOMA (Antwerp City Development Company), a municipal non-profit company responsible for managing supra local programmes, formed a Planning Team to coordinate the project. The Planning Team made an inventory and analysis of all relevant information[3] to prepare debates on the redevelopment. This resulted in an Initial Memorandum (containing information, legal and financial preconditions and ideas raised earlier) and a Discussion Memorandum, including a proposal about the course of the planning process and draft plans for the yard and its surroundings. In April 2001 the city government approved the proposed planning process.

This process consisted of three phases: 1) the city's view on the role of the area; 2) a feasibility study and negotiations with the owner; 3) plan and project development, including an international urban design competition. The Planning Team was commissioned to organise the first phase. To that end three brainstorming days were organised with representatives of municipal departments, city councillors and external experts. At the same time two workshops were organised for residents to make their wishes known. This resulted in a detailed Consensus Report. This important document presented a straightforward vision about the desired development of the railway yard together with a number of principles and a strict timetable.

Initially there was resistance against the idea of a park on this location. Every party had its own ideas and several proposals were put forward. However, the clear vision of the Consensus Report to provide green space where it was tremendously needed, the strong support among residents and the availability of a substantial budget that could only be used for deprived area upgrading, resulted in the acceptance of a park as the new purpose of the railway yard.

The Consensus Report formed the basis of negotiations between the City and the landowners NMBS/EIS (EIS: Euro Immo Star NV, the real estate branch of NMBS, formally owner of the site). The NMBS had not been involved in the vision-making process. They were, according to our interviewees, suddenly confronted with a press conference during which the City launched the intention that a park should be developed on their site. NMBS was not amused. A first reaction was that this was impossible: a city government cannot expropriate land from a federal organisation. NMBS tried to take advantage of the political fragmentation by starting consultations with individual parties about possible uses of their site. They had in mind taking over the lead of the project from the City.

After months of (not very smooth[4]) negotiations a compromise was reached with NMBS/EIS and a policy agreement signed. The strong support among residents was an especially powerful factor in the negotiations. The principles of the Consensus Report remained intact with one exception: the floor space in the commercial zone was raised from 60,000 to 192,000 square metres. In exchange for the ownership of the 18 hectare park land and the cleaning of the polluted site NMBS/EIS got rights to develop a mix of office, residential and public functions. A public limited company (ANNEIS) was set up in December 2001 as a financial and legal instrument for developing the project jointly.

PLANNING PHASE (2002–5)

In early 2002 the City Council decided to organise an international urban design competition. Twenty-six Belgian and foreign offices submitted themselves as candidates. A jury selected five projects to explore their proposals more concretely. Early in 2003 the project 'Villages and Metropolis' was chosen and the winning team[5] was invited to draw up a definitive design scheme.

In February 2003 a first element was opened to the public (a temporary sports hall). In March 2005, the final design was approved by the City Council, which completed the planning stage. In April 2005 sanitation works started. One year later construction works began.

EXECUTION PHASE (2005–9)

To keep up the momentum, during this stage a range of activities were organised like an annual street run, the opening of the renovated Dam railway station (a local railway station bordering PSN, headquarters of the Planning Team), an interactive exhibition *Proper!* (Clean!) about the soil decontamination works, a photo exhibition *Park in beeld* (Park in the picture) and several other events to keep the attention alive.

The festive opening of the western part of the park, where most of the leisure installations are situated, was held on 25 May 2008. The eastern part, primarily lawns and playgrounds, followed one year later. Meanwhile the park, the leisure installations and the renovation of most of the industrial heritage buildings have been completed. Construction activities in the commercial zone have started recently (2011).

PARK MANAGEMENT

After the opening in 2008 AG Stadsplanning established a park management team, led by a park coordinator. This team had to take care of all necessary preconditions for a clean, safe and attractive park. The management team organised the regular maintenance, including the 'social' management, such as striving for a balance as regards to activities and visitors, or trying to involve residents. After one and a half years the management team was abolished. The regular maintenance tasks were allotted to the municipal departments already active in the surrounding neighbourhoods. The current situation implies that the park management is formally a task of the city department Samen Leven (Living Together), but for all kinds of park problems this department has to involve others. This situation makes it not easy to act promptly and decisively like in the past. This new structure seems not to be well thought out and is not appreciated by everyone. The question nowadays is whether the park coordination should be restored.

4.2.2 Actors

CITY OF ANTWERP

The City of Antwerp is initiator, coordinator and leader of the project. The multidisciplinary Planning Team, appointed by the City, coordinated the entire project from the vision-making to the execution phase. They did the daily operational work. Within the team, the programme leader focused on the financial and juridical aspects and the project leader on buildings and architecture. Other members focused on landscape architecture, commissions and contracts, communication and park management. In 2008 the Planning Team was extended with a park coordinator and a park programmer. The park coordinator set up a management plan for the park in collaboration with local residents, the police and the city departments involved. The park programmer stimulates and coordinates local participation by supporting activities.

Next to (not above, because there are no hierarchical relations) the Planning Team were a Design Team (winning team plus external experts) and a Steering Committee with representatives of NMBS, EIS, the city administration (civil servants), city councillors, the chairperson of the OCMW (responsible for social policy) and the regional government. The Steering Committee met four times a year. A representative of the Planning Team (usually the coordinator) attended these meetings to report on progress.

A public company AG ANN was explicitly established to manage the project financing. Later on (in 2006), civil servants working for the project within the municipal organisation were transferred to AG ANN. Other municipal departments involved include Buurtsport (neighbourhood sport), VZW Antwerpen Stad (supports citizen initiatives) and city department Samen Leven (Living Together).

NMBS/EIS

NMBS, the national railway company, and its subsidiary Euro Immo Star NV, owned the site until 2012, when the ownership of the park part was formally transferred to the City of Antwerp for one symbolic euro. The railway yard was used for parking, repair and maintenance of trains and served as an educational centre for railway workers. Several reasons caused the NMBS to leave the area of Spoor Noord in 2001 and relocate in the industrial port area of Antwerp where there is much less impact on the environment and a more practical layout of the yard could be realised. EIS NV is responsible for the development of the 'head' of PSN, which has to fit within conditions laid down in the Bijzondere Plan van Aanleg (Land Use Plan).

RESIDENT GROUP

A resident group has played an important role. A group of around 15 well-educated persons living in the neighbourhood, led by a local physician, appeared to be very keen in lobbying towards the key players NMBS and the various government levels. Resident representatives were invited to the meetings of the Planning Team. Excursions with politicians to the Westergasfabriek (Amsterdam) and the Emscher Park (Ruhr area) were organised to get an idea of the possibilities, especially with regard to the reuse of the railway buildings. Their convincing ideas and strong basic principles (safety, reuse buildings, no splitting up, not too many provisions) were accepted as starting points for the final design. The same holds for some elements that were at first not fully embraced by the City, like water elements and the reuse of railway buildings.

FEDERAL GOVERNMENT

In 1999 the new national government established a Grootstedenbeleid (Federal Urban Policy). This policy was set up to support the largest cities to develop programmes to prevent urban flight by upgrading deprived neighbourhoods. A special budget was made available for projects submitted by the eligible cities of Antwerp, Ghent, Charleroi, Liège and a part of the Brussels Capital region. During the first years €30–40 million were made available; later this was raised to €67 million. The PSN was considered to be a project fitting very well into the objectives of the policy. The fact that NMBS as owner of the site is a federal organisation made it even easier. At the start of PSN, the federal state was closely involved in the project, for example as participant in the jury for selecting the design team. After some time the federal state withdrew. The implementation was left to local and regional authorities because tailor-made solutions were preferred instead of matching preconditions set by national government.

FLEMISH GOVERNMENT AND EUROPEAN INSTITUTIONS

Like the federal government, the Flemish regional government has subsidised PSN through its Sociaal Impuls Fonds programme (Social Impulse Fund). This fund, directed to the large Flemish cities, co-financed the budget allotted to PSN through the European ERDF programme. The Antwerp region was eligible for receiving a subsidy through the Objective 2 funds and the City successfully lobbied for a European grant for the project. The grant was co-financed by the Flemish regional government and the City of Antwerp.

OTHERS

Other actors that have played a role in the realisation of PSN are the design team, the Artesis university college, subcontractors and consultancies (the latter mainly for technical construction activities).

4.2.3 What made the project happen? Key enablers and organising capacity

POLITICAL SUPPORT COUPLED WITH A DEVELOPMENT OPPORTUNITY

The first factor is the ambition of local government to give political priority to urban regeneration. As an 'arrival city' Antwerp is confronted with severe integration problems. An investigation by the regional government into the presence of deprivation in Flanders (Kesteloot *et al.*, 1996) had shown that the neighbourhoods around PSN belonged to the most deprived of Flanders and Antwerp: high densities (2,200 per square kilometre, four times higher than in the rest of Antwerp), high migration pace (almost 25 per cent annually), almost no green spaces or playgrounds available within a reasonable distance, few leisure and cultural facilities, drugs and criminality nuisance, and a chaotic traffic and parking situation. The opportunity to create a green, open territory with recreation facilities in these neighbourhoods to support urban regeneration fitted well into all policy levels.

POLITICAL SUPPORT THROUGH GRANTS

An important factor were abundant budgets made available through the various government levels. In 1999 the new national government coming into office established (following national urban policy practices of France and the Netherlands) a federal urban policy for large cities. The PSN project fitted well into the objectives of this policy. According to the data submitted to the authors by the federal Dienst Grootstedenbeleid (Department for Federal Urban Policy) more than €28 million was set aside between 2000 and 2008 for PSN, of which €21 million were for 'physical' or spatial investments. Antwerp could also make use of regional funds in the framework of the SIF programme. In the late 1990s cities had the opportunity to formulate projects to tap into funds made available by European Regional Policy initiatives (such as URBAN, Interreg, Objective 2 of the ERDF). Since Antwerp belonged to the regions eligible for Objective 2 support the PSN project was successfully put forward by the city government for support.

SOCIETAL SUPPORT THROUGH CITIZEN PARTICIPATION

Citizen participation has been a crucial factor in the project's realisation. Residents were outspoken about the new function of the abandoned railway site: it had to become a green lung and nothing else. This group was keen in lobbying for the project. When the just appointed Belgian prime minister visited the area in 1999 he was given a leaflet with the idea produced by the resident group. During his visit he promised to contribute to improve the living circumstances in these neighbourhoods. Already in 2000 a budget of almost €4 million was made available for the project. During the development of the plan the resident participation became more refined on details of the project, like changing the structure of the paths, conservation of the railway buildings, construction of a playground and the provision of water elements.

LEADERSHIP

The quality of local organising capacity is a major factor in the success of PSN. Several things should be mentioned, like the innovative structure set up and the personal qualities of people involved. The multidisciplinary Planning Team, focused on one project only, got full competence to sign contracts. Thanks to its proactive approach, its strong link with the residents and its solid vision, the Planning Team took the lead in the cooperation with the local Antwerp coalition. If this had been unsuccessful a lot of money would have been lost (Winkel, 2011).

The set up of AG ANN was another innovation. Thanks to its successful operations AG ANN was in 2009 extended into a planning company for all urban projects in Antwerp (i.c. AG Stadsplanning Antwerpen). Until the late 1990s planning in Antwerp followed the classical approach, mainly restricted to set conditions leading to the approval or disapproval of projects initiated by (private) parties, primarily real estate developers. Urban planning had no priority; in 2001 about ten planners were employed by the City. Nowadays around 45 people are employed within AG Stadsplanning (AG Stadsplanning, 2012). The young, non-profit municipal company shows that proactive project planning in cooperation with private and public parties is an important innovation to Antwerp.

PSN was the City's first experience of strategic planning to initiate and develop a multi-functional project by itself. Since then several other large projects have followed the PSN example. Its approach must be considered an innovation for the City of Antwerp, as is its multidisciplinary approach.

Personal aspects were important too. The Planning Team kept the same composition during the whole development period. Strong, consequent and trustful leadership qualities can be mentioned as important ingredients. With respect to the negotiations with the site owners, the representative of Euro Immo Star NV was during our interviews explicitly mentioned for his constructive contribution to the negotiations.

PROJECT FINANCING

A decisive factor in the development of PSN were the grants linked to the project objectives by the state, Flanders and European institutions. During the period 2000–8 around €40 million were made available for spatial investments in the project. According to AG Stadsplanning Antwerpen (2011) about 21 million came from the federal government, about 14 million from Europe (including Flemish and local co-financing of ERDF) and an additional 4 million from the City. Next, the Artesis university college participated in the costs of the sports hall (1.75 million). According to information presented by the department Grootstedenbeleid of the federal government, grants devoted to PSN (exclusive investments in the surrounding neighbourhoods) amounted to about €28 million between 2000 and 2008 (Table 8.3).[6]

The investments in PSN are presented in Table 8.4.

Table 8.3 Federal GSB grants Park Railway North

Item	*Million €*
Spatial investments (buildings, park construction)	21.0
Process management (project coordination)	3.8
Vision development (policy vision)	1.8
Communication (mobilising programme)	1.5
Legal framework	0.1
Total	28.2

Source: Dienst Grootstedenbeleid, POD Maatschappelijke Integratie, Brussels

Table 8.4 Investments in Park Railway North

Item	€
Park construction	15,000,000
Skate bowl	845,000
Playgrounds	520,000
Water gardens	1,700,000
Parkloods	16,000,000
Platform	2,600,000
Boemel	2,250,000
Water towers	500,000
Demerstraat Tunnel	1,000,000
Park bridge	1,500,000
Total investments	41,415,000

Source: Antonoom Gemeentebedrijf Stadsplanning Antwerpen

The AG ANN was able to quickly and easily decide where and how the money had to be invested in the park. Permission from the City Council was not necessary. The budgets were set aside for the moment that the money was needed for implementation. Formally this was not allowed, as a budget has to be spent in the year that it becomes available. The Belgian Auditor's Office made remarks about this behaviour, but it was accepted. In their turn the Planning Team used the budgets as a means of pressure on the City Council to keep pace in the project. They pointed to the risk of losing the money if decisions were not be taken in time. This worked out very well, although this way of working is no longer allowed. Currently, grants are linked to timetables and deadlines, rather than to project objectives. A result of this approach is that many (good) projects lose their budget because the process is more time-consuming then expected (Winkel, 2011).

COMMUNICATION

The Planning Team worked with an open, transparent agenda, sharing each step with others. This created trust among the parties involved (Winkel, 2011). From the vision-making phase to the construction phase (2002–5) a lot of attention was given to participation and communication. Alongside providing basic information about the park (through websites, posters, flyers, articles in local newsletters), the communication strategy comprised the organisation of 'participation moments' in the form of workshops, debates, a *praatcafé* (debate pub), etc. Moreover, 'design days', such as the Trek je plan dag (Make Your Plan Day), were organised during which residents were invited to make a model of their ideal park.

Initially people could not imagine that this ugly, industrial part of the city could be transformed into an attractive environment. There was quite some scepticism about the plan. Conventional ways of disseminating information did not work, especially among foreign citizens. Therefore flyers were sometimes distributed in more than one language, but far more interesting was to show concrete results as soon as possible, because seeing is believing. This was done by organising events during the construction period, like walks in the park grounds under construction, the opening of the temporary Perrier neighbourhood sports hall in February 2003 and a seven-day indoor happening Park in zicht (Upcoming Park) in November–December 2003.

Communication costs, including the Wervend Programma (Mobilising Programme), were subsidised by the federal government and amounted to €1.5 million (about 3.5 per cent of total investment) which can be considered as quite substantial for a public project.

4.3 Results

In this section we try to answer the question whether the objectives of PSN have been met. What are results of the project? Did it develop into a popular spot for Antwerp's citizens and did it contribute to the social upgrading of the adjacent neighbourhoods? Did the PSN contribute to sustainable competitiveness?

A preliminary remark is that the park is not yet completely finished. Activities in the 'head' of Spoor Noord are in full swing (summer 2012) and the park was only opened in 2009. This makes a solid evaluation of effects on the neighbourhoods difficult, because some of these effects (like migration) are partly of a medium or long-term character.

SUCCESSFUL AS A NEW URBAN AMENITY

A first conclusion was already given above in Section 4.1.2 where it was stated that since its opening, the park can be considered a big success as a new urban amenity. Indeed, all information made available to us and all our collocutors confirm this statement. In 2011 a survey to evaluate how people experienced the park showed that citizens of Antwerp are very positive about PSN (Winkel, 2011). From its start, the park has attracted many visitors and users in a balanced mix of people with a Belgian or with an ethnical background, young and old, from the direct environment and from the rest of Antwerp and even from the region. People from Antwerp are proud of this new amenity. According to the project leader the special design (incorporating vast green spaces, which was a powerful wish by the local residents) has contributed to this success. The initial fear by some groups that the park could have been dominated by sub-groups or could have suffered from nuisance (drug addicts, prostitutes or other unwanted people) or would only be used as a place for dog walkers did not come true. From the opening, well-organised management, spontaneous social control and of course the layout and design have prevented negative developments. In this sense – as recreational amenity – the project is a great success.

NEIGHBOURHOOD UPGRADING

PSN's major objective was to create new, high-quality, urban green space that should function as leverage for the upgrading of the adjacent neighbourhoods. Since the start of the vision-making process in 2000, characteristics of the neighbourhoods such as socio-demographic and local needs data have been mapped by field recording on the level of individual buildings. This zero measurement in the frame of the federal urban policy allows the effects and emanation of the PSN on the neighbourhoods to be measured. These effects have been measured with the help of geographical information systems (GIS). Based on quantitative data and with the help of spatial techniques the physical and social dimensions of the project can be sketched.

Although an up-to-date evaluation is not yet available, there are some first impressions of effects. The park option was underpinned by the lack of green in this part of the city. For the year 2000 the GIS maps show that only a very limited part of housing (10 per cent at most) met the standard for district green[7] (AG Stadsplanning, 2011). After the realisation of PSN, the whole area plus parts of the adjacent neighbourhoods fulfil the district green criterion. Another interesting figure are house renovation permissions. Eleven renovations were authorised in 2000; in 2009 this figure was 47. Another factor are vacancies. Since 2000 the degree of vacancy has decreased substantially, again based on GIS maps. A study by the City of Antwerp confirms that between 1995 and 2009 real estate prices around PSN increased more than the Antwerp average prices, albeit that this increase is in line with the increase in other inner city districts. Convincing data with respect to the course of real estate prices do not exist. Moreover, data are based on a limited number of transactions. According to one of our collocutors, prices of apartments looking out on the park have quadrupled in price, but the effects on more remote real estate are moderate. Another factor is the change of functions. GIS maps show that the residential function is increasing due to the conversion of former industrial and professional buildings into residential use. These transformations have increased the living quality in those areas (AG Stadsplanning, 2011). Finally, probably the most important of all indicators is the index of deprivation. This index is made up of three variables[8] that can be considered representative to measure deprivation. It appears that in 2001 as well as in 2010 the three neighbourhoods around PSN belong to the most deprived in the city. The observation by AG Stadsplanning is that the concentration of deprivation in this part of Antwerp was the *raison d'être* to invest here, but it appears that it has not been enough. This should be described as a disappointing result, given the park's prime objective. The conclusion drawn is that investing in this area should be of a permanent character (AG Stadsplanning, 2011).

On the other hand, gentrification effects, as they were feared by some parties, seem to be moderate and concentrate in the streets that have front sight on the park only. Social displacement, the negative effect of gentrification, has not been recorded yet. Some groups consider this as a positive result.

Moreover, the project is not yet complete. The arrival of 3,000 students in the area in 2014 (of which a part will reside in the new student housing in the 'head' of Spoor Noord and probably also in the neighbourhoods) will bring in another public. Their presence could work as a catalyst for the area.

RETURN ON INVESTMENT

The (spatial) investments in the project amounted to over €40 million. What has been the return on investment? It is not possible to answer this question. Choosing a park was opting for more social return and less economic return. An economic effect has never been an explicit objective, at most a side effect as soft location factor. Expressing the benefits in monetary terms has not been attempted. These benefits are that the quality of life has improved and that the park (and its amenities like the new sports hall and the skate and BMX bowl) has produced a new positive feeling among citizens. A survey among citizens confirms that the (internal) image of the city has improved, even among those who had not yet visited the park. Some variables (renovation permits, vacancies) show positive developments, but these are minor effects compared with the investments. The 'head' of Spoor Noord offers an attractive new location on a former derelict site that contributed to the negative image of the area. This part, owned by the NMBS, is a welcome addition to Antwerp's working and living environment.

INNOVATIONS

Finally, innovations for the City of Antwerp that have been dealt with in this case study are:

- Integrated proactive urban planning, replacing the classical approach of approving or disapproving development initiatives by others. PSN was the first large integrated urban development scheme for Antwerp.
- Establishment of an autonomous municipal development company, AG Stadsplanning Antwerpen, successor of the AG ANN, explicitly established for PSN. This young modern company has proved to be able to act promptly, effectively and efficiently in an entrepreneurial way.
- Innovative form of citizen participation, resulting in broad societal and political support.
- Innovative convincing communication strategy to maintain the momentum of the project during the preparation and construction phases.

5 Conclusions

This chapter explored the case of the Park Railway North in Antwerp: an abandoned railway yard that has been turned into an urban park and a commercial zone. In 2000, the Belgian national railway company NMBS left its marshalling site, situated between deprived and densely populated residential quarters. After thorough research and consultation with all parties involved, the City Council decided to use the site as a driving force for upgrading the quality of living in this problematic part of the city.

The factors that contributed to make the project happen are related to finance (European, federal and regional) becoming available to tackle social problems, active residential lobbying and participation, leadership and the formation of relevant networks. Negotiations between the City and the landowner NMBS resulted in a compromise of a large urban park and a commercial zone (192,000 square metres of floor space) to be developed by NMBS.

Since its opening in 2009 the park can be considered a big success as a new amenity. All information available confirms this statement. The park attracts many visitors and users in a balanced mix of people. PSN and the neighbouring areas are back on the citizens' mental maps. As to neighbourhood upgrading – a major objective – some early impressions of effects can be given, as a more comprehensive evaluation is not yet available. The adjacent neighbourhoods now meet the internationally accepted district green criterion, the number of private house renovations quadrupled (until 2009), the degree of vacancies decreased substantially, former industrial and professional buildings have been transformed into residential functions and real estate prices of residences with a view on the park increased considerably.

Displacements – a feared consequence of gentrification – have been limited. Perhaps the most important indicator, the index of deprivation, showed in 2010 that the neighbourhoods around PSN still belong to the most deprived. This outcome demonstrates that, despite the big success of the park, it did not really influence deprivation. It is not possible to present clear statements about the return on the investment of more than €40 million, because expressing the benefits of the park in monetary terms has not been attempted. Choosing a park was opting for more social and less economic return. An economic effect has never been an explicit objective, at most a side effect.

Notes

1 This section is for an important part based on a case study about Antwerp's characteristics and dynamics conducted by Steven Sterkx (2014).
2 For the description of the process the publication *Antwerp Spoor Noord. A City Park Off the Beaten Tracks* by De Wever and Lamberts (2003) has been an important source of information.
3 Extensive use was made of geographic information systems (GIS) techniques to show the characteristics of the neighbourhoods on maps, helping decision-taking by the municipal government and for monitoring the effects.
4 In contrast to the NMBS the City was in favour of public functions on the six- hectare 'commercial' NMBS location. A mediator between the City and NMBS had to be appointed to work out the details of the contract.
5 The winning design team consisted of Bernardo Secchi and Paola Viganò (Italy) and Pieter Kromwijk (Netherlands).
6 Of which €21 million were for spatial investments.
7 Internationally accepted standards are that neighbourhood green should be available within a distance of 400 metres from a house and district green within 800 metres.
8 These are percentages of long-term unemployed, persons on welfare and disposable incomes below €10,000. These three are selected from a list of 23 variables as most representative to measure deprivation.

References

AG Stadsplanning Antwerpen (2011). Park Spoor Noord. Van idee tot park. Antwerp: AG Stadsplanning Antwerpen.

De Wever, H. and Lamberts, E. (2003). *Antwerp Spoor Noord. A City Park Off the Beaten Tracks*. Antwerp: Ludion.

Kesteloot, C., Vandenbroecke, H., van der Haegen, H., Vanneste, D. and van Hecke, E. (1996) Atlas van de achtergestelde buurten in Vlaanderen en Brussel. Brussels: Ministerie van de Vlaamse Gemeenschap.

Stad Antwerpen (2009). *Urban Development in Antwerp. Designing Antwerp*. Antwerp: Stad Antwerpen.

Stad Antwerpen (2011). Indicatorenmonitor 2011. Antwerp: Stad Antwerpen.

Sterkx, S. (2014). Antwerp. In L. van den Berg, J. van der Meer and L. Carvalho (eds), *Cities as Engines of Sustainable Competitiveness: European Urban Policy in Practice*. Aldershot: Ashgate.

Van den Berg, L., van der Meer, J. and Pol, P. (2003). *Social Challenges and Organising Capacity in Cities*. Aldershot: Ashgate.

Vandermotten, C., van Hamme, G. and Marissal, P. (2010). *La production des espaces économiques*. Brussels: Éditions de l'ULB.

Van Hamme, G. and Strale, S. (2011). Port gateways in globalization: the case of Antwerp. *Regional Science Policy and Practice*, 4, 83–94.

Verhetsel, A. and Sel, S. (2009). World maritime cities: From which cities do container shipping companies make decisions? *Transport Policy*, 16, 240–50.

Winkel, E. (2011). Urban MLG Profile Belgium. In M. Grisel and F. van de Waart (eds), *Multilevel Urban Governance or the Art of Working Together*. The Hague: European Urban Knowledge Network.

9 Phoenix project (Dortmund)

Erwin van Tuijl and Leo van den Berg

1 Introduction

This case study discusses the development of the Phoenix project, a large brownfield redevelopment programme in the city of Dortmund. This project consists of the transformation of two large and connected industrial sites of the iron and steel industry, that were closed to society for one and half centuries, into a location for the post-industrial 'New Dortmund' that combines modern industries with housing and leisure. Among other developments, the latter is achieved by the development of a large lake, the 'Phoenix See'. The project also deals with the redevelopment of the Hörde district – a run-down working class area that was squeezed by the two industrial sites and railway tracks – and the environmental and ecological development needed to restore the damage created in the industrial times.

Even though there are many brownfield redevelopment projects in the Ruhr area, Phoenix is a special one for a number of reasons. Firstly, because of its size: the total project covers more than 200 hectares, which is comparable to 300 football pitches. Secondly, a large number of stakeholders are involved in the project, including private ones as well as authorities of different layers, ranging from district level to the European level. Integrating the different interests of all these stakeholders as well as various funding sources is a major challenge. Thirdly, the project requires an integrated approach to make the transformation from a mono-functional industrial site into a mixed-use urban area. It attempts to combine all pillars of sustainability (economic, social and environmental development) and integrates various disciplines, including hydraulic engineering, landscaping, urban development and mobility management.

This chapter is structured as follows. Section 2 analyses the structure (and recent dynamics) of Dortmund across a number of dimensions or 'capitals' for sustainable competitiveness. We particularly pay attention to the industrial history, the negative effects of the de-industrialisation, the assets the city has for recovery and the possible role of the Phoenix project in this recovery. Section 3 sketches the city-wide economic vision and strategy, and discusses the role of the Phoenix project in this wider strategy. Section 4 is the core of the chapter and explores Phoenix in depth, namely its start-up history, vision and content (4.1), the key features and enablers that 'made it happen' (4.2) and a number of preliminary results of the project (4.3). Section 5 concludes raising lessons and challenges ahead.

2 City profile: capitals for sustainable competitiveness

2.1 Performance

With more than 580,000 inhabitants, Dortmund is among the ten largest cities in Germany and forms the largest city in the Ruhr area. The Ruhr area itself, with about 5.2 million

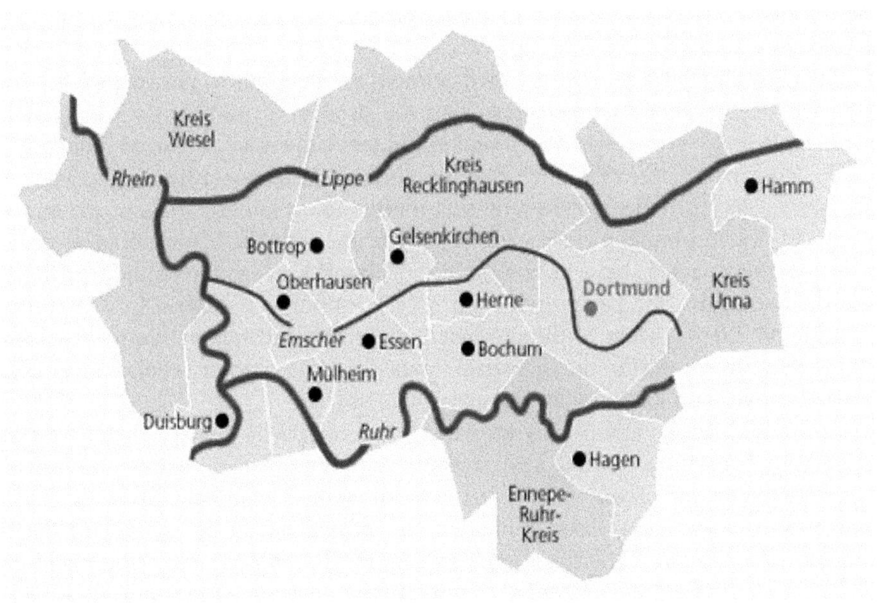

Figure 9.1 Dortmund in the German Ruhr area

Source: Blanke-Bartz (2011)

inhabitants, is one of Europe's largest polycentric urban areas (see Figure 9.1). During the last decade, the city's population has declined from 590,213 in 1999 to 580,956 in 2011 (Stadt Dortmund, 2013), and further demographic change (e.g. migration and aging population) is a serious challenge for Dortmund as well as for Germany as a whole (Ellwein and Mai, 2014).

Dortmund's economic development between 2002 and 2012 can be seen as positive. The GDP per capita has increased since the middle of the 1990s to €32,637 in 2008, which is above the German average (€30,392) and other cities in the Ruhr area, with the exception of Essen (€38,429) (Ellwein and Mai, 2014). The city's export rate has grown rapidly from 25.6 per cent in 1997 to about 50 per cent in 2012. The number of firms has increased considerably from 25,189 in 1998 to 35,708 in 2012 and also the number of employees has increased from 278,100 in 2000 to 304,200 in 2010. The unemployment rate has declined from 17.5 per cent in 2005 to 13.1 per cent in 2012 (Stadt Dortmund, 2013), but remains among the highest in the Ruhr area. It is a major challenge to reduce structural unemployment, as about 40 per cent of those who are unemployed are jobless for more than two years (Ellwein and Mai, 2014). It concerns a large pool of old industrial workers who have difficulties in learning new skills and applying for jobs in other sectors.

2.2 Economic structure and knowledge

Dortmund's economic structure has changed considerably during the transition from an industrial economy towards a modern service economy (see also van den Berg *et al.*, 2005). The three dominant manufacturing industries – mining, steel and brewing – have declined strongly since the 1960s resulting in the closure of the last mine in 1987 and the last steel plant on the Phoenix site in 2001. Also the last two breweries (in the past the city had eight breweries producing 10 per cent of German beer production) were bought by a multinational outside Dortmund (Ellwein and Mai, 2014).

Nowadays, the city has a service-oriented economy with about 80 per cent of the workers in the service sector. This does not mean that the production sector is not important for Dortmund, since many services in Dortmund might serve the manufacturing industries elsewhere in the Ruhr area where manufacturing remains a large sector (Ellwein and Mai, 2014). The City has successfully developed new sectors including IT, logistics, micro and nanotechnology (MST) and biotech (these are all part of the so-called 'dortmund-project', see Section 3). All these sectors have grown rapidly (Table 9.1). Also the manufacturing sector seems on the rise again, and has grown from 25,136 employees in 2005 to 27,920 in 2012 (Stadt Dortmund, 2013). Most of the producers are small and innovative niche players in high-tech sectors. In contrast to the industrial times, the contemporary economic structure is dominated by small- and medium-sized enterprises (SMEs).

The successful development of new growth sectors can partly be explained by the strong knowledge base consisting of advanced research and competence centres and knowledge institutes, both public and private. The decision to set up the Technical University Dortmund (TUD) in 1968 in a working-class dominated city turned out to be crucial for the future development of Dortmund. TUD forms the basis for the Technology Park (founded in 1984) and the Technology Centre Dortmund (Technologie Zentrum Dortmund, TZD) (1988), which are based in direct geographical proximity of the university. Apart from TUD, Dortmund has a University of Applied Sciences and Arts and another 50 research institutes, including the Frauenhofer Institute for Material Flows and Logistics and the Max Planck Institute for Molecular Physiology (Ellwein and Mai, 2014).

2.3 Social conditions

Despite the strong knowledge base that supplies highly qualified professionals, balancing the labour market is a major challenge in Dortmund. The labour market is characterised by a qualitative mismatch. On the one hand, there is a shortage of high-skilled labour in the rapidly growing new sectors. Dortmund has many students, but has difficulties keeping them after study due to competition from other cities with more attractive job opportunities.

On the other hand, there is a large group of low-skilled old industrial workers (Ellwein and Mai, 2014). The latter group has limited skills to move to other sectors and it is hard to make this group decline, also due to the inflow of new low-skilled migrants. Nevertheless, the total share of migrants is equal to the German average (15 per cent) and is not seen as a

Table 9.1 Employees and firms in new key sectors

	Employees		*Firms*	
	2005	*2012*	*2005*	*2012*
IT	11,600	12,954	720	864
MST	1,865	2,373	30	46
Logistics	24,150	26,984	662	923
Biotech	67	332	5	27

Source: Own elaboration, data from Stadt Dortmund (2013)

major issue. The city does have the challenge of socio-economic polarisation and unequal income distribution, very visible in a number of poor and run down neighbourhoods, especially in the north of the city (Ellwein and Mai, 2014).

2.4 Environment and accessibility

Besides the socio-economic challenge caused by de-industralisation, the industrial past has both positive and negative impacts on the physical and environmental development of the city. On the positive side, due to the various industrial brownfields, the city has space for new urban functions inside the city borders (van Winden *et al.*, 2010). In total, the city has about 600 hectares of development areas, of which the Phoenix site is the largest with 200 hectares (van Winden *et al.*, 2010). On the negative side, many brownfields and other parts of the city are polluted and need to be cleaned. Illustrative is the Emscher River that was used as a sewer. Cleaning this river is part of various state-funded programmes to improve environmental quality and develop new green spaces in the state of North Rhine-Westphalia – NRW (Ellwein and Mai, 2014).

The city has many green spaces in the city borders, with the Hoeschpark, Westfalenpark and the Rombergpark as some major examples. Outside the city, the hills of the Sauerland and the green Münsterland give further possibilities for recreational purposes. In addition, the city has various urban amenities that citizens can use for recreational and leisure purposes. Yet, missing in the city is a large, central recreational spot and water. The city has some waterfronts, but these are occupied due to port and industrial activities.

2.5 Image

The city's industrial past also has implications for its image. The long tradition of mining and heavy industries is associated with pollution and a 'working-class' city. Therefore, the City is challenged to promote the 'New Dortmund' on various spatial levels as a modern R&D location with a pleasant working and living environment (Ellwein and Mai, 2014). Progress seems to be made, but the perception change is a difficult and long-term process. For instance, the city was one of the hosting cities of the World Cup in 2006 and was part of the European Capital of Culture in the Ruhr area in 2010. These events have been used to show the city in a national and international scene and seem to have contributed to an improvement of the city's reputation, although long-term results are still hard to assess. The city is also increasingly known as a high-tech location thanks to among others TUD, Technology Park and TZD.

2.6 Institutional conditions

In order to understand the role of different actors and their relations in large redevelopment processes such as the Phoenix project it is important to discuss some features of Dortmund's institutional and governance system. First of all, Germany has a complex multi-level government system consisting of the federal government, 16 federal states (Dortmund is part of North-Rhine Westphalia, NRW) and lower governments, including municipalities and districts. Municipalities have the right to regulate all issues for the local community within a legal frame and are tasked with the implementation of laws set by higher governments. As such, municipalities have a lot of freedom, although financial sources are often limited (Ellwein and Mai, 2014).

Second, various authorities have independent planning and development firms, such as NRW.Urban (a planning agency of the state of NRW) and DSW21 (a multi-utility and infrastructure development agency of the City of Dortmund). Such firms are primarily owned by public authorities, but operate as independent firms. This makes them less vulnerable to political change, which is important in long-term projects such as Phoenix. In addition, the authorities have less financial risks and responsibilities are outsourced to the development firms.

Third, the economic department of Dortmund is different from many other municipalities in Germany. The economic department is a direct part of the city government (in other cities local economic development agencies are more independent) and covers a broad range of activities to make the transition to a modern service industry, including cluster management, locational marketing, start-up support, human resource development and development of industrial sites. It is also worth mentioning that the economic department increasingly cooperates with other municipalities in the economic development agency for the Ruhr area (Wirtschaftsförderung metropoleruhr) (Ellwein and Mai, 2014).

Fourth, in general, cooperation on the regional level has improved considerably. The different municipalities jointly organised the European Capital of Culture in 2010 and have a stake in the regional authority, Regionalverband Ruhr (RVR), which has a formal parliament with 70 members who are representatives of the different municipalities. Metropoleruhr is a joint business development agency operating under this regional authority and it deals with marketing of the region, infrastructure development, environmental issues and regional planning.

2.7 Conclusions: main issues

Dortmund is making considerable progress in the long and difficult transition from an industrial economy based on three pillar industries ('coal, steel and beer') towards the 'New Dortmund', a knowledge-intensive service city. It has developed new growth sectors that are linked with its strong knowledge base with TUD, Technology Park and TZD as key assets. However, there are also remaining challenges, especially countering the socio-economic legacy of the de-industralisation in the form of structural unemployment and social polarisation. Other challenges include further improvement of the image of the city and cleaning the polluted sites.

3 City-wide vision and strategy

The major guide to steer Dortmund's transition from an industrial economy to a service and knowledge economy is the so-called 'dortmund-project', a best practice of an integrated urban development strategy and balanced public–private partnership. The project was a direct response to the decision of Thyssen-Krupp AG to close down the steel factory on the Phoenix site. Thyssen-Krupp AG and the City of Dortmund initiated the dortmund-project and with a private consultancy's support the concept was developed. Labelled the 'New Dortmund', the main goal of the dortmund-project was to 'create 70,000 new jobs from 2000 until the year 2010, 34,000 of them in the IT-industry, 16,000 in the microstructure industry, 10,000 in logistics and 10,000 in the other core branches' (Bömer, 2005, p.23). The number of 70,000 new jobs does not come out of the blue, but equals the total number of jobs lost in the three old pillar industries of coal mining, steel and beer (Ellwein and Mai, 2014).

The project was formally accepted by the City Council in 2000 and responsibility for it fell directly under the Mayor's Office. It focuses on three fields of action: 1) people and skills (education and training programmes); 2) future industries (the three initial focus industries include logistics, MST, IT and in 2007 bio-tech was added as a fourth); and 3) future locations for new economic functions and industries. Further fields of action include location marketing on the national and international level and the introduction of monitoring systems for all activities in the project (Ellwein and Mai, 2014). Even though the ambitious (and highly criticised) goal to generate 70,000 new jobs was not reached, the dortmund-project can be perceived as a success. Many new firms have been created in the new fields, the redevelopment of many locations for new functions has taken off and there are many indications that the perception of Dortmund has improved (Ellwein and Mai, 2014).

Because of the success of the project, the City Council decided in 2008 to continue with the programme and extend it until 2018. The follow-up strategy is linked with the vision 'to compete for global markets, in regard of economic globalisation and demographic change' (Ellwein and Mai, 2014). The ambitions for 2018 include the development of: 1) an internationally acknowledged business location; 2) a creative environment; and 3) high-quality living and working environments. An important tool for realising the ambitions is cross-clustering. The City's idea is to link a number of key clusters (the four new key clusters, but also healthcare, production technologies and creative industries) and to explore the following new markets: simulations, mobility, energy and resources.

The Phoenix site plays a central role in the 'dortmund-project' and the follow-up strategy. The closure of the steel plant on the site is the direct cause of the ambition to develop the 'New Dortmund'. The site is one of the major future locations in the programme. In the following section we explore the development process of the Phoenix site in more detail and analyse to what extent it indeed contributes to the development of the 'New Dortmund' project and city development.

4 The project

4.1 Vision and content

The industrial history of the Phoenix site dates back to 1841 when entrepreneur Hermann Dietrich Piepenstock established a steel rolling mill on the site of the former Hörde Castle on the eastern side of the district of Hörde (by that time still an independent municipality). Later, on the other side of the Hörde district (Phoenix West) a blast furnace plant was opened, and steel production, processing and pig iron production were combined. It resulted in one of the major industrial sites in the City of Dortmund that was closed to society. However, the decision of ThyssenKrupp, the owner of the industrial sites at the end of the 1990s, to close down the blast furnace (Phoenix West) in 1998 and the steel mill (Phoenix East) in 2001 meant the end of the area's industrial era. The City Council's idea was to give the site back to society and to transform it into a multi-functional area for the 'New Dortmund'.

The integrated master plan for the redevelopment covers four parts (Table 9.2). The first (see Figure 9.2) is the transformation of Phoenix West into a technology park targeted to production technologies (hi-tech production), and IT and MST industries, as well as general offices for other business and for leisure functions. In order to stimulate business in the mentioned key industries, two competence centres have been set up in Phoenix West: the MST Factory and Centre for Production Technologies (Zentrum für Productionstechnologie Dortmund, ZfP). The idea is to attract new (high-tech) production firms to the site in order to keep the history alive. This has also been done by the conservation of industrial heritage that have been used as landmarks for the area and will be transformed for other functions

Table 9.2 Details of the Phoenix project

Part	*Planning targets*	*Management and ownership*	*Funding*	*Size*
Phoenix West	High-quality sites for industrial sectors High-standard service quality for business Cluster development MST industry Commercial leisure facilities Conservation blast furnace	NRW.Urban (state)	EU State City	110 ha
Phoenix East	High-quality housing Leisure and recreational facilities with public activities Offices and services	Phoenix See Development Agency, PSDA (City)	City State	98 ha
Emscher Landscape Park and River	Making the river area accessible Ecological revaluation High-water management Leisure and recreation	Emschergenossenschaft (Water Management Association of the State)	State	N.A.*
Hörde	Integration with Phoenix East and revaluation of the district city centre	City	City	N.A.

Source: Own elaboration, data from Blanke-Bartz (2009)

*The Emscher Landscape Park is one large environmental project dealing with the development and connection of green spaces along the 84 kilometre Emscher River that flows through the Ruhr area.

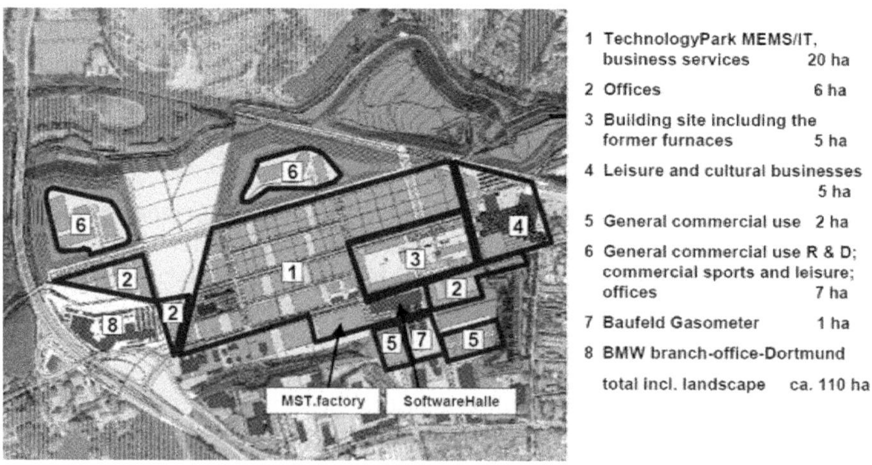

Figure 9.2 Plan for Phoenix West

Source: Blanke-Bartz (2009)

(see Table 9.3). Finally, the area is targeted to leisure and recreational functions including an event hall and green spaces. Especially the northern part has been converted into green spaces as this was the most polluted and limited other possibilities for redevelopment other than covering the contaminated soil.

The second part is the creation of high-quality housing, offices and recreational facilities on the Phoenix East site (Figure 9.3). The core of the plan is the creation of a lake

Table 9.3 Industrial heritage and re-use industrial buildings

Industrial heritage	*(Planned) new function*
Blast furnace	Landmark Other plans in concept stage
Gas tank	Concept stage
Warehouse Phoenix Halle	Events
Factory plant	Business and office centre for IT firms
Schalthouse 101 – Factory workshop	N.A.
Ammoniakhalle – Former warehouse	Office-business centre
Laboratory	Offices
Lavatory building	Offices and information centre

Source: Own elaboration, data from Blanke-Bartz (2009)

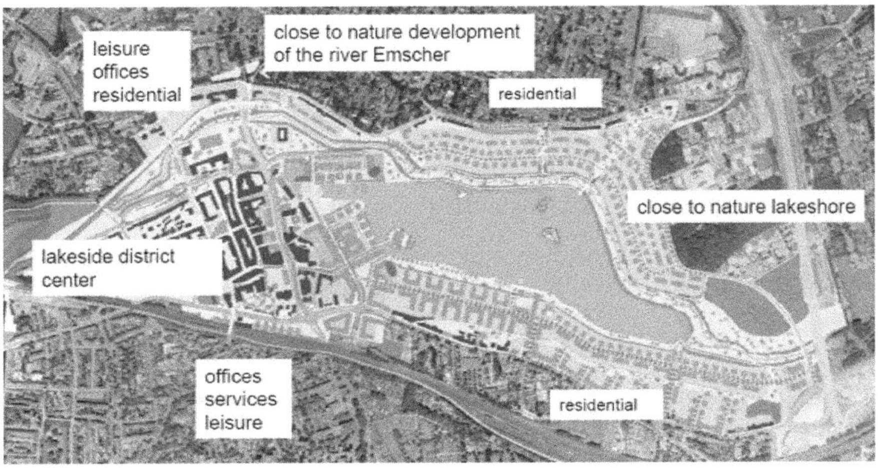

Figure 9.3 Plan for Phoenix East

Source: Blanke-Bartz (2009)

(Phoenix Lake, in German 'Phoenix See') that covers about 24 per cent of the surface area. There are some reasons behind the idea to develop a lake on this former industrial site. First of all, the lake could function as a major recreational spot for water activities that was missing in the city of Dortmund. Second, the lake could differentiate Dortmund from other (competing) cities. There are many brownfields in the Ruhr area, but none of them have been converted into a lake. Third, from an historical point of view, the water element has been brought back to the area. In pre-industrial times Phoenix East was a valley for the Emscher River, but in order to construct the steel plant the Emscher was covered and was no longer visible on the surface. High-quality houses have been constructed on the northern and the southern shores of the lake, covering about a quarter of the Phoenix East area, while the western part, bordering with the centre of the Hörde district, is a mixed-use area with offices, restaurants, bars and apartments. The remaining parts of the area are used as green spaces and for infrastructure.

The third part of the master plan is the revitalisation of the centre of the Hörde district. The district is squeezed by the two former industrial sites in the west and east and by a railway track in the north. The area is a low-income working-class neighbourhood with social problems and limited services. The idea is to connect the area with Phoenix East and to attract new investments in commercial functions.

The fourth part includes the development of the Emscher Landscape Park and cleaning and 'restoring' the Emscher River. In the industrial times, the Emscher was used as a sewer and polluted heavily. The idea is to bring the Emscher back to the surface, clean it and separate it from the sewage system. The project covers the development of a complex water management system as the Emscher cannot be connected to the lake for ecological reasons, but both are used as spaces to collect rainwater. In addition, the banks of the river are converted into green spaces for recreational purposes and are part of the larger Emscher Park project that covers the entire flow of the Emscher River through the Ruhr area.

Together, the parts form an innovative environment that aims to: 1) structurally improve the eastern Ruhr area; 2) improve competitiveness; 3) develop about 15,000 workstations; and 4) upgrade the Hörde district. Note that the four aspects are partly overlapping, especially the Emscher Landscape Park and River plan which covers both the Phoenix West as well as the Phoenix East sites.

The implementation and development of the master plan is complex as there are different owners with their own interests, funding sources, functions and target groups. In the remainder of this section we analyse this development process in more detail – focusing on leadership and organisation, finance and communication – as well as the first results of the plan.

4.2 Implementation

4.2.1 Leadership and organisation

The organisational model for the Phoenix project is composed of various stakeholders with complementary tasks. As said, the project is supervised by the City of Dortmund and falls directly under the responsibility of the Mayor's Office. The daily coordination is done by the Economic Department, while the master plan has been set up by the Planning Department. The individual parts are steered by different coordinators: Hörde centre by the City; Phoenix West by NRW.Urban; Phoenix East by the Phoenix See Entwicklungsgesellschaft (Phoenix See Development Agency, PSDA) and the Emscher River and Park development has been the responsibility of the Water Management Association of the State. Overall, the organisation seems to work well with limited conflicts between the stakeholders and the project is running smoothly, despite the complexity and long-term horizon of the project. This success does not come by chance, but can be explained by a number of (often related) factors that made it happen.

First of all, the project started at the right time. The closure of the steel plant was another shock for the city that was already suffering heavily from the pain of de-industralisation due to the closure of the mines, other steel plants and reduced beer production. The overall mood was low – there was a feeling that the future could never get worse than the current situation and all ideas are good; that helped to bring stakeholders together and to get support for new ideas.

Second, the project has a clear vision and goals that were jointly set up by various stakeholders. After the announcement of the closure of the Phoenix West site, the City of Dortmund, NRW.Urban and ThyssenKrupp organised a series of workshops in order to discuss the future of the area. The variety of participants (from different departments of the City, ThyssenKrupp, the state, consultants, architects, planners, housing agencies and knowledge institutes) helped to integrate the project into the wider dortmund-project and the general city planning. The information gained through the workshops had been used as input for the planning process of the area. Even though the workshops and initial planning were mainly targeted to Phoenix West, the City's visionary Planning Department already had the idea of the Phoenix See in mind. Later on, this turned out to be crucial in integrating the different parts into one master plan and distinguishing the areas from each other as well as from other areas in the city.

Third, the project had clear leadership at various levels. At the highest level, and as a 'face to the outside world', the project was steered by the Mayor. The Mayor believed in the project and had the charisma and skills to convince the society of its long-term benefits. The Mayor created societal support not only at the start but also during the project, tackled problems by creating a 'let's do it feeling' and convinced the stakeholders to continue with the project. At a lower level, a project team within the Economic Department of the City managed the total project on a daily basis. This project team had the right skills to constantly interact with many stakeholders, listen to them and make sure that all interests were covered as far as possible. The project team has limited the amount of conflict in the development process and keeps the project running. This is also possible due to stability in the project organisation (e.g. the same people), regular meetings between the stakeholders, the small size of the project team (enabling decisions to be taken quickly) and the clear goals set up beforehand.

Fourth, as the site is too large to be developed by one actor, the project has been split up into different parts with different coordinators. This decision is partly driven by politics; it cannot entirely be developed by NRW.Urban as it would create an imbalance in the total assets of NRW.Urban – Dortmund would get too much support compared to other areas in NRW. But this has other advantages as well: by splitting the project into different parts, various (financial) resources could be accessed and combined.

PHOENIX EAST

After negotiations with ThyssenKrupp, who owned the site, the city government bought the land of Phoenix East for €27 million. This was considered a win-win deal for both sides. The steel giant did not only transfer the land to the City, but also the responsibility to clean it, although it is worth mentioning that the land on the eastern part is relatively clean compared to the western site. The City in its turn sold the land to the agency tasked to redevelop it and to then sell it back to the City.

The use of PSDA as an 'independent' developer has a number of advantages. First, as a specific development agency, PSDA is able to borrow additional money, which the City cannot do by law. Second, PSDA can connect the different City departments that are involved in the plan (e.g. finance, economics, planning, infrastructure, etc.). It is easier to connect the different departments from the outside than from the inside. Third, PSDA already has experience in brownfield redevelopment (e.g. it transformed former military sites into new urban areas) and knows the rules and how to play the game.

Although PSDA is responsible for the redevelopment, it works in close cooperation and under control of the City. This is arranged via strict development contracts. For instance, regarding housing development, the City Planning Department set strict regulations for each house in order to guarantee a certain quality.

PHOENIX WEST

NRW.Urban is the main coordinator of Phoenix West and has various tasks, including cleaning the site and preparing it for redevelopment (e.g. removing the pollution and industrial structures), finance and marketing. Just like PSDA, NRW.Urban has a large amount of experience in brownfield redevelopment. Since the 1970s when it was set up as an independent state-owned redevelopment agency it has transformed many smaller and larger brownfields throughout the entire Ruhr area. This broad and long experience has been used in the Phoenix site. NRW.Urban has also learned from the redevelopment of the Phoenix project. In general, the cooperation between the City and NRW.Urban seems to work well and is arranged in a cooperation contract in which the City is tasked to do the general planning, control the retail of real estate and develop all infrastructure. NRW.Urban deals with the economic and physical development of the area and with applications for EU funding (Blanke-Bartz, 2009).

ORGANISATIONAL CHALLENGES

Despite the good organisational structure, planning, leadership and clear division of tasks, the Phoenix project has faced some challenges in terms of the organisation. First, compared to Phoenix West, cooperation with the project team responsible for the Eastern part has not always been without tensions. This might be caused by the weaker financial situation of PSDA (by comparison, NRW.Urban has many projects and is able to cross-fund Phoenix with other projects) and because PSDA is still relatively close to the city government, despite its independent position. Second, in both Phoenix West as well as East there are conflicts between the developers and construction companies which delay the redevelopment process. It is illustrative that certain parts of Phoenix East needed new pathways due to construction mistakes in an earlier version that had just been constructed.

4.2.2 Finance

A good organisational model does not work without funding and in the Phoenix project there are at least important linkages between these two 'enablers'. First of all, and also from a financial perspective, the project started at the right time. Despite the pain of the de-industrialisation at the start of the project, the City, and probably other stakeholders as well, were in a better financial position than nowadays due to budget cuts taken in order to tackle the global financial crisis. In addition, at the start of the project, the City received an additional income of €75 million in a public transport project that could be fully used in order to fund the Phoenix Lake. Second, as mentioned, by splitting the project into different parts, different financial sources could be accessed and used. The revitalisation of the Hörde centre has been largely financed by a fund of €30 million from the City Hall; the work on the Emscher Landscape Park and River has been made possible by funding from the Water authority of the State of NRW. The Phoenix East and West sites have been funded by various sources as we discuss in more detail in the remainder of this section.

PHOENIX EAST

PSDA has two important financial sources in order to fund the transformation of the former steel production site into a vibrant urban district that combines living, recreational and business functions. First, it received a grant from the City, in addition to the grants it received to buy the land. This budget is used for the general development of the area, including the construction of the lake. Second, it received various grants from different departments of the state of NRW. These grants are targeted to ecological development in the area and the canalisation of the Emscher River. The latter is part of a larger ecological programme in the Ruhr area and is done under the responsibility of the Emschergenossenschaft (EG), a public institution.

Initially, the PSDA also received a grant co-funded by the EU, but this budget was too complex to use and possibilities for development were limited. The EU grant only allowed projects with over-regional effects, and therefore, such EU-funded projects are often limited to the development of industrial and advanced service functions. The EU grant could not be used to realise the investments in the mixed-use functions on the western side of the lake (e.g. restaurants, houses and offices), which was the plan of the City and PSDA. These functions are important for the local to regional economy, but have a limited (direct) effect on higher scale levels. Therefore, it was decided not to spend the grant and to return it to the EU. The 'loss' could be compensated for by extra income from plots around the lake which could be sold for higher prices than was originally planned. However, the relation between the two variables is not clear. Similarly, it is unclear why there is a change in focus from some social housing as was initially planned to mainly luxury houses.

PHOENIX WEST

The Phoenix West site is funded by a combination of sources from the EU, NRW and City. It has a financial security on the medium to run due to two reasons. First, NRW.Urban has many projects and is able to cross-fund financially well-performing projects with weaker ones. Second, NRW.Urban and the City managed to get funding for seven years from the EU, which is an exception as normally EU funding is available for only 3–4 years. This financial security might be seen as a benefit in economic weaker times – like the current economic crisis – as the developer can 'wait' for better times.

On the other hand, the project has limited flexibility, especially the central part that is targeted to firms in the MST industry and technological production. Other firms that do not fit into these narrow fields cannot move to the area. Similarly, the sites and buildings cannot be temporarily used for other purposes. Changing the plan and paying back the EU grant as happened in Phoenix East is not an option as the dependency on the EU money is too large (the site, buildings and the restoration of industrial heritage are all co-funded by the EU) and there are no (unexpected) other sources of income.

FINANCIAL CHALLENGES

All in all, the project seems to be healthy in financial terms, despite the high dependency of Phoenix West on EU funding which gives limited flexibility for changes in the plan. There are also two other relatively small financial challenges. First, even though there is budget to develop the infrastructure in both sites, there is as yet no budget to maintain the infrastructure and public spaces. Second, again in line with the organisational aspects, conflicts with construction companies not only delay the process but also have financial implications. For instance, it is estimated that NRW.Urban has incurred about €8 million in extra costs in order to deal with legal affairs.

4.2.3 Communication

As Phoenix used to be a closed site that was separated from the rest of the city by huge walls, it is a major challenge to reintegrate the area into the urban fabric. Therefore, communication with society is crucial, not only to create awareness but also to get support for the redevelopment plans. This was especially important for the idea to create a lake, which was initially perceived as a 'crazy and unrealistic idea' that received negative media attention. Worth repeating here is the convincing role of the Mayor who actively supported and promoted the idea of the lake and the general plan via field visits and press conferences.

Apart from the Mayor's important role, some other specific measures increase awareness about the site. First, an events strategy has been used to bring people to it. Various larger and smaller events have been organised at Phoenix West, including the Theatre Festival Ruhr (2002); an open-air cinema (2003 and 2004); World Soccer Games Media Night (2006); an art exhibition and an exhibition dealing with the 'New Dortmund'. So there are a variety of events, combining indoor, outdoor and general events as well as events directly linked with the Phoenix site (such as the exhibition of the 'New Dortmund' and the industrial history events). Overall, the events strategy seems to work as many visitors attended the events at the site, although it is difficult, if not impossible, to measure to what extent there is an increased awareness of the site.

Second, the industrial heritage and past of Phoenix West have been used as historical landmarks to promote the site and to distinguish it from others. The industrial heritage has an historical value as a memory of the industrial times, but it is also used as a bridge to the future referring to the use of the site as a location for production technologies. The blast furnace is an especially important landmark in the area that gets extra attention during events when it is put in the spotlight. It is worth noting that both the developers as well as the existing users (competence centres and firms) acknowledge the historical value of the area which has also been used to promote the individual actors on the site.

Third, from the early beginning of the redevelopment process – when the old industrial structure was still visible – NRW.Urban has regularly organised field trips to the Phoenix West site for inhabitants. This has helped to give them insights into the past of the 'forbidden area' and they can follow the redevelopment through step-by-step visits during different development stages. This strategy seems to work well, as nowadays inhabitants frequently play sports using the public part of the site. In addition, this 'field trip strategy', which was applied by NRW.Urban for the first time at Phoenix West, is nowadays also used in other projects of the state-owned development agency.

Fourth, for the development and promotion of houses on the Phoenix East site a special online competition was organised to assess consumer demand. PSDA invited 70–80 architects to design proposals for houses that fit within the quality guidelines set by the City's Planning Department. The best proposals (selected by the Planning Department and PSDA) were collected in a catalogue that was published online. Visitors' responses were ultimately used to develop the houses, which in the end were based on the requirements of the City and on market demand.

Finally, more common communication and promotional channels have been used to promote the area, including three information centres (in Phoenix East, West and in the centre of Hörde), brochures and websites. In addition, PSDA informs inhabitants via a specific newspaper – *Phoenix See Nachrichten* – which covers the development of Phoenix East and the redevelopment of the centre of Hörde.

4.3 *Results*

4.3.1 *Phoenix East*

The major goals for Phoenix East – development of high-quality housing and public recreational facilities – seem to be realised, along with some further positive effects. The core of the project – the Phoenix See – is finished and can be regarded as a 'new pearl for the city'. It is very popular and widely visited for recreational purposes and as such the lake has successfully filled the gap in available leisure areas which were missing in the city. It is also well connected via cycling and walking lanes with other recreational areas, such as the Westfalenpark in the north and Romnbergpark in the south and is part of a large green corridor in the former industrial city. In addition, the lake functions as a new 'landmark' for the city and can be regarded as the physical image of the 'New Dortmund'. All interviewees agreed that the lake has contributed to a better image of the city.

The lake also seems to have a positive effect on the development and sales of the real estate around it as nearly all houses, offices and other real estate have been sold at relatively high prices. Housing development started with a small plot on the northern side of the lake with a price of €300 per square metre. Despite the relatively high price – by comparison the average real estate price in Dortmund is €200 per square metre – the plot sold out quickly. The prices of the houses on the southern side and the real estate in the mixed-use zone on the west side of the lake are higher, but these plots have also nearly all been developed and sold. The extra income earned from the higher real estate prices has not been used as profit – which is not allowed as PSDA is a public non-commercial company – but is used to cover the costs of infrastructure and public spaces and to pay back the EU grant.

Yet, due to the relatively high prices, the houses and apartments around the lake are mainly attractive for high-income groups. There is no social housing in the Phoenix area. Regarding retail, the strategy is to allow only small commercial spaces in the mixed-use area in order to stimulate local entrepreneurship and to avoid large chains, but it is questionable to what extent this strategy helps lower income groups. Due to the high prices, the area seems to be mainly attractive to advanced service providers or luxury shops that earn enough to cover the high costs. So, as such, Phoenix East has no direct effect on social development.

Final points of attention at Phoenix East include the financial situation of PSDA and the Hörde Castle. Despite the high returns on the real estate, for PSDA it seems to be impossible to cover the costs meaning large financial losses. However, if the non-monetary and indirect effects have been taken into account (e.g. the improved image and new public recreational facilities) the project as a whole is beneficial. The Hörde Castle is the former office of the steel works built in the style of the castle, which was based in the area in the pre-industrial times. This historical building is a protected monument and as a consequence limited changes are allowed, making it hard to transform it for other functions. For instance, the rooms are perceived as being too small for a hotel. The Hörde Castle is owned by the City, which is challenged in its search for investors.

4.3.2 *Phoenix West*

The major targets for Phoenix West are development of (high-tech) business and commercial leisure functions. How far have these goals been achieved? Concerning the business function, Phoenix West can indeed be considered as an important location for the 'New Dortmund' that is different, and thus complementary, to other locations in the city. It has some important location factors.

First of all, all the interviewees mentioned the presence of space as the key asset of Phoenix West, making it suitable for production firms (as is also mentioned in the plan). In this sense it is clearly complementary to Technology Park Dortmund that is nearly fully occupied and in which firms have limited space to expand their business. An illustrative example is Raith, a high-tech company that combines an emergent technology field (nanotechnology) with a traditional one (machinery). The firm started as an incubated company, then moved to Technology Park Dortmund and ended in Phoenix.

Second, knowledge and business support facilities are well developed. The two planned incubators have already been realised and work well, not only as incubators, but also as magnets for other firms. The MST Factory is completely full and opened a second unit in 2007. Similarly, ZfP was opened in 2007 when it had six tenants, while currently (June 2013) it is home to 13 firms and is in negotiation with four others. In total, ZfP will have about 20 tenants, depending on the size of the firms. The two competence centres also function as knowledge/discussion partners of other firms and give a physical face to the location for high-tech production and MST industries. Moreover, the site has a favourable geographical location and good accessibility. The airport can be reached quickly (important for international operating firms), just like the city centre and TUD and the Technology Park. TUD is only 5 kilometres from Phoenix West and can be reached quickly.

Despite these important location factors, there are some challenges as a well. In general, the development of the site as a business location is slower than expected (see Table 9.4). Currently 31 firms are based on Phoenix West, although the data available do not give insights into whether this includes or excludes the firms in the two competence centres. The relatively low number of firms also has implications for further development of the site, as currently there is no central electricity system yet and the individual buildings receive power from single generators. Similarly, there are no basic catering facilities in place yet due to a lack of demand.

Most firms in the area nowadays fit into the target industries: MST (8) and production technologies and machinery industries (12), although this is also the major current challenge

Table 9.4 Firms at Phoenix West

Field	*Number of firms*
Chemical and environmental technologies	4
Energy and electronic technology	4
Leisure, media and culture	2
IT	2
Machinery	6
Medical and pharmaceutical technology	3
Micro and nanotechnology (MST)	8
Production technology	6
Business services	23
Total	31*

Source: Own elaboration, data from http://www.phoenixdortmund.de/de/firmenindex/ (accessed 26 June 2013)

*Some firms are counted in more than one field. Therefore the total number of firms is lower than the sum of the firms in the specific fields.

and subject of discussion. In general, it is hard to attract firms, also due to the financial crisis. For small niches, such as MST and production technologies it is even harder to attract firms. In addition, competition has increased. For instance, the concept of TZDO was innovative when it started, but now many more cities develop similar competence centres. Various firms from other fields have shown interest in moving to the site, but are not allowed to as they do not fit the profile and temporarily renting spaces to them to avoid vacancy is not part of the strategy. Another challenge identified is the knowledge infrastructure of the MST cluster. At the start of the dortmund-project, MST was seen as one of the new major fields of TUD, but TUD changed its focus and MST is not as important as expected. Therefore, the MST Factory and major MST firms in Dortmund search for other knowledge partners in this field elsewhere in the region.

Regarding the development of the (commercial) leisure functions, the results are mixed. On the one hand, the site has many green spaces, which are intensively used for sports and recreation. On the other hand, due to the crisis and the low demand, there are as yet no investments in commercial leisure facilities. Similarly, and despite innovative ideas, there are no concrete business plans and investors for the restoration and transformation of most of the historical heritage. Their function as landmarks is important, but commercial use of the heritage is required in order to cover maintenance costs. Currently, the historical industrial infrastructure is protected against demolishment by law for 15 years. It is important to realise the goal of conserving the blast furnace, as currently the only possible strategy is to search for investors.

4.3.3 Hörde

The results concerning the ambitions to revitalise the centre of the Hörde district and to connect the district with the Phoenix East site are mixed. On the one hand, public investments have been followed by further private investments. In particular the renovated Hörde station and shopping centre, Einkaufscity Hörde, have attracted several new shops, including large chains such as Burger King and C&A, but also small entrepreneurs have opened new shops. In addition, the cultural supply in the area has been enriched with a new theatre and a new bookshop. The latter has also been used for seminars and courses (e.g. reading or writing) and includes events for all ages (PSDA, 2012). As such, the bookshop seems to have an important social function in terms of education and community building.

On the other hand, in the interviews it was mentioned that the effects in Hörde are not as large as was hoped and it is still too early to analyse the long-term effects, just like the spillover effects of the Phoenix East site. Remaining questions include to what extent do community projects work in the long run; do the new shops generate jobs for the local community; to what extent is gentrification playing out, etc.

4.3.4 Emscher Landscape Park and River

The activities concerning the Emscher Landscape Park and River are part of a large and long-term project and it might be too early (the master plan dates from 2010) to analyse the results. However, the project seems to develop smoothly and natural improvement is clearly visible. The heavily polluted part of Phoenix West has been covered and has been transformed into the new Phoenix Park which offers new recreational possibilities. Interestingly, between the natural areas, there are some industrial heritage to explore, including the remains of an old railway bridge and pipelines. The park is part of an integrated larger green corridor running from Phoenix East to the Westfalenpark. As such, the Phoenix site is integrated well with the rest of the city.

5 Conclusions

5.1 Broad assessment

The Phoenix project can be regarded as an example of good practice of a complex and integrated brownfield redevelopment project, tackling the challenge of transforming a monofunctional industrial site into a mixed-use future location. The transformation process from a closed industrial area into a location for the 'New Dortmund' overall went smoothly and most of the development goals have been met. The project is well integrated into the larger economic strategy (the 'dortmund-project') and is different, and therefore complementary, to other major locations in the city. A key asset of the Phoenix West site is the availability of space, making it attractive for firms that want to expand their business, and as such it is a follow-up location to Technology Park Dortmund. In addition, the industrial heritage gives the site a specific character, although it is a remaining challenge to find the investors needed to convert these historical assets into other functions. The two competence centres – MST Factory and ZfP – have already been realised and function well.

The initially widely discussed Phoenix See on Phoenix East has turned out to be a 'pearl for New Dortmund'. It contributes positively to a new image of the city and functions as a popular recreational spot. Newly created green spaces on both Phoenix West as well as East offer additional recreational spaces and are well connected with each other as well as with other green spaces in Dortmund. Also, the complex water management plan dealing with the Emscher River seems to develop according to plan. Finally, new luxury houses, offices and leisure facilities have been constructed and are largely sold.

Overall, the Phoenix project has important positive effects in terms of environmental and economical development. The third pillar of sustainability – social development – seems to be harder to realise. Housing in Phoenix East is mainly targeted to the higher segments and the revitalisation of the Hörde district is slow, despite early positive signs such as private investments following the public ones.

5.2 Critical factors

The success of the Phoenix project does not come out of the blue, but is related to a number of factors and organising capacities of the stakeholders involved. First of all, and as a contextual factor that cannot be transferred to other locations, is the timing of the project. The project started during the deep pain of the de-industralisation. The end of the industrial activities on Phoenix symbolised this de-industralisation and the deep crisis brought different stakeholders together to jointly develop a new future translated into the 'dortmund-project', with the redevelopment of the Phoenix site as a major component of it.

Second, and related to the above, the clear vision and concrete goals that have been jointly set have functioned as the right recipe to guide the redevelopment process. This recipe is not only based on global trends (e.g. new growth sectors), but also on local needs (e.g. a lake for recreational purposes and as a new landmark) and on historical local assets (historical heritage on Phoenix West and the idea to keep the location for high-tech production).

Third, good leadership at the top and on the ground reduced conflict and kept the machine running. The charismatic Mayor convinced the society of the need for the project and made high level deals with important stakeholders, while project teams in the Planning Departments and Economic Department supervised the quality of the project and continuously interacted with all stakeholders in order to gain trust. The latter was possible due to small and constant project teams, but also via contractual arrangements.

Fourth, splitting up the site into different parts allowed access to various funding sources, enabled the involvement of various experienced coordinators (e.g. NRW.Urban and PSDA) and the development of a variety of functions. As the total control is in hands of the City, the different parts are integrated into one plan. Depending on the institutional context, such a strategy may also work in other countries.

Finally, apart from more common communication and marketing strategies (e.g. information centres, websites and brochures), some specific strategies have been used at Phoenix, including field trips to the sites at various stages of the redevelopment, development of houses based on input of the City and potential buyers (via an online catalogue), events and historical heritage. Apart from the historical heritage, which gives the site a relatively unique character, most of these strategies can also be applied to other sites. The field trip strategy seems to be especially crucial in gaining societal support and has already been applied elsewhere by NRW.Urban.

It is also worth noting that even large projects like Phoenix that tackle many issues at the same time (e.g. new industries development, image improvement, housing, etc.) are not a recipe for killing all of a city's diseases. For instance, the project will have a limited effect on the large pool of structural unemployed workers – the city's largest challenge – which requires continuous investments in other (educational) programmes.

References

Blanke-Bartz, P. (2009). Phoenix Dortmund. Accessed at http://nirov.platform31.nl/upload/media/projecten/Forum_Stedelijke_Regios/091127_PaulBlankePhoenixproject.pdf (7 June 2013).

Blanke-Bartz, P. (2011). Economic Development and Structural Change in Dortmund PHOENIX Brownfield Redevelopment. Presentation, September.

Bömer, H. (2005). *Modern Local and Regional Development Policies in Times of Mass-unemployment – The Example of Dortmund* (2nd edition). Dortmund: Institute für Raumplanung Universität Dortmund.

Ellwein, H. and Mai, H. (2014). Dortmund. In L. van den Berg, J. van der Meer and L. Carvalho (eds), *Cities as Engines of Sustainable Competitiveness: European Urban Policy in Practice*. Aldershot: Ashgate.

PSDA – Phoenix See Development Agency (2012). Endlich wieder eine Buchhandlung in Hörde! *Phoenix See Nachrichten*, September.

Stadt Dortmund (2013). Wirtschaftsstandort Dortmund 2013. Stadt Dortmund Wirtschaftsförderung.

Van den Berg, L., Pol, P., van Winden, W. and Woets, P. (2005). *European Cities in the Knowledge Economy – The Cases of Amsterdam, Dortmund, Eindhoven, Helsinki, Manchester, Munich, Münster, Rotterdam and Zaragoza*. Aldershot: Ashgate.

Van Winden, W., van den Berg, L., Carvalho, L. and van Tuijl, E. (2010). *Manufacturing in the New Urban Economy*. Abingdon: Routledge.

10 RDM Campus (Rotterdam)

Erwin van Tuijl and Alexander Otgaar

1 Introduction

This chapter discusses the transformation project of the former RDM shipyard in Rotterdam into a new business and knowledge location, the RDM Campus. The idea of this new campus is to put the 'golden triangle of business, education and research' into practice by putting start-ups, firms, educational and knowledge institutes together in renovated historical industrial buildings. Parts of the halls and the docks are targeted as experimental spaces, where students, researchers and firms can develop climate proofing-related solutions and prototypes. Another ambition of the project is to re-integrate the campus with the village of Heijplaat that was constructed for the workers of the shipyard, but currently faces major socio-economic challenges due to the closure of the shipyard. The campus should give new blood to the shrinking village surrounded by industrial and port activities on all sides.

The development of the RDM Campus and renewal of Heijplaat are part of the larger Stadshavens (City Ports) development strategy. This strategy concerns the long-term transition of port and industrial sites into new urban and industrial functions. The project encompasses 16,000 hectares; it is the second largest inner-city redevelopment project in Europe after the London Gateway. It is part of a political deal in 2003 that guaranteed the corporatisation of the Port Authority and financial sources for the development of Maasvlakte II, a new port extension into the sea (Daamen and Vries, 2013). In the present case study we will analyse the RDM project in relation to other parts of the Stadshavens area and the development of Rotterdam in general.

The RDM Campus is worth analysing for a number of reasons. First, from a management perspective the project can be regarded as a 'joint adventure', as all key players – educational institutes, Port Authority Rotterdam and a housing agency – are 'playing outside their core business'. None of them have large experience in integrated urban development and they all face the challenge to sell the project internally and to learn new activities, such as integrated area-based marketing. Second, the new campus concept is set up by visionary leaders of educational institutes who also take the lead in the implementation of the project. Third, a flexible planning approach is used, meaning space for temporary functions and changing directions of the project's content, depending on new opportunities and challenges that could not be foreseen at the start of the project yet fit its overall vision.

This chapter is structured as follows. Section 2 analyses the structure (and recent dynamics) of Rotterdam across a number of dimensions or 'capitals' for sustainable competitiveness; it particularly analyses the dynamics of the port and the development of the city. Section 3 briefly sketches the wider strategies and visions of the City of Rotterdam and the Port Authority, namely the overall Stadshavens development strategy. We discuss the links between these strategies and explore the role of the RDM Campus in them. Section 4 is the core of the chapter and explores the RDM Campus in depth, namely its start-up history, vision and content. Section 5 presents the key features and enablers that made it happen, while Section 6 presents preliminary results of the project. Section 7 concludes raising lessons and challenges ahead.

2 City profile: capitals for sustainable competitiveness

2.1 Performance

Rotterdam, with 618,793 inhabitants (1 January 2014), is the second largest city in the Netherlands after the capital Amsterdam. In comparison with the Dutch average and the other large cities, the population growth is modest. Between 2000 and 2013 the population increased with 4 per cent in Rotterdam versus 5.7 per cent on the national level and 9 per cent, 14.7 per cent and 37.7 per cent respectively in Amsterdam, The Hague and Utrecht (CBS Statline, n.d.). An important explanation for the success of other cities has been their ability to acquire land from neighbouring municipalities and to develop new areas such as Leidsche Rijn (Utrecht) and Leidsche Veen (The Hague). Rotterdam has much more land available within its municipal boundaries, such as in the Stadshavens (city ports) area.

Rotterdam had an increasing GDP per household in the 2005–11 period; yet, that growth is lower than in the other large cities. Similarly, the unemployment rate of the port city is 13.9 per cent higher compared to the other large cities – 10.0 per cent in Amsterdam, 11.1 per cent in The Hague and 8.3 per cent in Utrecht which equals the country's average (CBS, 2014). Furthermore, a mismatch can be observed in the regional labour market: there is a relatively large pool of job seekers (including low-skilled former port workers) but also a shortage of skilled workers, notably of people with a technical background. This can be explained by a general limited interest among students for technical professions, but also due to the negative image of the manufacturing and port-related sectors (van Winden *et al.*, 2010). In the near future, the need to find new skilled workers for jobs in the port is expected to be even higher due to the aging workforce that will soon retire.

2.2 Economic structure and knowledge

The development of Rotterdam is strongly linked with the development of the port. In particular, the opening of a canal to the sea (Nieuwe Waterweg) in 1872 caused a spectacular growth of the city and the port. The port expanded with new sites in the direction of the sea, while the city was a magnet for job seekers from the countryside (Braun, 2008). Individual firms played a key role in the development of the port. So-called Port Barons (Havenbaronnen) helped Rotterdam to become the world's largest port (Nijdam, 2010), a position the city kept until 2004.

After the Second World War, which destroyed a large part of the city centre and the port, priority was given to the reconstruction of the port and the city. New large port sites were developed such as the Botlek (1955–66), Europoort (1957–70) and Maasvlakte I and Maasvlakte II (1967–now). The city itself was redeveloped and the reconstruction of the port led to growth of the port and the city with peaks in 1965 when the city had a population of 732,000 inhabitants while the port and industrial complex provided 116,000 jobs (Braun, 2008).

Over time the city and port have gradually become less integrated and mutually dependent: not only in physical terms, but also from a socio-economic and organisational point of view. Physically, the harbours have moved away from the city towards the sea mainly due to an increasing scale in port operations and the lack of space in the city. This migration process of port functions has made available land within the city's boundaries for new urban and industrial functions. Economically, the relevance of the port in terms of direct employment has declined strongly due to automatisation and containerisation. Moreover, in organisational terms, the port has become more independent, especially with the corporatisation of the Port Authority as an independent company in 2004, although the city still is the largest shareholder.

While the port has moved away to the sea and has become less relevant as an engine for jobs, other industries developed in the city. Between 2001 and 2010, employment has grown particularly in healthcare, social services, public utilities, business services, real estate and construction (OBR, 2011). The number of jobs in trade has remained on the same level, while employment in industry and financial services has decreased. If we look at the contribution to added value in the same period, a similar picture emerges though with some minor differences. Industry and trade seem to have become less labour intensive and more productive as the contribution to regional GDP has grown while the number of jobs remained equal (trade) or even declined (industry) (CBS Statline, n.d.).

The Rotterdam economy can also be analysed by looking at the most important (sometimes overlapping) clusters. A key cluster is the port and industrial complex, which is still a major engine of the regional and also national economy, as becomes clear from the key data in Table 10.1. This heterogeneous cluster consists of various economic activities, such as offshore, chemical and oil production and trade and logistics. These activities still dominate various parts of the Stadshavens area, notably the Waal-Eemhaven that surrounds the RDM Campus.

Another relevant cluster for Stadshaven in general and RDM in particular is 'cleantech', covering the bio-based and energy sector, hydraulics and maritime engineering (including offshore), infrastructure and mobility, and construction and integrated area development. These industries have in common that they look for new technologies which can be applied to become more efficient in the (re-)use of energy and resources, new climate adaptation solutions and delta design. This cluster covers a larger geographical area stretching from between Delft in the north to the Drechtsteden in the south. The port-industrial and the cleantech cluster are both connected to the expertise of several knowledge institutes in the region, including Technical University Delft (TUD), Erasmus University Rotterdam, Hogeschool Rotterdam (university of applied research), Albeda College (vocational level), and TNO and

Table 10.1 Key data Port of Rotterdam, 2010

Direct value added	€11.6 billion
Indirect value added	€7.3 billion
Share Direct and indirect value added as share Dutch GDP	3.2 per cent
Direct employment	87,000 employees
Indirect employment	50,000 employees
Average private investments	€1.5 billion

Source: Nijdam *et al.* (2012)

Deltares (independent public research organisations). Hogeschool Rotterdam and Albeda College are both based at the RDM Campus and can be regarded as the pioneers and founding fathers of the new campus concept (see Section 4).

2.3 Social conditions

Rotterdam differs from other large cities in the Netherlands by having a relatively young population and a high share of immigrants. Due to its role as a port city it has attracted migrants from all over the world. The industrialisation firstly attracted farmers from the southern provinces of the Netherlands, followed by migrants from China, Southern Europe, the Mediterranean, the Netherlands Antilles and Surinam and more recently from Eastern Europe. This has resulted in a multi-cultural society with nationalities from 166 countries all over the world (Azaoum and De Lange, 2014).

The city is characterised by a relatively high share of households with a low income and living under the social minimum. Social divisions have a physical shape. Roughly, the northern bank of the New Meuse River, especially Kralingen-Crooswijk and the city centre, has a population with a relatively high income and attracts most of the students and business. The southern bank is faced with socio-economic challenges, including unemployment, poverty and crime and is therefore also part of major social policy programmes, such as Pact op Zuid (Azaoum and De Lange, 2014).

One of the areas in Rotterdam South is the village of Heijplaat, which was built for the workers of the RDM shipping wharf. After the closure of the wharf, the village entered a period of decay, with population decline, unemployment growth and a loss of amenities. In this period Heijplaat mainly attracted low incomes looking for affordable housing. The only services that remained are a small primary school (with less than 100 pupils), a supermarket and a bakery. It remains to be seen what role the RDM Campus can play in changing the tide for Heijplaat.

2.4 Environment and accessibility

Namely due to its port and logistic functions, the Rotterdam region suffers from more pollution and congestion than most other regions in The Netherlands. The City of Rotterdam and the Port Authority are aware of the environmental and accessibility challenges and take various measures to reduce air pollution and congestion, as also becomes clear from the strategies of these authorities (see the next section). According to its strategy, the Port Authority aims to optimise existing infrastructure via pro-active traffic management and to develop green spaces between the port and the city. The measures taken to tackle the environmental and traffic challenges offer new opportunities as well, for example in the cleantech domain.

2.5 Institutional conditions

The Netherlands has a decentralised governance system with powerful city governments that can initiate and coordinate large projects, often (co-)funded with national sources. Cities do not only lobby for resources, but can also influence regulation. Illustrative is the so-called 'Rotterdam Law' (in Dutch: 'Rotterdam wet') that enables city governments to set minimum income requirements to new inhabitants of low-income areas. This tool that aims to counter poverty was accepted by the state after lobbying by the City of Rotterdam.

On the local level in Rotterdam, there are two important players: the City of Rotterdam and the Port Authority. The City of Rotterdam is responsible for general development of the city, covering a wide variety of fields. A key player within the municipal government is the City Development department (Stadsontwikkeling) that deals with economy, spatial planning, infrastructure development and real estate, and within large projects it is responsible for quality safeguarding and acts as public developer as well. The Port Authority works as an independent agency (with the City of Rotterdam as the major shareholder, having about 70 per cent of the shares, while the remaining part is owned by the state) and is responsible for development of the port and industrial complex, among others via a role as a landlord of industrial sites and as developer of port infrastructure.

Port and city development do not always go hand-in-hand leading to shared challenges for (and conflicts between) the Port Authority and the City Government, like tackling negative externalities of port functions, including congestion, noise and pollution. The largest joint initiative of the two authorities is the Stadshavens development strategy, which the RDM Campus is part of (see the next section).

Besides the national and local governments, the Netherlands has three intermediate authorities: the water boards (responsible for water management), city regions (dealing with spatial development, mobility, living, working, environment and youth care) and provinces. With the exception of the water boards, the other authorities are regarded as less powerful compared to higher and lower governments, although provinces formally supervise lower governments.

The City Council (directly elected by citizens) as well as the city region (not elected) increasingly search for partnerships with other authorities to increase competitiveness by increasing the critical mass, especially in the so-called Metropole Region Rotterdam-The Hague. Besides new governance structures under development on the larger metropolitan level, there are various reorganisations within the City of Rotterdam. On the one hand, this is a response to the economic crisis requiring lay-offs and outsourcing of municipal services to cut costs. On the other hand it includes reorganisations to establish more strategic ways of governance. Some recent examples of reorganisations include: the establishment of the City Planning department (merger between the Development Agency, OBR and Infrastructure and City Development, dS+V) and the abolishment of district governments. In Section 4, we analyse the impact of the restructuring and budget cuts on the development of the RDM Campus and Heijplaat.

2.6 Conclusions: main issues

The development of the city of Rotterdam is traditionally linked with the development of the port. Nowadays, they seem to grow apart, as the port is moving away from the city centre and is managed by an independent authority, while the city has developed new growth sectors that are not related with the port. Nevertheless, this is only a seemingly separation. The port and industrial complex is still a major engine of the city's economy; it is linked with innovative new clusters (e.g. cleantech) and causes negative externalities leading to common challenges of the city and port, such as congestion and pollution. Moreover, the city and port authorities are dealing with socio-economic challenges, such as a relatively large share of inhabitants with low income, a high unemployment rate and a mismatch on the labour market. Many opportunities and challenges of the link between the port and the city come together in the Stadshavens, the central strategy linked with the physical migration of port and industrial function away from the city. This strategy is discussed in more detail in the next section.

3 City-wide vision and strategy

3.1 City Vision and Port Vision

The City Vision of Rotterdam 2030 was published in 2008 and refers to the 'development of a strong economy and an attractive city to live' (Gemeente Rotterdam, 2008). The 'development of a strong economy' refers to an increase in employment, while the attractive city part refers to the development of a more balanced population (e.g. especially keeping and attracting higher educated citizens). In order to realise these two major goals, the vision includes 13 so-called 'VIP' areas, being major (re)development areas that receive priority. Stadshavens is one of these VIP areas, although it only covers the sub-areas of RDM (our case study), Waalhaven-Oost Rijn and Maashaven.[1]

Just like the City Vision, the Port Vision refers to the year of 2030 (Port of Rotterdam, 2011). It was set up by the Port Authority and approved by the City Council only in 2011, three years after the City Vision. The Port Vision covers two parts: 1) 'Global Hub Vision' (development of a leading European hub for global and intra-European cargo flows of containers, fuel and energy) and 2) 'European Industrial Cluster' (development of an integrated industrial and energy cluster with Antwerp). These visions are translated into a strategic agenda with actions in the following fields: 1) realisation of the European industrial cluster; 2) realisation of the global hub; 3) improvement in accessibility; 4) improvement of quality of life; 5) technical and social innovation.

Stadshavens is one of the key projects in the current Port Vision and was already on the agenda in the Port Visions 2010 and 2020. Stadshavens is explicitly described in the fourth field – improving living environment. It concerns a major exception as the improvement of the living environment is normally done by the development of green buffer zones, working as a 'green wall' limiting noise hindrance and offering environmental space for leisure purposes. These buffer zones cannot be used as housing construction space. Stadshavens is different from other areas in Rotterdam, as the idea is that the port and city jointly develop new functions allowing housing at the border of port and industrial functions. It is used as experimental space for new regulations for noise and pollution and therefore, it is also mentioned in the fifth field of action.

3.2 Stadshavens development strategy

The Stadshavens is a large and long-term development strategy for former port areas in the city (e.g. Daamen, 2010; Daamen and Vries, 2013; Schoo, 2013). The formal history of Stadshavens dates back to the early twentieth century and is part of a political deal to create an independent Port Authority and the development of Maasvlakte II, and extension of the port (Daamen and Vries, 2013). A first concept development strategy dates back to 2004, while in 2011 a detailed structure vision was published that was formally approved by the authorities of the city and port.

This structure vision uses the label 'Creating on the edge' referring to the spatial development task of the transition of port areas into new urban functions. In total, the Stadshavens area covers 1,600 hectares including the river, or 1,400 hectares excluding the river, covering 600 dry land and 800 wet docks. The project can be divided in four areas (see Figure 10.1 and Table 10.2) and covers restructuring of port areas to urban areas (especially in the Rijn-Maashaven and Merwe-Vierhavens) as well as intensifying of port functions (mainly in Waal and Eemhaven).

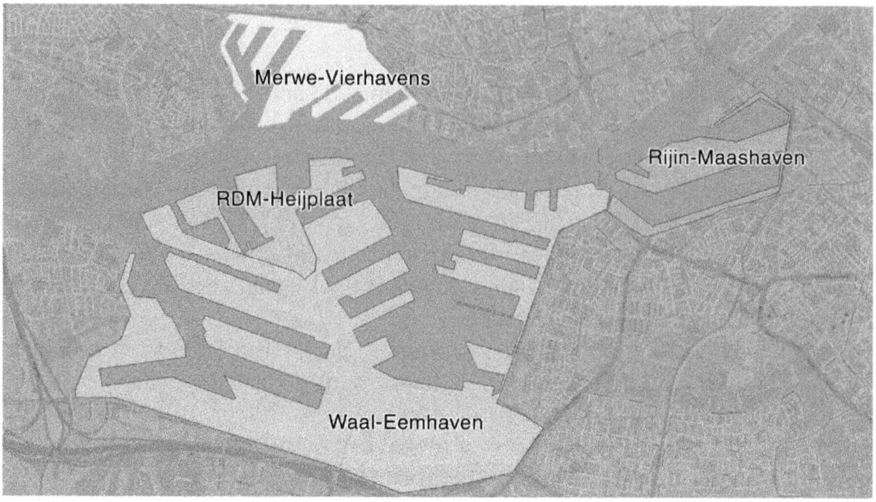

Figure 10.1 Areas of the Stadshavens strategy

Source: Own elaboration

Table 10.2 Areas of the Stadshavens strategy

Area	*Plan*
Rijn-Maashaven	Showcase for cleantech cluster: events and floating objects, including pavilion and houses ('floating city') Leisure (e.g. cruise vessel as tourist attraction) and housing
Merwe-Vierhavens	Housing (4,500–6,500 houses) and business (in fields of cleantech cluster)
Waal-Eemhaven	Renewal/modernisation fruit and container terminals; Development maritime service cluster ('port valley')
RDM-Heijplaat	RDM Campus: linking education research and business Re-integration of RDM Campus with village of Heijplaat

Source: Own elaboration, based on Projectbureau Stadshavens, 2011

The RDM Campus and Heijplaat cover both elements. The main goals of the Stadshavens development strategy include strengthening the economy of the city and port and the development of attractive and high-quality living and working environments (Scheepmakers, n.d.), hence following the general vision of the city. It also follows the City's climate agenda (e.g. Rotterdam Climate Initiative), which is not despite the environmental challenges the city is confronted with, but because of the challenges (Daamen, 2010), which also leads to new opportunities such as development of the cleantech cluster. The climate agenda is translated into two goals for the Stadshavens area: 1) to make the area energy neutral and 2) to make it resilient to climate change. As such, it is indeed a major part of the cleantech cluster, with the RDM Campus and Merwe-Vierhavens important hotspots for innovation in these fields. Similarly, the Rijn-and Maashaven was selected as the demonstration area to showcase new energy and climate concepts to the general public, taking place via events and demonstration projects, including floating houses. This area was selected as the demonstration area due to its proximity to the city centre, generating a higher number of visitors and (media) attention than more peripheral locations (Projectbureau Stadshavens, 2011).

The Stadshavens development strategy is linked with other major projects in the city, in particular to Maasvlakte II, as was already mentioned in the Port Strategy of 2020. The original thought was that Maasvlakte II could be used to relocate industrial and port activities from the Stadshavens area, giving new space for urban functions. However, so far there is limited evidence of the migration of port functions from the Stadshavens area to Maasvlakte II as becomes clear from Daamen's study (2010) and as is also acknowledged by the involved authorities. This opens possibilities for migration of port and industrial functions within the Stadshavens area, and thus to keep industrial functions in the area. This latter is an important adaption of the original plans done after a major organisational change: the shift from the Development Agency Stadshavens (Ontwikkelingsmaatschappij Stadshavens Rotterdam, OMSR) to a small joint project bureau of the City and Port Authority that is tasked with branding of the Stadshavens area and facilitating, while the two 'parents' – City of Rotterdam and Port Authority – get the responsibility to develop the various areas. This change is also known as the so-called 'north-south deal', roughly speaking giving the Port Authority control of the southern areas, with priority on the development of port and industrial functions, while in the north the City is taking the lead, with a focus on urban functions in the first place.

As the whole project covers a long period with many uncertainties (e.g. economic and political cycles), the project is divided in three time periods. The short run covers the period until the year 2015, which equals the timeline of a major state funding source (Randstad Urgent), which is an important enabler for the Stadshavens development strategy. The medium-term period covers the period until 2025, the year until the two parents of the strategy have formalised cooperation via a cooperation contract. Finally, 2040 is the end of the long-term period, running parallel to the vision for the Randstad area (Randstad 2040). For each period, concrete goals and projects have been identified. The development of the RDM Campus is mainly part of the short-term aspect of the project (Projectbureau Stadshavens, 2011). Besides the division into different timeframes, the strategy allows a flexible approach through adaptations of the plans based on new opportunities and challenges. One way of doing so is by the use of temporary rents of business locations, a strategy that has been used in the RDM Campus as well as in the Merwe-Vierhavens area.

In order to realise the goals set, the project covers five strategic lines (Projectbureau Stadshavens, 2011):

- 'Re-inventing delta technology': development of the cleantech cluster, focusing on energy transitions and water management.
- 'Volume and value': growth of port and related services, in volume and value (e.g. growth of short sea shipping; attracting offices – in particular headquarters – of maritime service providers; etc.).
- 'Crossing borders': connecting the port and city via the development of innovative and attractive living and working environments.
- 'Floating communities': the development of floating houses, waterfront redevelopment and the opening of the SS Rotterdam – a historical cruise ship as a new tourist attraction and seminar space.
- 'Sustainable mobility along various transport modes': covering optimisation of current connections as well as development of new modes and connections (e.g. public transport over water).

The strategies are implemented at various places in the Stadshavens area. As we describe in more detail in the next section, the RDM Campus covers all strategies.

4 The project

4.1 Vision and origins of the concept

The RDM campus aims to connect research, multi-level education (vocational and higher level) and business – the 'golden triangle' – by clustering start-ups, educational and knowledge institutes and firms on the former Rotterdamsche Droogdok Maatschappij (RDM, Rotterdam dry dock corporation) shipyard. RDM has been redefined as 'Research, Design and Manufacturing' with the following, partly related, goals (Hooijer, n.d.).

The first aim is to develop a cluster for innovative production industries in the fields of 'building, moving and powering', thus bringing manufacturing industries back to the city. The large industrial buildings allow large-scale production and the involvement of educational and knowledge institutes opens roads for more advanced types of production. This ambition is clearly related with two other goals of the campus: to facilitate crossovers between creative industries and technology and to promote knowledge exchange between research/educational institutes and business. This all should lead to new economic activities: new firms and the introduction of new products and services. More particularly, RDM intends to set up a framework for cooperation between small- and medium-sized enterprises (SMEs) and education and research institutes, reasoning that SMEs generally have limited capabilities to do this on their own. This framework goes beyond offering internships for students, and companies are closely involved in educational programmes (e.g. they can offer business cases and guest lectures) and can set up joint (applied) research projects. This latter is also driven by the need of Hogeschool Rotterdam – a university of applied research – to do more research and to move beyond education. The final goal is the development of inspiring and attractive surroundings for work, (temporary) living, studying, leisure and (short) stay.

The goals are based on the City and Port Visions 2030 and Stadshavens' general development strategy (Table 10.3). In addition, the plan takes into account the dynamics between the different parts of the Stadshavens area. For instance, spaces for inland shipping will move from the Rijn-Maashaven (where less space will be available due to new urban functions,

Table 10.3 Stadshavens strategic lines and RDM Campus

Stadshavens Strategy	*Translation for RDM Campus*
'Re-inventing delta technology'	Education and business in these fields – especially offshore Application space in halls and on water new (floating) concepts Application of new energy solutions, solar and wind technologies
'Volume and value'	More space for 'break bulk' New jobs in technological firms New space for inland shipping (relocation from Rijnhaven)
'Crossing borders'	Re-open link between Heijplaat and RDM site New facilities for village
'Floating communities'	Floating constructions built and tested
'Sustainable mobility'	Reduce car traffic; improve public transport; waterbus connection

Source: Own elaboration, based on Projectbureau Stadshavens (2011) and fieldwork

including floating houses) to the Heijehaven in the RDM area. Another idea is to move 'fruit terminals' (companies dealing with logistics of fruit) from the Merwe-Vierhavens to the RDM area, giving the Merwe-Vierhavens space for new urban functions, while RDM remains also an area for port and industrial functions.

The goals will be realised in the various parts of the RDM site as is depicted in Table 10.4. Here, the mixed-use character of the area becomes clear with space for business (including start-ups), leisure, living and education/research and innovation. A key asset to realise the 'golden triangle' is the RDM Innovation Dock, a renovated production hall that is targeted to educational institutes, an incubator centre and firms.

The campus concept and the redevelopment of the RDM site started through the action of educational institutes. The Albeda College, a school on the vocational level, was already on the RDM site with a welding school dating back to the times of the shipyard. In addition, Hogeschool Rotterdam was one of the first new tenants of the area. It searched for a new location for its automotive department and moved to a former production hall of the RDM complex as it offered a suitable working space for the large equipment (e.g. vehicle lifts) that did not fit in the other locations of the educational institute. Inspired by notions such as open innovation and the role of informal 'communities of practice', and field visits to other renovated historical industrial sites, such as the Zollverein in Essen, the directors of the two schools jointly developed the 'campus idea' and 'sold' the campus concept to the Port Authority. The Port Authority supported the concept as it saw it as a tool to attract other tenants to the area, although the campus concept is a new approach, outside its core business.

4.2 Content and key domains

Moreover, the concept is an important solution to tackle a major challenge in the port: the labour shortage of especially technical workers. To deal with this issue, the Port Authority has already had intensive contact and various programmes with educational institutes, and the campus concept on the historical shipyard is seen as a new locus to train and attract workers. More concretely, regarding the education function, the plan is to keep a port and maritime atmosphere in the area and to use the former shipyard for training and production. This combination of the history of the area and the new training equipment is used to attract youths and give them an impression about the port as work environment.

The original fields of the campus –'moving' (mobility); 'building' (construction); and 'powering' (application of 'sustainable energy') – refer to education, research/innovation and business functions and are based on the major economic clusters of the city and the Stadshavens area (for instance, cleantech).[2] In time, 'logistics' and 'water' (including offshore and maritime industries) were added to the core fields.[3] The new plan is to create a campus that is strongly linked with economic profile (and image) of the city, that is, the port. The Port Authority (in agreement with the educational institutes) saw especially offshore and maritime industries as an important new field for two reasons. First, it fits well into the historical profile and atmosphere of the area; second, offshore is an important growth sector – especially in times of rising oil prices, which drive the need for more advanced ways of getting energy sources, such as new deep offshore oil fields – making it an important target group for the business location. Attracting offshore firms could prevent vacancy and kick-start the site as a new business location.[4]

As such, flexibility and adaptation to change are key dimensions of the location's concept, yet keeping a link with the context of the area (via existing players and past functions and

Table 10.4 Various parts of the RDM site

Building/area	*Previous function*	*Plan*	*Status January 2013*
RDM Droogdok / Middenstrip	Headquarters	Offices (e.g. Stadshavens; congress centre) Knowledge and education institutes (e.g. Aqua Dock Water Centre; Academy of Construction; Cityport Academy)	Renovated, in use
Dokkantoor	Office	Final goal to be decided	Renovated and available for temporary rent
RDM Dokhaven	Harbour	Exposition and research space; possible terraces and parking (also on water); floating house and garage Space for vessels (e.g. for embarking)	Available and partly in use
RDM Village	Living (housing and leisure)	Renewal of Heijplaat and re-integration with RDM Campus Short stay Leisure facilities Concept house village	New facilities of RDM Campus available Two concept houses constructed and in use; two others in construction and four others in concept stage Some new buildings constructed, but general renewal delayed due to financial problems
RDM Innovation Dock	Production hall	Albeda College and Hogeschool Rotterdam Firms and incubators Work, educational, office and research spaces	Renovated and in use Largely rented
Scheepbouwloods	Production hall	Workspace; office; general services (dressing rooms)	Renovated for and in use by Franklin Offshore
Onderzeebootloods	Subway production halls	Exhibition space and company location	Exhibition space in use Renovated for Ampelmann
Aqua Dock	Harbour	Floating office/work/ experiment place	Concept
Heysehaven	Harbour	Harbour Inland shipping; leisure boating; yacht construction	Partly renovated and (still) in use by firms and educational institutes, including SAIO – off shore training centre
Quarantaine terrein	Site to threat sailors with infectious diseases and to disinfect vessels and cargo	Creative business in natural surroundings	National heritage buildings and nature used as living and working space for artists Recreational area (e.g. beach and parks)

Source: Own elaboration, based on fieldwork

atmospheres). The same flexibility can be observed regarding the incubating facilities. Initially, an incubator programme 'Dnamo' offered support for firms in the pre-incubation and incubation space.[5] However, the concept was regarded as being too similar to other incubator centres, and, therefore, it changed into 'RDM Makerspace' that focuses on fast prototyping. Concretely, this means that the machinery of Hogeschool Rotterdam in the RDM Campus (including a 3D printer) is available for rent to third parties (see Box 10.1).

The last goal of the area – creating an attractive working and living environment – is linked with the renewal of Heijplaat. This was not included in the initial plans that were set up by the educational institutes and the Port Authority. The initiative to integrate the village with the RDM Campus came from housing agency Woonbron that is responsible for renewal of the village. Initially, it had the idea of using the RDM site for new housing and as an extension area of the village. However, after the north-south deal, the site was targeted for (new) port and industrial functions, making housing on the RDM site not possible due to safety regulations. As a consequence, renewal of the village is limited to refurbishment of existing housing stock (especially rental housing) and new housing construction (mainly for sale to attract higher income groups) within the borders of the village. Therefore, Woonbron developed a new vision that was later integrated into the RDM site plan, including the plan to use the RDM Campus as flywheel for renewal by increasing the critical mass of new urban functions in the village (e.g. shops). In addition, the village can also be used for new activities of the educational institutes and firms at the RDM Campus. One concrete example is Concept House Village, a number of houses where (temporary) users can test all kinds of new living concepts. Inhabitants of these houses are monitored in order to analyse their reactions to new concepts in order to adapt the technologies.

The latter illustrates the innovation function of the area. The RDM Campus attempts to put concepts of 'open innovation' into practice, and different (competing) firms and educational institutes develop a shared labour pool and work on joint development projects

Box 10.1 RDM Makerspace

RDM Makerspace offers machinery for its members according to a 'sports school model'. Members include consumers (e.g. hobbyists), start-ups and existing firms which do not have their own equipment. The machinery offered includes 3D printing facilities and more 'classical' timber and metal processing equipment. In addition to the machinery, members can make use of classrooms, office spaces, storage spaces (e.g. for raw materials of produced prototypes) and general catering facilities. Moreover, RDM Makerspace offers courses, which are partly mandatory for users in order to use the equipment safely. There are also specific courses for children in order to encourage them to be enthusiasts for technology and manufacturing. An advantage of the concept (which emerged originally in the USA) is that machinery is used outside working hours – during the day it is used by Hogeschool Rotterdam – and it provides expensive equipment for third parties who cannot afford the high investment costs or the low frequency of usage. Moreover, it may increase interaction between various actors, like consumers or start-ups that use the equipment, and Hogeschool professors.

Source: Fieldwork and http://www.rdmmakerspace.nl/ (accessed 27 March 2014)

and experimentation. This is possible due to the large halls and water spaces that can be used as working and experimentation spaces where new concepts can be developed and tested – for example, the development and experimentation of floating houses. For instance, the RDM Innovation Dock includes a large working space with (partly shared) equipment (e.g. robots, welding machines) that can be used by firms and educational institutes. Furthermore, the RDM Innovation Dock houses a major incubator centre, and as such, the site attempts to form a springboard for entrepreneurship.

Finally, the plan includes a new waterbus connection between the RDM Campus and the city centre to improve the accessibility of the area (including Heijplaat) and to move towards the ambition of being an energy neutral area. As such, it is not only a location for new energy business, but also for trying to implement (new) clean energy solutions, such as locally generated wind and solar energy.

5 Implementation

5.1 Leadership and organisation

From an organisational point of view, the development of the RDM Campus can be split into two periods. The first period (running from about 2004 to 2013) was led by the leaders of the two educational institutes – Hogeschool Rotterdam and Albeda College – which can be regarded as the 'founding fathers' of the campus. They introduced the campus concept and had the skills and patience to introduce it to others, ultimately leading to a vision that is shared with the Port Authority and later Woonbron as well. In the end, there was a 'we feeling' among the core project partners ('Kwartiermakers' in Dutch), being the two educational institutes, the Port Authority and to a lesser extent Woonbron.

This shared belief refers to various organisational levels. At the top, the directors of the Port Authority and the educational institutes promoted the Campus Concept among other players and had the power to start projects. For instance, a delay of the Waterbus connection was prevented due to a proactive role and fast decision making by the mentioned leaders who also had close contacts with the Mayor. On a more operational level, the managers of the two educational institutes and the Port Authority founded a development agency, RDM Campus BV, dealing with the coordination of the area transformation process, including marketing, acquisition, renovation of buildings and new construction, accommodation and facilitation. The core project partners are all based (at least part-time) at the RDM site in an office of the former headquarters.

In addition to the actors organised in the RDM Campus BV, there is close cooperation with Woonbron (the housing agency), namely for the re-integration of the former shipyard and the village. This is translated into a formal cooperation agreement that describes the major aims of the renewal of the village including the tasks and responsibilities of all actors. Even though the agreement is not a strict contract, it sets directions for development and is targeted at preventing conflicts between the stakeholders. The Development Agency of Stadshavens (OMSR), Woonbron and Hogeschool Rotterdam signed a first cooperation agreement in 2005, while much later, in 2012, a second version was signed by more stakeholders,[6] which contains a number of concrete projects including the development of a ring road and a park. It also includes the ambition of developing an energy neutral village via local generation of sustainable energy.

As said, for all actors, the project is seen as a 'joint adventure', as none of them has experience in integrated mixed-use area development, combining education, business, research and living functions, and the marketing of all that. For example, Woonbron has experience with communication with inhabitants, but lacks skills to promote business areas, while for the Port Authority it is the other way around. Similarly, they are dealing with new functions outside their core business, including running an event and congress bureau – used as a strategy to generate income and to attract people to the campus – and the Innovation Dock that puts joint research and education into practice. Due to the novelty of functions outside the key actors' business, the core project partners need to 'sell' new initiatives internally – within the own organisations – to get the required resources.

To overcome such issues, a new organisation has been set up. This department obtains the required skills via new talent coming from other organisations as well as via specific training courses for existing employees. For example, to find tenants for the Innovation Dock, the Port Authority hired a new commercial manager who had the right skills and networks that are not covered by the other commercial managers who are specialised in large industrial clients. Moreover, experience has been built via learning-by-doing during the redevelopment process.

Despite a shared vision and high level of trust, especially among the core project partners, there are a number of factors that make cooperation less straightforward and easy than in the past, leading to the second development period, RDM Campus 2.0 (2014 onwards). For instance, the Port Authority and the Hogeschool Rotterdam have new directors and it is unclear to what extent they support the RDM Campus. In addition, Woonbron and the City of Rotterdam are both in an internal reorganisation process and seem to have less capacity to develop the RDM Campus and the renewal of Heijplaat (see also the next section). In order to guarantee stable cooperation in the future, the actors search for new ways to cooperate and another way to divide tasks.[7]

5.2 Finance

The RDM Campus and renewal of Heijplaat have been funded by investments by the Port Authority, the educational institutes, Woonbron and the City of Rotterdam. These players also have access to national and European subsidies, including €1.9 million of EU funding (Rutten, 2012). The main investors in the RDM Campus and the renewal of Heijplaat differ, also leading to differences in (re)development progress.

The largest investor of the area is the Port Authority (about €100 million), covering renovation of buildings and new infrastructure development, such as a new road to the RDM Campus that prevents hindrance for the inhabitants of Heijplaat who live on the old route to the RDM site (Port of Rotterdam, 2011). The Port Authority is a financially healthy player with leeway for large investments and can prevent developments from coming to a standstill. A major example is the Waterbus connection, which was pre-financed by the Port Authority.[8] The investment was crucial for further development of the RDM Campus, as it reduces travel time between the city and the RDM Campus and the village of Heijplaat. Similarly, the realisation of the Innovation Dock, largely funded by the Port Authority, but with the Hogeschool Rotterdam funding the building's interior design, has not only created trust among the stakeholders, but it also used as a selling point attracting other tenants to the area.

The Port Authority uses its 'traditional' funding model for the development of the RDM Campus. As a landowner and port manager, it actively searches for tenants, and after long-term contracts have been signed, halls are renovated and adapted to the wishes of clients. In addition, it provides empty plots on the RDM site that can be used for new construction. As such, it has limited financial risks, and income generated by long-term contracts can be used to cross-fund other parts of the area, including the incubator centre in the Innovation Dock that works with flexible contracts. Note that this is a traditional model for the Port Authority, but a new approach for area development which normally works with a strict master plan that has been pre-financed with large investments, while the Port Authority's approach is more flexible and includes less financial risk. As such, the RDM Campus is regarded as an example of 'integrated area development new style' (van Leeuwen, 2013).

The renewal of Heijplaat is another story. The renewal process moves slowly (at the time of writing, 2014) due to a low demand for new houses as a consequence of the financial crisis. Moreover, the largest investor in this part of the project, Woonbron, is in a large reorganisation process and is obliged to cut its budget from €10 million to €3.5 million (interview data). As a consequence, it needs to transfer the land development rights to other actors. It is expected that private developers will pick up this task, but it will take time and the renewal process has slowed down.

With the exception of the Port Authority, other players have less funding available, making it harder to start projects in the area. The two educational institutes benefited from subsidies and could move to the area and contribute to its redevelopment and the Campus concept. However, the time periods of these subsidies are over, and consequently the leaders are challenged to find new funding sources to implement the Campus concept further. Similarly, the City of Rotterdam is in a reorganisation process and needs to postpone or cancel projects; this also delays parts of the RDM Campus in which the city is involved. For instance, Aqua Dock, a new innovative pilot for floating construction techniques, only started in March 2014, later then the core project partners hoped for.

5.3 Communication

As the RDM Campus was a closed area for many years, it is a challenge to promote the area to new tenants and to reintegrate it with the village of Heijplaat. This happens in various ways and by the various key actors involved.

First, an event strategy has been used to bring people to the site and to create awareness for the new location among various target groups, including inhabitants of Heijplaat, business visitors and general audience. Small-scale events have been organised, such as movie evenings and small music events (e.g. jazz). Not only the core project partners organise events for inhabitants, but also firms. An important example is the offshore company Franklin, which organised a 'welcome party' for the village after it opened its new European Headquarters and a large production hall at the RDM Campus. Business visitors are attracted to the conference centre that was purposely opened in the former headquarters building. Finally, the general public has been attracted via larger events, including exhibitions by Museum Boijmans (a museum in the city centre), a dance performance ('Conny Janssen danst') and 'Blown away', a unique music performance by the Rotterdam Philharmonic orchestra combined with electro music. Adding to this, guided tours through the area have been organised, targeting potential new inhabitants and business.

Second, an active media strategy has been devised to constantly generate news from the area. This concerns large happenings that are broadcasted in the national media – such as the opening of the Waterbus Connection and the symbolic re-opening of the RDM Campus – as well as small announcements such as a new tenant moving to the area. An important part of this strategy is the use of famous people as 'ambassadors' of the area (such as former ministers and aldermen). The rationale is that when such a high-level person says something about

Box 10.2 Ampelmann – 'Born at Technical University Delft and raised at the RDM Campus'

Ampelmann is high-tech offshore company that develops, constructs and leases new types of offshore access solutions ('making offshore access as easy as crossing the street'). It was founded in 2008 as a spin-off of Technical University Delft. It searched for a suitable place for its business and found the right facilities on the RDM Campus: large production spaces, cranes to lift new installations and water access to try new floating concepts. It benefited from the flexible lease contracts in the RDM Innovation Dock and started as a small company with the lease of two lots, but grew in time to three and later even to five lots. Nowadays, it will even get its own production hall on the RDM Campus; the former 'submarine construction hall' will be converted to accommodate further growth of this key niche player. This example is often used to show that (and how) the flexible lease contracts work, but it also gives a 'new story' in line with the profile of the former shipyard that has been used for business promotion and may have helped in attracting other offshore sector firms.
Source: Interview data and http://www.ampelmann.nl/ (accessed 27 March 2014).

the RDM Campus it has a much wider impact than announcements made by the core project partners themselves. In addition, key people are invited to major happenings to increase media attention. For instance, the RDM Campus was formally opened in 2009 by Prince Willem Alexander (currently the King of the Netherlands). Moreover, the innovative drive of the area and the focus on new concept development and prototyping – e.g. new floating solutions (e.g. Aqua Dock) or the Concept House Village – is also strongly promoted in order to attract companies and inhabitants to the area. Key assets of the area are also used as selling points, including the history of the area, the Innovation Dock and a number of new 'success stories' (see Box 10.2).

Finally, more regular communication channels have been used, including a website, magazine, newsletter and brochures. In addition, new modes of communication with inhabitants have been established. For instance, initially, the communication with inhabitants of Heijplaat was mainly about informing them, with limited space for feedback. However, after several tensions in the past (mainly concerning new industrial functions), a new communication platform – Heijplaat Vitaal – was set up. At this platform, the key actors meet inhabitants on a monthly basis and not only inform them but also leave space for feedback; as such, inhabitants are involved in the development process. This new platform strategy has turned out to be a success and is also used by the Port Authority in other port areas that are close to the housing areas, such as the village of Pernis that is surrounded by petrochemical industry sites.

6 Results

6.1 The golden triangle: business, education and research

The transformation of the shipyard into a new business, research and study location seems to progress well. Many historical buildings have been converted as new business and educational spaces, and new halls have been built on empty plots. Most developments are done in the central and western part of the RDM site, while the development of the eastern part is regarded as more difficult due to the presence of existing tenants with long-term contracts and a relatively large distance to the core of the campus where the educational facilities and general services (e.g. restaurant and congress centre) are based. The former headquarters building, Middenstrip, has been fully occupied with new tenants and also the Innovation Dock has been largely filled with new firms and educational institutes. The Innovation Dock had a slow start in 2011 due to the financial crisis, but benefited from growth in the offshore industry that was added as a new field.

Most of the attracted firms are active in the offshore, energy and construction sectors, in line with the target fields, which are partly defined based on initial tenants and new growth sectors (see Appendix 1). Besides, and also partly due to the success of Ampelmann (Box 10.2), the RDM Campus managed to attract most of the giants of the offshore sector who opened new large production facilities (such as Franklin) or rent a space in the Innovation Dock (e.g. IHC Merwede, Imtech Marine, Huisman and Damen). With major exceptions, such as, again Ampelmann, developing a location for start-ups is harder to realise. First of all, start-ups had less priority for the Port Authority compared to large players that can guarantee a higher income. Now, after attracting the offshore giants, income can be used to cross-fund start-ups. Second, the Port Authority has less experience in supporting start-ups and it takes time to develop such skills and networks. Finally, due to the change of Dnamo to RDM Makerspace, there is less support for firms in the pre-incubation and incubation stages.

Regarding the education function, the Campus seems to develop well. The Hogeschool Rotterdam (including Rotterdam Academy of Architecture and Urban Design) and Albeda College started their first courses at the campus, and the Maintenance College (a joint initiative by STC, Zadkine and Albeda College) opened a demonstration centre at the former shipyard. In addition, SAIO, which offers professional safety training (for the transport sector, chemical industry and offshore sector among others), is based at the other side of the Heijsehaven. Education at the RDM Campus is in general limited to the practical part (e.g. the third and fourth years of the courses offered by the Hogeschool Rotterdam). It remains a challenge to attract students and teachers to the peripheral location of the RDM Campus, despite the fast waterbus connection. As such, it is illustrative that Hogeschool Rotterdam offers its Industrial Design Course again at a central location in the city, after a 'try out' at the RDM Campus, which did not attract sufficient students. In order to increase the critical mass as an educational location and to reduce dependency on Hogeschool and Albeda College, the Port Authority tries to attract other educational institutes as well.

Concerning the RDM Campus as an innovation location, first steps have been taken. Various pilots have been realised in the area, including the new types of Ampelmann's offshore solutions (tested and constructed at the RDM site) and the Floating Pavilion, a congress centre constructed at the RDM Campus and currently in use in the Rijnhaven. Also, a number of pilothouses in the Concept House Village are already in use (Box 10.3). Moreover, the Aqua Dock formally started in March 2014, with the launch of a pilot of a floating tree (an artwork that is part of a larger innovation project, the floating forest) as a formal kick-off for the project. In addition, a floating road is ready to be tested as the first large part of the Aqua Dock project. However, Aqua Dock started later than expected, and many other initiatives have not begun due to difficulties in making a business case, often caused by a shortage of funding or due to bureaucratic barriers (including tendering procedures for new floating constructions).

Box 10.3 Concept House Village

Concept House Village is a user-oriented testing environment for new building and living concepts, including new construction materials (e.g. wall isolation made of recycled paper) and new energy concepts. The project is initiated by CityPorts Academy Rotterdam and is executed by Hogeschool Rotterdam, Technical University Delft and Woonbron. Initially, the plan was to realise the Concept House Village at the RDM Campus, but this was not possible due to resistance by the Port Authority that wants to prevent conflicts between housing and port functions. Therefore, the project started at Heijplaat, and as such, it could be seen as a new link between the RDM Campus and the village. Moreover, the project is targeted to attract new temporary inhabitants who want to participate in studies around experimental living, and would normally not move to Heijplaat.

The plan covers the construction of eight experimental houses. So far, four pilot houses have been constructed and are in use. The first concept house is part of an international research project of Technical University Delft (in cooperation with partners in Gothenburg and London) and the second is a project by Hogeschool Rotterdam. Studies in the houses have a long time period (four to five years) and cover the entire building cycle: construction, use and demolition.

Source: Fieldwork.

Finally, the concept of the 'golden triangle' seems to work, but is also regarded as a long process. There are various joint pilots between educational institutes and firms; firms hire students and provide business cases, but more structural cooperation between firms and the educational institutes requires a change in working attitude. It is argued that both workers and researchers from firms as well as teachers and researchers from educational institutes need to get used to involving each other and to setting up joint initiatives. The concept of 'Communities of Practice' (CoP) has also been applied in the RDM Campus, namely in the CoP Floating Construction (CoP Drijvend Bouwen).[9] This network deals with collecting and sharing of information in the field of floating constructions, covering many dimensions of this new field, such as legal issues, economic feasibility and social acceptance. Students' theses form an important source of information for this network.

6.2 Integration and renewal of Heijplaat

The integration plan of the RDM Campus and Heijplaat aims to increase the critical mass for certain new facilities at the Campus that can also be used by the inhabitants of Heijplaat, or to keep facilities in Heijplaat due to new demand from users (workers and students) of the Campus. The best realised example is the fast Waterbus connection between the RDM site and the city centre. It was not only set up to bring workers and students to the RDM Campus, but also to give the inhabitants of Heijplaat a fast and direct public transport link to the city centre. Another examples are basic services such as ATMs and restaurants – these functions were not present in Heijplaat as the demand from the village was too low to operate them efficiently.

The RDM Campus also offers new business, working and educational possibilities for the inhabitants of Heijplaat. One example is the ESFA, an assembly plant for electric scooters that hires students, mostly youths with social problems coming from Heijplaat. They have the possibility of obtaining working skills and experience instead of living on the street. An entrepreneur from Heijplaat set up this social project in cooperation with educational institutes (including Albeda College and Hogeschool Rotterdam), the City of Rotterdam and QWIC (the scooter brand). The City provides subsidies to this project.

Despite these positive examples, the village remains challenged with social problems and decline, and the general socio-economic integration with the RDM Campus offers space for improvement. Due to low market demand for new houses and Woonbron's financial problems, the construction of new housing is limited to two flats so far, while further realisation of other projects has been postponed. This does not only lower the societal support from the existing inhabitants, who are disappointed due to the delay, but it also postpones inflow of new inhabitants required to increase the critical mass for provision of services in the village. For instance, the last supermarket is facing bankruptcy due to the lack of sufficient clients.

Finally, it remains a challenge to balance industrial development and development of the village, despite new communication strategies such as Heijplaat Vitaal. For instance, there were long discussions about new projects, such as the new ring road and a new production hall at the edge of the RDM site next to the village. Such tensions seem unavoidable in this rather unique place where industry and living functions are located so close to each other.

7 Conclusions

7.1 Broad assessment

The RDM project can be seen as a front-running practice of a transformation process of a single industrial site into a new location that combines and integrates business, education and research ('the golden triangle'), making use of new innovative planning concepts. To date, the RDM Campus seems to be progressing well and many buildings have already been transformed. Many of them focus on 'cleantech', one of the new key transversal clusters of the city, namely offering possibilities for developing, prototyping and testing new solutions. Examples include the Floating Pavilion, Amplelmann's new concepts and the start of Aqua Docks. Similarly, the Concept House Village is developing well. Finally, the RDM Campus has contributed to create new demand and brought new facilities for the inhabitants of Heijplaat.

The project is well integrated into the city and the port's long-term strategy and is part of their joint Stadshavens development strategy. As such, there are linkages between the RDM Campus and other parts of the Stadshavens area, e.g. new (floating) concepts have been tested and produced in the RDM area and are used in the Rijnhaven. Similarly, there are plans to apply the campus concept in other city locations as well. As such, the experiences obtained via 'learning-by-doing' are a major result of the project. A concrete example is the communication platform with inhabitants that the Port Authority now also applies in other port areas in direct geographical proximity of living areas.

7.2 Critical factors

The success of the RDM Campus so far does not come out of the blue, but is related to a number of enabling factors and organising capacities.

First, the new campus concept was introduced by visionary leaders of educational institutes with the skills and patience to create trust among other core project partners. The mutual trust and shared vision are also the result of shared challenges, including the labour shortage of technically skilled port workers and Heijplaat's large socio-economic problems, which were crucial to gain a constituency for the new campus concept. The project has been supported and managed on various levels (decision-making and operational), although it is still a challenge for the core project partners to sell new initiatives and projects internally as they are acting outside their core business.

Second, the project benefits from the organisational and financial power of the owner of the area (the Port Authority), which employs a flexible development approach. By developing 'building-by-building' and 'plot-by-plot', based on the requirements of tenants with long-term contracts, it limits financial risks and creates opportunities to cross-fund general facilities and spaces for tenants with short-term contracts such as incubators and smaller companies.

Third, a certain degree of flexibility becomes clear from the target innovation fields that are 'not too narrow nor too broad'. They are based on the history of the area and the profile and needs of the city, but can also be slightly adapted to new opportunities and challenges. For instance, adding the growing offshore sector to the key fields of the campus has resulted in a large concentration of offshore firms on the RDM site, while still offering the right facilities to new types of start-ups, including the flexible lease system of the Innovation Dock, enabling new firms to rent more space in times of growth and less in less favourable economic times.

Appendix

Table A10.5 Tenants in the RDM Campus, March 2014

Tenant	*Field*	*Location*
Aimabel	Furniture design	Dokloodsen
Alternative Energiewinkel	Energy	Dokloodsen
Ampelmann	Offshore	Innovation Dock and Onderzeebootsloods (reconstruction)
Archimedes	Energy	Innovation Dock
Damen	Offshore	Innovation Dock
Doet	Mobility	Innovation Dock
EICB	Mobility	Innovation Dock
Energy Floors	Energy	Innovation Dock
Esfa	Mobility	Innovation Dock
Franklin Offshore	Offshore	New production hall
Ingenieursbureau Gemeente Rotterdam	Construction	Innovation Dock
Jules Dock	Offshore	Innovation Dock
HeatMatrix	Energy	Dokkantoor
Huisman	Offshore	Innovation Dock
Hyperbody / TU Delft	Construction	Innovation Dock
IHC Merwede	Offshore	Innovation Dock
Imtech Marine	Offshore	Innovation Dock
Innovatie Centrum Duurzaam Bouwen	Construction	Centraal Magazijn
Innovatiecentrum Kunststoffen	Metal-electro, supplier to offshore firms	Innovation Dock
Itanks	Logistics	Scheepsbouwloods

Table A10.5 Tenants in the RDM Campus, March 2014 (Continued)

Tenant	*Field*	*Location*
Mobiel Erfgoed Centrum / TU Delft	Offshore	Innovation Dock
Mr. Elevator	Offshore	Innovation Dock
Perlex	Mobility	Innovation Dock
Port-able	Media/Communication	Innovation Dock
Public Domain Architects	Construction	Innovation Dock
RDM Makerspace	Business services	Innovation Dock
Skilledin	Human resources/training/ education	Innovation Dock
Van Gansewinkel	Waste management	Innovation Dock
Vrijhof Ankers BV	Offshore	New production hall (under construction)
Yumm Concepts	Food	Innovation Dock
Albeda College	Education: Maintenance and Metal engineering (Vocational level)	Innovation Dock
Hogeschool Rotterdam	Education: Automotive engineering and Academy of Architecture and Urban Design	Innovation Dock and Middenstrip/RDM Droogdok
SAIO	Professional: safety training	Quarantaineweg

Source: Own elaboration, data from http://www.rdmcampus.nl/bedrijvengids (accessed 26 March 2014)

Notes

1 In time, the vision has been revised and updated. For instance, the 13 VIP areas are rearranged, putting priority on the implementation of (parts of) three programmes: central city, Rotterdam South and the Stadshavens area. Within the Stadshavens area, the focus is on certain parts and not the whole 1,600 hectares. Priority is on further development of RDM Campus and Heijplaat, Rijnhaven and the start of parts of the Merwe-Vierhavens area.

2 However, it is also based on fields of some pioneer tenants who moved to the area before the campus concept was introduced, such as Hogeschool Rotterdam (automotive, moving), BAM (a construction company) and Formula Zero (a start-up dealing with zero emission cars, hence moving and powering).

3 This was done by the Port Authority who gained more control in the area after the 'north-south deal' (see Section 3) and argued that the original fields did not cover the port sufficiently.

4 At the same time, other niches get less attention. For instance, initially the focus was on luxury yacht construction, which was in line with the history of the RDM, and with Merwelands, which since 2008 has been based in the eastern part of the RDM site, it has already one player in this niche. Whereas this niche was booming at that time (2008), later demand declined quickly and the focus was put on the offshore sector.

5 This programme was founded by Enviu, Technical University Delft (TUD), Yes Delft (the incubator service of TUD), Hogeschool Rotterdam, the Port Authority and Rabobank.

6 These stakeholders include the City of Rotterdam, Woonbron, the Port Authority, Wijkvereniging Heijplaat (a community association representing the interests of the inhabitants of Heijplaat), Eneco (a utility company), Stedin Netbeheer (an electricity grid company) and RDM Campus BV/ Hogeschool Rotterdam.

7 Some changes have already been made. For instance, the key actors stopped with RDM Campus BV, and this company's tasks will be carried out by a project organisation of the Port Authority that hires the managers from the other partners. In addition, the responsibility for the congress bureau

will shift from the Hogeschool Rotterdam to the Port Authority. As such, the Port Authority gains power in the further transformation process, and the role of others is as yet unclear.

8 Organisational uncertainties and the large number of players made it hard to divide the financial burden, and there was a risk that the project would be delayed. In order to prevent this, the Port Authority decided to pre-finance the project and the Waterbus connection that was needed to start the first educational course in 2009 was realised on time.

9 This is a network of the City of Rotterdam, Hogeschool Rotterdam, Public Domain Architects, Platform Drijvend Bouwen, Flexbase, Hogeschool InHolland Rotterdam, Stichting Stad op het water and RDM Campus.

References

Azaoum, K. and De Lange, C. (2014). Rotterdam. In L. van den Berg, J. van der Meer and L. Carvalho (eds), *Cities as Engines of Sustainable Competitiveness: European Urban Policy in Practice.* Aldershot: Ashgate.

Braun, E. (2008). *City Marketing: Towards an Integrative Approach.* Rotterdam: Erasmus Research Institute of Management.

CBS – Centraal Bureau voor de Statistiek (2014). Hoogste werkloosheid in Flevoland, *Webmagazine*, 24 February.

CBS Statline (n.d.). Online statistics. Accessed at http://www.cbs.nl/nl-NL/menu/cijfers/statline/zelf-tabellen-maken/default.htm (5 February 2014).

Daamen, T. (2010). *Strategy as Force: Towards Effective Strategies for Urban Development Projects: The Case of Rotterdam City Ports.* Amsterdam: IOS Press.

Daamen, T. and Vries, I. (2013). Governing the European port–city interface: institutional impacts on spatial projects between city and port. *Journal of Transport Geography*, 27, 4–13.

Gemeente Rotterdam (2008). *Stadsvisie Rotterdam – Samenvatting Ruimtelijke Ontwikkelingsstrategie 2030.* Rotterdam: Gemeente Rotterdam.

Hooijer, B. (n.d.). RDM Campus, online presentation. Accessed at http://www.transurban.nl/media/bestanden/stad per cent20water/Conferentie per cent20Rotterdam per cent20Waterstad/Bert per cent20Hooijer per cent20RDM per cent20Campus.pdf (5 February 2014).

Nijdam, M. (2010). *Leader Firms: The Value of Companies for the Competitiveness of the Rotterdam Seaport Cluster.* Rotterdam: Erasmus Research Institute of Management.

Nijdam, M., Bakker, D. and van der Lugt, L. (2012). *De economische betekenis van Nederlandse zeehavens.* Rotterdam: Erasmus University Rotterdam.

OBR – OntwikkelingsBedrijf Rotterdam (2011). Werkgelegenheidmonitor Rotterdam – Winter 2011. Rotterdam: OntwikkelingsBedrijf Rotterdam.

Port of Rotterdam (2011). *Port Vision 2030: Port Compass.* Rotterdam: Port of Rotterdam.

Projectbureau Stadshavens (2011). Stadshavens Rotterdam: Structuurvisie. Accessed at http://stadshavensrotterdam.nl/wp-content/uploads/2011/10/11442-Structuurvisie-Stadshavens-2011-08-LOWRES.pdf (5 February 2014).

Rutten, J. (2012). EU 2014–2020: van subsidies naar revolving funds, Zorg dat je erbij bent. *Building Business*, September.

Scheepmakers, H. (n.d.). Stadshavens: Organisatie, Strategie, Uitvoeringsprogramma. PowerPoint presentation. Accessed at http://www.transurban.nl/media/bestanden/stad%20water/Conferentie%20 Rotterdam%20Waterstad/Hans%20Scheepmakers%20Stadshavens.pdf (11 February 2013).

Schoo, T. W. (2013). Nieuwe tijden aan het waterfront, Op zoek naar een nieuwe vorm van publiek-private samenwerking voor de herontwikkeling van Stadshavens. Master thesis, Utrecht University.

Van Leeuwen, M. (2013). Gebiedsontwikkeling nieuwe stijl standaard voor RDM Heijplaat: Op maat herontwikkelen voor gecontracteerde eindgebruikers, Vastgoedmarkt, April 2013.

Van Winden, W., van den Berg, L., Carvalho, L. and van Tuijl, E. (2010). *Manufacturing in the New Urban Economy.* Abingdon: Routledge.

11 Water management strategy (Singapore)

Sarah Lidé

1 Introduction

This case study describes Singapore's water management story and strategy and how it has evolved from being an issue of national security to a source of economic value and competitive advantage for the tiny city-state. All across the globe, there is a growing concern about the increasing likelihood and implications of a decline in water quality and availability worldwide. In fact, water supply crisis was one of the top five global risks identified by the World Economic Forum in its 2014 Global Risks report (World Economic Forum, 2014). However, water (or lack thereof) has been a critical issue for Singapore since the country gained independence from the British in 1965. With a limited land area for collecting and storing rainwater, imported water has always played an important role in meeting Singapore's water needs. Singapore has been importing water from across the Strait of Johor for the past five decades or so, with the signing of two long-term supply deals in 1961 and 1962 ensuring water supply to Singapore until 2011 and 2061 respectively. However, Singapore has made it a key priority to become truly self-sustainable in terms of its water.

The unique aspect of Singapore's water management journey is not its success in nearly reaching complete self-sustainability, but also that it has managed to convert vulnerability into opportunity. A far cry from its water rationing days in the 1960s, Singapore is today positioning itself as a 'Global Hydrohub', with approximately 130 water companies and 26 research institutes operating in a vibrant ecosystem (PUB *et al.*, 2014). Singapore has been host to the Singapore International Water Week (SIWW) since 2008. As an early adopter of water management solutions such as NEWater (wastewater reclamation) and the Deep Tunnel Sewerage System, Singapore is leveraging on and exporting locally developed water and wastewater treatment technology and expertise overseas. Leading global industry players, such as Black & Veatch and CH2M HILL, are employing expertise gained in Singapore to their other projects worldwide. Home-grown giant Hyflux has developed two seawater reverse osmosis desalination plants in Singapore and one of the world's largest membrane-based seawater desalination plants in Algeria. These success stories are building towards Singapore's stated goal of growing value-added contribution from its water sector while contributing to the sustainability of water provision in the city-state.

This chapter is structured as follows. Section 2 analyses the structure (and recent dynamics) of Singapore across a number of dimensions or 'capitals' for sustainable competitiveness, including its multi-agency approach for integrated urban planning. Section 3 sets forth the city-state's vision and strategy, and how its water management strategy fits into the overarching national strategy. Section 4 explores Singapore's water management strategy in detail, illustrating the key features and organising capacities leading to Singapore's success in achieving water sustainability and economic value from new solutions. Section 5 concludes with identifying some critical factors to Singapore's success.

2 City profile: capitals for sustainable competitiveness

2.1 Performance

Since its independency in 1965 and within one generation, the city-state of Singapore became one of the most developed economies in the world. It has one of the highest GDP per capita worldwide, and was one of the fastest growing economies in the world for the first half of 2010 with a 17.9 per cent growth rate (Adam, 2010). Singapore shows up on many global rankings, for instance, it topped the World Bank's Doing Business report on the ease of doing business for eight years running. Its unemployment rate remains low at 1.8 per cent (Ranasinghe, 2014), driven primarily by public infrastructure projects, which have boosted growth in the construction sector to offset the slowdown in services. The Singapore economy in turn grew by 3.7 per cent in 2013 (Chen, 2013).

2.2 Economic structure and knowledge

Singapore has been intentional in its targeting of industries to focus on over the years. In the early years of its independence, it aimed at attracting labour-intensive foreign manufacturing in order to provide employment for its then low-skilled workforce. As Singapore's efforts at raising the educational attainment of its people bore fruit, it shifted to developing capital- and skills-intensive industries, particularly precision engineering, in the 1970s and 1980s, with a corresponding emphasis on technical fields. With the formulation of the Strategic Economic Plan in 1991, this was expanded from manufacturing to include services, and today, Singapore is positioning itself as a leading player in the global knowledge economy, with a focus on R&D and knowledge-intensive industries and higher value-added services such as professional services.

Regardless, manufacturing remains a key pillar in Singapore's economy, contributing between 20 and 30 per cent to Singapore's GDP annually, and the sector showed resilience in 2011 despite the global downturn, thanks primarily to the pharmaceutical and biomedical sectors (Hawksford, n.d.). Other major manufacturing industries include electronics (such as semiconductors, precision engineering, etc.), pharmaceuticals, chemicals, construction and ship-building, and its main export industries comprise machinery and equipment, electronics, consumer goods, pharmaceuticals, chemicals, and other mineral fuels (Hawksford, n.d.). Many economists believe that the biomedical, water and medical technology sectors could prove to be bright spots in Singapore's near future (Economic Development Board, 2012).

Over the years, Singapore has adopted a developmental strategy, namely to support industries where it is either already a globally leading player, or has the potential to become one. This strategy has also a very strong educational dimension. From the early days of putting in place a bilingual policy (where students learnt both English as well as their native tongue) and a national system of public education bringing its multiracial population under one roof, today, Singapore's education system is often upheld as a role model for many countries. Singapore emerged top in Asia for higher education; its universities have significantly risen up in many rankings and are strongly connected to other excellence centres worldwide (Toh, 2012; Sedghi, 2012; Magnanti, 2012). The emphasis on R&D is boosted by strong support from the government, particularly from the government agency A*Star (Agency for Science, Technology and Research). A*Star offers substantial funding for research and seeks to attract top scientists and scientific companies to Singapore. In large part due to the government's efforts, approximately 1 million foreigners with scientific, technical or managerial skills have been encouraged to work and live in Singapore (OECD, 2010).[1]

2.3 Institutional conditions

The Singapore government is considered a highly efficient meritocracy with a strong focus on integrated strategic planning and detailed execution. The country's small size and political stability has enabled it to be versatile in responding to rapidly changing environments, and its ability for swift policy implementation is facilitated in part by the strength of its civil service. Singapore has invested a considerable amount of resources in its government agencies, with the civil service showing a willingness to pay market rates in order to attract and retain senior staff. The relatively flat hierarchy and highly disciplined nature of the civil service has also enabled it to make rapid technological change and implement cross-government systems.

Government agencies adopt a whole-of-government approach – all the public agencies (ministries and statutory boards) serve the people as one (e.g. the 'no wrong door' philosophy) even though each agency has autonomy to set directions and build its own capacity. So while each agency is accountable for its defined and focused role, cross-agency issues are addressed as a networked-government to provide solutions from the citizens' perspectives (Ho, 2014). In addition, the civil service, in line with recent calls for increased engagement with the public, is moving beyond passive policy feedback from citizens to an active engagement of citizens in policy formulation. For instance, the 'National Conversation on our Singapore' was launched in August 2012 in a serious effort to engage citizens regarding the future of Singapore.

Specifically, given Singapore's limited land area and lack of hinterland, urban planning becomes an important tool in ensuring that competing land needs are met. As such, perhaps by necessity, Singapore has taken an integrated and holistic approach to managing its land use. This is achieved through a Concept Plan, which maps out the long-term directions for Singapore's land use and transportation plans over the next 40 to 50 years. It takes into account Singapore's land use demands including housing, industry and commerce, community, recreation, water catchment, transport and infrastructure needs, and safeguards land for key growth sectors to ensure there is enough land to meet all development needs even in the long term. The Concept Plan thus goes beyond a mere physical land use plan and strategically adopts a macro view of land use requirement to ensure sufficiency of land amidst competing needs and in turn enhance investor and business confidence. While the review of the Concept Plan is led by the Urban Redevelopment Authority of Singapore (URA), it is a multi-agency effort involving the participation and cooperation of many other government agencies, and also involves public consultation in the review process.[2]

2.4 Conclusions: main issues

Singapore has come a long way since its early days of independence to become a well-regarded player in the global knowledge economy. However, the fact remains that its challenges have always been associated with its lack of natural resources, such as land constraints and its dependency on its neighbours for water. In particular, its ever growing population and economy means that water catchments continually compete with activities such as housing, commerce, transport and schools for space on the small 700 square kilometre island. This limited catchment size to collect and store rainwater, as well as high evaporative losses and lack of natural groundwater, led it to being ranked 170th among 190 countries in terms of fresh water availability (UNESCO, 2002).

Further, given its tiny size and geographical position, it is vulnerable to regional forces and its dependency on exports ties its fate closely to global upswings and downturns. As such, Singapore is continually looking to stay one step ahead in finding areas of competitive advantage that yield economic value to the nation, while simultaneously addressing the fundamentals of survival.

This thus presents itself in Singapore's water management strategy, which, if to be appreciated, must be seen as part of a larger whole comprising Singapore's vision in becoming a global player in, among others, water management. Its 'Four National Taps' long-term water supply strategy, coupled with its 'Active, Beautiful, Clean Waters' (ABC Waters) programme which aims to integrate water bodies with the surrounding environment in a holistic way, do not just look to enhancing Singapore's water security, but also play an important role in the nation's economic trajectory and global positioning.

3 City-wide vision and strategy

3.1 Vision

Singapore's founding father and first Prime Minister, Lee Kuan Yew, identified the strategy of developing Singapore into a 'first world oasis in a third world region' (e.g. Yew, 2000). This entailed establishing first world standards in public and personal security, health, education, telecommunications, transportation and services, eventually making Singapore a base for entrepreneurs and professionals wishing to do business in the region. The focus was on building confidence among and consequently attracting overseas investors by making Singapore better organised and more efficient than other countries despite the lack of a domestic market and natural resources (Lee, 2000).

This vision was encapsulated in Singapore's Strategic Economic Plan (SEP), which was launched in 1991 by the Economic Planning Committee under the Ministry of Trade and Industry. One of the SEP's eight strategic thrusts was that of reducing vulnerability, which Singapore's water management strategy clearly falls under in terms of achieving self-sustainability in its water supply. However, its economic value-add to the nation speaks to its alignment with other strategic thrusts set forth by the SEP, particularly that of spearheading economic redevelopment and maintaining international competitiveness, linked to innovation and the development of clusters.

Today, Singapore is preparing itself to maximise opportunities in a new world environment by exploiting its Global-Asia position – 'connected to the world, located in Asia' (Lee, 2013). This vision is elaborated in the set of recommendations put forth by the Economic Strategies Committee (ESC), which was established by current Prime Minister Lee Hsien Loong in 2009 to develop strategies for Singapore to achieve 'sustained and inclusive growth'. These recommendations are captured under three broad priorities: 1) boost skills in every job; 2) deepen capabilities among Singapore companies to seize opportunities in Asia; and 3) make Singapore a distinctive global city (Economic Strategies Committee, 2010).

3.2 Strategies

In line with the aforementioned three broad priorities, the ESC sets forth seven strategies (see Figure 11.1) that should be achieved before 2020 in order to sustain economic growth and enable broad-based improvements in the living standards of Singapore. In particular, the strategy that seeks to anchor Singapore as a Global-Asia hub identifies water management as one of the areas that could establish Singapore as the location for 'future-ready' urban solutions. This would entail fostering collaboration amongst Singapore-based companies to provide solutions, which could be in turn scaled up and exported.

This Global-Asia positioning is also reflected in the Economic Development Board's (EDB) 'Home' strategy for Singapore, which seeks to help businesses by leveraging on Singapore's value proposition and growing their top line through establishing and deepening strategic activities in Singapore (Economic Development Board, 2014). Specifically, one of the aims is for companies to tap into Singapore as a 'living laboratory' to collaboratively conceptualise, co-create and test-bed new solutions for commercialisation in Asia and globally.[3] Clearly, a red thread running through all of Singapore's policies is the desire to achieve and maintain sustained economic growth, and this continued focus on extracting economic value in all areas is in turn manifested in Singapore's water management strategy, thus contributing to enhancing the nation's sustained competitive advantage, as the following sections will describe.

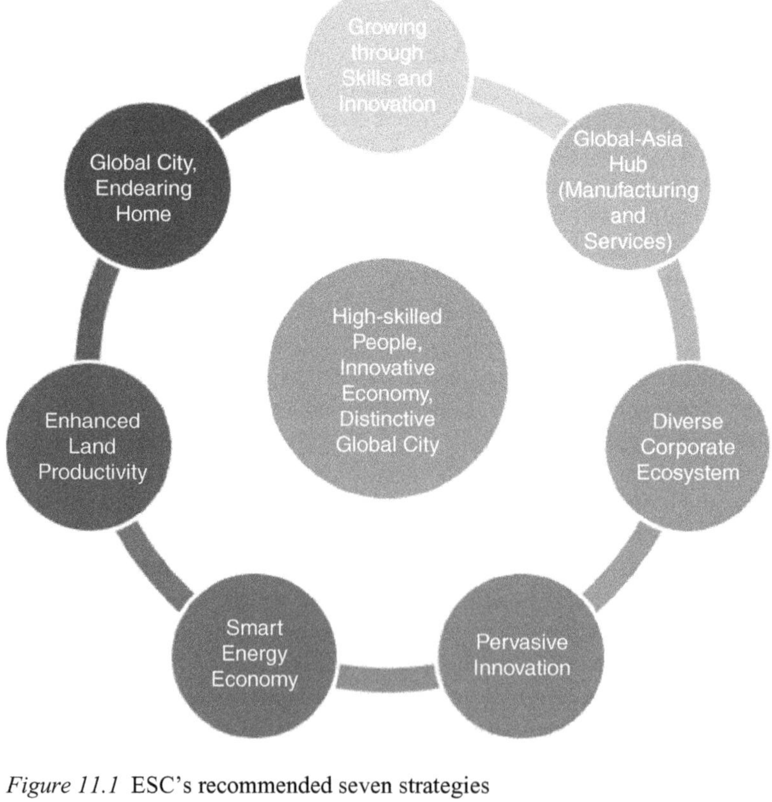

Figure 11.1 ESC's recommended seven strategies

Source: Economic Strategies Committee, 2010

4 Singapore's water management strategy

4.1 Vision and content

In line with the aforementioned strategic thrust of reducing vulnerability is the aim to achieve water sustainability. When Singapore first became independent, one of the first things then Prime Minister Lee Kuan Yew did was to set up a unit in his office to coordinate the prioritisation of water at the whole-of-government level. As he aptly recalls during a dialogue session at the inaugural Singapore International Water Week in June 2008, 'This [water] dominated every other policy. Every other policy had to bend at the knees for water survival.'

4.1.1 Achieving self-sustainability in terms of water

This desire to achieve sustainability in terms of water led to the formulation of Singapore's long-term water supply strategy called the 'Four National Taps' (PUB, 2014), in which Singapore's water supply comprises four sources:

- *Local catchment water*: Rainwater is collected through protected catchment areas as well as urbanised catchment areas, through a comprehensive network of rivers, drains, canals, storm-water collection ponds and reservoirs before being treated for potable use. In fact, Singapore is one of the few countries in the world to harvest urban storm-water on a large scale for its water supply. It is estimated that as of 2011, two-thirds of Singapore's land acts as water catchment areas, and this is set to increase to 90 per cent in the long term with the use of technology to harness yet untapped water from the remaining streams and rivulets near Singapore's shoreline.
- *Imported water*: Singapore has had two bilateral agreements with Malaysia to import water from the neighbouring Malaysian state of Johor. The first agreement expired in August 2011, while the second agreement under the 1962 water agreement will continue to allow Singapore to draw water from Johor River up to 2061. The aim is for Singapore to be completely self-sufficient in terms of water supply by the time the second agreement expires.
- *Desalinated water*: Singapore hosts one of Asia's largest seawater reverse-osmosis plants. Producing 30 million gallons of water a day (mgd), it meets about 10 per cent of Singapore's water demand. A second and larger desalination plant with a capacity of 70 mgd of desalinated water opened in September 2013. Today, the two desalination plants can meet up to 25 per cent of Singapore's current water demand. Singapore aims to continue to increase desalination capacity by 2060 such that desalinated water will meet up to 25 per cent of water demand in the long term.
- *Highly-purified reclaimed water*, i.e. NEWater: One of Singapore's most recognised success stories in terms of water technology, NEWater is an ultra-clean, high-grade reclaimed water produced from treating used water that is further purified using advanced membrane technologies, namely microfiltration, reverse osmosis and ultra-violet disinfection. To date, NEWater meets about 30 per cent of the nation's water needs, and there are plans to triple current NEWater capacity by 2060 to meet up to 55 per cent of future water demand.

Alongside these 'four taps' are multiple water projects initiated to ensure a sustainable water supply for Singapore. This includes cleaning up the Singapore River, building the Marina Barrage and creating the Deep Tunnel Sewerage System (DTSS).

4.1.2 Engaging and educating the community

In 2002, URA, together with the National Parks Board (NParks), launched a new Parks and Waterbodies Plan under the Singapore Master Plan. The plan's aim is to transform Singapore beyond a 'Garden City' into a 'City in a Garden', towards the vision of creating 'a city that is nestled in an environment of trees, flowers, parks and rich bio-diversity' (Ministry of National Development, 2002). This meant providing for an island-wide network of green spaces and creating better access to waterways and water bodies for recreation. In particular, the waterways had the potential to act as focal points for community activities given their proximity to residential heartlands, and would be further enhanced by the presence of park connectors which increase accessibility to these sites.

Promoting the enjoyment and usage of the waterways would also be crucial in engaging the community and educating the public on the need to keep the waters clean and have greater ownership of the nation's water resources. This paved the way for the Active, Beautiful and Clean Waters Programme ('ABC Waters'), which has the following objectives (PUB *et al.*, 2013):

- *Active*: providing new community spaces and bringing people closer to water through recreational activities;
- *Beautiful*: transforming reservoirs and waterways into aesthetically pleasing lifestyle attractions that integrate well with parks, estates and even commercial developments; and
- *Clean*: improving water quality through holistic management of water resources in slowing down runoff, keeping water clean at source, while beautifying the landscape. Aims also to minimise pollution in the waterways through public education and by building people–water relationships.

The ABC Waters Programme captures the Singapore government's desire to go beyond achieving adequate and affordable water supply, to also obtain public buy-in and increase ownership of water resources. As of January 2014, 23 projects have been completed by Singapore's national water agency, PUB.[4] By 2030, more than 1,000 potential projects will be implemented across the island. One such project, achieved through the partnership of PUB with URA and NParks, was creating access to selected waterways and water bodies. Simultaneously, PUB worked together with various sports associations to introduce activities such as sailing, dragon boating and canoeing in Singapore's reservoirs. Today, a variety of recreational activities take place in and around most of Singapore's 17 reservoirs.

4.1.3 Water management as a source of economic value

While not an explicitly stated goal of Singapore's water management strategy, Singapore has recognised environment and water as an essential sector and has backed it up with heavy investment in R&D and technology in the area. It has a goal to grow value-added contribution from the water sector from S$0.5 billion (0.3 per cent of GDP) in 2003 to S$1.7 billion (0.6 per cent of GDP) by 2015, with a doubling of jobs in the sector to about 11,000 in 2015 (PUB, 2014).

4.2 Implementation

4.2.1 Long-term, integrated planning

Singapore's implementation of its vision towards water self-sustainability relies much on long-term and integrated planning. This long-term perspective towards achieving its vision has allowed it to make decisions which may incur short-term costs, such as heavy infrastructural expenditure from the provision of proper sanitation and sufficient water, but which has long-term pay-offs.

For example, the creation of urbanised water catchments was an innovative solution borne out of the need to combine multiple uses of land for housing, commerce, transport and industry in land-scarce Singapore. Even today, in the global context, urbanised catchments are the exception rather than the norm, let alone back in the 1980s, when the notion of tapping water from urbanised catchments was considered risky given that the water run-off from urbanised areas was subject to many potential sources of pollution and contamination such as household waste and industrial discharge. Nevertheless, Singapore decided to plow ahead with urbanised catchments given the lack of feasible alternatives at that point in time, and its success can be attributed to the inter-agency coordination required to safeguard the quality of the water run-off from urbanised catchments. For instance, in the case of the Bedok catchment area, Bedok Reservoir, which was constructed from a sand quarry, was to be filled with water collected in storm water abstraction ponds constructed in the highly urbanised 'New Towns' of Bedok and Tampines. Because Bedok was a greenfield site, the URA played a key role in stipulating development and land-use guidelines from the outset, ensuring that pollutive developments were kept away from the water sources from as early as 10 to 15 years prior to the construction of the reservoir scheme. Further, the Housing and Development Board (HDB), which used the original sand quarry for its development programme, made sure to excavate the sand it required for its future projects and create a stockpile at Tampines to ensure the timely conversion of the quarry into Bedok Reservoir.

Another fruit borne out of Singapore's approach to integrated urban planning is the Marina Barrage. Completed in 2008 to form a reservoir in the heart of the city, it is an outcome of cleaning up the Singapore River, which was a process that started nearly 30 years ago in 1977. The Marina Barrage, which is a dam, was built across the Marina Channel to create the Marina Reservoir. As envisioned, today it is a three-in-one project providing water storage, flood control and lifestyle attractions. With a catchment area of 10,000 hectares, or one-sixth the size of Singapore, the Marina catchment is the island's largest and most urbanised catchment. The flagship project of the ABC Waters programme, the Marina Barrage serves as a tidal barrier to prevent high tides from advancing further inland and flooding low-lying city areas in the Marina Catchment. The Marina Reservoir in turn has become an integrated water sports hub and tourist attraction, and is conducive to water-related recreational activities such as sailing, canoeing, dragon boating and so on. Its close proximity to and synergy with other key attractions in Singapore, such as Gardens By the Bay, Marina Sands Integrated Resort, and Downtown Marina, also serves to bring vibrancy to the city centre.

Further, the integrated approach to formulating and implementing the water management strategy has allowed all government agencies to collaborate in identifying a clear vision and shared outcomes, and coordinating efforts to achieve these goals. This creates the platform for objective discussion and deliberation of trade-offs, with decisions based on overall national interest (Tan *et al.*, 2009). This is particularly important in terms of land-use planning given Singapore's limited land area and ensures that Singapore's land is optimised without compromising the nation's development needs. This is illustrated in the Water Catchment Policy, put in place in 1983, which aimed to control developments within unprotected catchment areas. The total land area that can be developed within water catchments (urbanisation cap) was capped at 34.1 per cent and a population density limit of 198 dwelling units per hectare was set on anticipated developments up to 2005. The resulting lowered intensity of development, together with strict pollution measures, allowed Singapore to collect good quality water even from these unprotected water catchments. In turn, PUB was able to adopt advanced water treatment technology to upgrade treatment plants for water from increasingly urbanised and unprotected areas. Subsequently, the urbanisation cap and population density limit were lifted in 1999, subject to continued strict water pollution control measures.

The above example highlights how planners and engineers were able to effectively collaborate in reviewing and improving policy as technology and pollution management practices evolved. The integrated approach allowed each agency to understand the issues and challenges faced by its counterparts and provided the flexibility for review and tweaking of its own plans for Singapore's overall benefit (Tan *et al.*, 2009). In the same way, the successful implementation of the urbanised catchments in obtaining good quality water can be attributed to a number of factors – understanding the nature and challenges of urbanised catchments, putting in place bold policies and long-term plans to minimise or eliminate sources of pollution, harnessing technology, and perhaps most crucially, a common vision and shared determination to succeed by various government agencies willing to work together to make things happen (Tan *et al.*, 2009).

At the same time, it is important to note that the government was careful to address water management from all parts of the value chain, not just from the supply perspective. This meant that potable water is priced with four components – the water tariff which was set to recover the full cost of production and supply; a Water Conservation Tax imposed as a percentage of total consumption which reflects the marginal cost of water supply since water is a scarce resource; a Sanitary Appliance Fee (SAF); and a Waterborne Fee (WBF). The SAF and WBF are levied to offset the cost of treating used water and for operating and maintaining the used water network. The SAF is a fixed component based on the number of sanitary fittings in each premises. The WBF is charged based on the volume of water supplied to premises, regardless of the location and how the water is used or discharged. Both charges are imposed as a tax contribution to Singapore's national used water system. This pricing strategy served to manage the demand for water and promote conservation among citizens and businesses.

In turn, the effective development of the ABC Waters programme called for urban planning, environmental engineering and landscape design, skill sets foreign to PUB engineers. To overcome this challenge, PUB engaged the skills and knowledge of the public and private sectors through Watershed Managers, who worked in consultation with other planning agencies like URA, HDB, JTC Corporation and NParks.

4.2.2 Close collaboration with the private sector to develop innovative water technologies

Given that Singapore had limited primary sources of natural water, it turned to water reclamation to supplement existing water supplies. However, used water reclamation also had its fair share of challenges. When the use of reclaimed water (also known as NEWater) was considered in the 1970s, concerns were raised relating to the reliability and costs of the technology, public health risks, safety assurances, and other technical issues, such as the presence of microbial pathogens, trace organics (e.g. pharmaceuticals and hormones) and toxic waste (e.g. cyanide) in used water. To address these questions, the PUB and the then Ministry of the Environment (ENV) conducted a comprehensive and extensive two-year study from 2000 to 2002, with an independent expert panel providing advice on the study. The exercise revealed that NEWater quality was well within the drinking water standards, and as such was safe for potable use. In fact, it was established that NEWater is of higher quality than normal tap water – it was clearer and more sparkling (due to the presence of lower levels of minerals and organic substances as compared to water from river or reservoir sources). Because of the higher purity of NEWater, NEWater has actually helped industry users, for example wafer fabrication plants increase operational savings, since the wafer fabs could save on spending money and chemicals treating the water to reach ultra pure water standards.

However, the wafer fabrication industry's acceptance of NEWater did not happen overnight. The reason was simple – at the time, no other wafer fabrication facility in the world had ever used recycled water for its processes. To address these concerns, PUB built a 'plug and play' pilot plant that could simulate 'real' conditions at the wafer fabs, and over the 18 months when the plant was running, the ultra pure water produced from the plant was tested for its quality. At the end of this period, the wafer fabs were satisfied with the reliability and security of NEWater. In fact, the wafer fabs found that they achieved 20 per cent savings in chemical costs for UPW production by using NEWater instead of potable water (Tan and Seah, n.d.).

While the demand risk for NEWater was then addressed, there remained a certain level of technological risk given the infancy of the NEWater production process. To address this, PUB took the lead in developing Singapore's first NEWater plants. This served to facilitate collaboration with global water companies (e.g. GE Water, Veolia Water and Siemens Water Technologies) as well as local companies (e.g. Hyflux) to implement various technologies at the NEWater plants. By working with PUB onsite, the private sector was able to gain confidence in the new technologies, as well as build capability in operating a NEWater plant. This subsequently paved the way for private sector participation in NEWater supply via a Public-Private Partnership (PPP) approach. The private sector was allowed to design, build, own and operate (DBOO) subsequent NEWater plants, and would produce and supply NEWater to PUB based on a tendered price. This not only freed PUB to acquire NEWater at a stipulated level of quality and at the most cost-effective price, but also served to encourage innovation in the private sector towards greater production and operational efficiency, thus facilitating the long-term development of the local water industry into one which was innovative, vibrant and export-oriented.

4.2.3 Positioning Singapore as a global hydrohub

Given Singapore's success at strengthening its water supply through comprehensive policies and the application of new technologies, the country is increasingly seen as a model of sustainable water management. This has facilitated its positioning as a 'global hydrohub' with a thriving water ecosystem. The efforts of building Singapore as a global hydrohub are spearheaded by the Environment and Water Industry Programme Office (EWI) since the government identified the environment and water industry as a strategic growth area in 2006. With funding of S$470 million from the National Research Foundation (NRF), the interagency setup led by PUB and involving many other government agencies such as EDB, SPRING Singapore, and

International Enterprise (IE) Singapore, has put forth a number of initiatives into three strategic thrusts, namely Cluster Development, Capability Development and Internationalisation.

Cluster Development focuses on bringing in overseas companies, growing local water 'champions' as well as proliferating start-ups to create a vibrant water ecosystem, which will in turn directly contribute to the economic value added and employment of the water sector. Singapore's water industry has experienced a rapid growth from a small cluster of around 50 water companies in 2006 to over 130 international and local water companies.[5] The number of public and private R&D centres has also increased from 3 to 26 over the same period. There is in place a strong ecosystem of water management players comprised of public sector agencies, the private sector, academia, hybrid organisations such as PUB's Waterhub, and citizens who play an important role in water conservation. While there have been both push (Singapore's small and limited market) and pull factors (demand from overseas) resulting in Singapore's exporting of water management expertise outside its borders, this would not have been successful without the close collaboration and partnerships existing between these players.

In particular, the relationship between the private sector and academia, though relatively new in the Singapore context, is surprisingly cohesive as well. Companies such as Hyflux and GE Water work closely with academia and their research counterparts in developing applications and test-bedding, providing input into curriculum, jointly organising and running seminars, and offering internships to students, among others. For instance, Mr Foo Hee Kiang, Hyflux's Advisor to the CEO, sits on the programme advisory committee of NTU's (Nanyang Technological University) School of Civil and Environmental Engineering, while GE Singapore Water Technology Centre, together with NERI (NUS Environmental Research Institute), has jointly launched the Visiting Scientist Scheme that aims to invite international water experts to NUS (National University of Singapore) to share their expertise, knowledge and experience (NERI, 2014). Under this scheme, a prominent scientist is nominated by NUS and sponsored by GE to spend a week at NUS where both the NUS research community and GE colleagues would have the opportunity to interact with the visiting scientist for learning and exchange. In turn, the universities and research institutions are committed to maintaining a regular dialogue with industry players in relation to pipeline projects and research with a view to commercialise. Though these academic stakeholders note the common challenges associated with academia–private sector partnerships (e.g. differing goals of publishing scientific papers versus profitable products; different expectations in terms of time-to-market), they are actively seeking to find a win-win solution that aligns the interests of both sectors.

Capability Development is the key growth driver aimed at developing innovative and commercially viable technologies, building up research capabilities and grooming talent in the local water ecosystem. A series of Funding Initiatives (FIs) have been put in place to support R&D activities from upstream basic research to downstream commercialisation and adoption. To date, more than 100 R&D projects have been funded, including the development of breakthrough technologies such as biomimetic and biomimicry membranes, and over 140 test-bedding projects involving over 70 companies have been hosted at PUB installations which cements Singapore's position as the 'living-lab' for innovative water solutions. In line with this strategic thrust, the government has been increasingly providing funding support for research. For example, NERI, together with PUB, is jointly developing the New Smart Water Assessment Network (NUSwan), a water quality-monitoring tool that builds upon aesthetics (coming in the shape of a swan) and recreation (NERI, 2014). This is an idea-to-market concept that has already attracted some overseas water companies, which are interested in performing the technology transfer and customising the tool for overseas water environments.

Internationalisation helps Singapore-based companies secure their foothold and expand their businesses overseas. The government, in particular, has been named by stakeholders as critical in opening doors to foreign markets. For example, government agencies such as PUB often introduce private sector companies to foreign delegates who come to Singapore to learn about water management policies and strategies, and representatives from the private sector also accompany local ministers during trade missions overseas to lend credibility and expertise to the team. In another instance, when local water company Hyflux completed its first desalination plant SingSpring in 2005, its launch coincided with the International Desalination Association (IDA) World Congress, which was jointly organised by PUB and IDA and held in Singapore. At the same time, the government has been careful to keep the market competitive by taking a merit-based approach and providing a level playing field for both local and global companies, understanding that this is necessary in order for Singapore to remain competitive.

Another initiative aimed at sharing knowledge and profiling Singapore's capabilities is the Singapore International Water Week (SIWW), the global platform for sharing and co-creation of innovative solutions. Launched in 2008, it is an annual gathering where public officials, industry leaders, water experts and practitioners meet to engage in dialogue regarding policies, business solutions and water technologies. It aims to be a global platform for water solutions by showcasing best practices in water management and successful application of water technologies. During the event, the Lee Kuan Yew Water Prize, a prestigious international award, is presented to individuals or organisations in recognition of outstanding contributions in solving water-related challenges.

4.3 Results

4.3.1 NEWater paving the way for an integrated water cycle

Since NEWater was launched in 2003, many industrial and commercial facilities have replaced potable water with NEWater for operational usage (e.g. cooling towers, general washing, processes, etc.) and this has served to release PUB potable water supply for potable use. Further, to encourage companies to look to manage their water demand through recycling more efficiently, the 2007 Water Efficiency Fund was put in place to provide financial incentives as well as support in defraying part of the initial capital outlay required and obtaining commercial funding. As a result, wafer fabrication plants, one of the larger consumers of water, have managed to cut their water demand by more than half through increased efficiency in recycling water.

With the entrance of the private sector, NEWater production expanded quickly, with economies of scale, productivity gains and more effective treatment technologies bringing down the cost of NEWater. With the recycling of used water to produce NEWater, PUB has effectively closed the water loop – NEWater IPU can not only be increased when needed, but is also independent of rainfall fluctuations, and can be used to top up Singapore's reservoirs during dry months, thus boosting Singapore's resilience in its water supply against periods of low rainfall or drought. During the prolonged dry spell which hit the nation in mid-January 2014, about 35 mgd of NEWater was injected into the reservoirs to maintain stock levels. The NEWater is blended with the raw water before undergoing conventional treatment at the waterworks for supply to the public. The closed water loop means that PUB could now manage all aspects of Singapore's water cycle in an integrated manner.

4.3.2 Sharing and exporting expertise overseas

Singapore's success in utilising new water technologies such as water reclamation (NEWater) and desalination to boost its water supply has gained international recognition. For instance, Harry Seah, PUB's Chief Technology Officer, serves as an advisor on the Queensland Expert Advisory Panel in Australia. Singapore also assisted the Maldives in its recovery efforts in the aftermath of the tsunami of 2004. To create a sustainable supply of drinking water for the Maldives, PUB closely collaborated with private Singapore water companies such as Keppel Integrated Engineering, GrahamTek and Hyflux to send water desalination plants and water generators to the country.

Further, Singapore's positioning as a global hydrohub means that the country is now home to approximately 130 local and international water companies. Many of these Singapore-based water companies have successfully exported water expertise to key overseas markets such as China and the Middle East, and have secured more than 100 international projects worth close to S$10 billion between 2006 and 2012. Investments by water companies during this period, when fully realised, are expected to contribute S$590 million of annual value added to the Singapore economy and generate 2,300 professional and skilled jobs. In turn, S$470 million has been committed to fund water research (Reuters, 2013).

Singapore's research institutions in the area of water management such as NERI and NEWRI are also becoming leading players in the academic world, with the National University of Singapore being ranked by Lux Research as being the top water research institute worldwide in 2013 (NUS, 2013). The university's focus on desalination, reuse and membranes has also seen it topping these areas of specialisation globally. This positioning of local universities as academic leaders on an international level has led to the attraction of many foreign companies and academic institutions to Singapore to leverage on the knowledge and expertise present in the city-state, thus further enhancing the growth and prominence of the local water sector. For instance, in 2009, Toray Industries, Inc., a multinational corporation headquartered in Japan and the world's largest producer of carbon fibre, announced the establishment of Toray Water Technology Laboratory in Singapore to conduct research and development of water treatment technologies, with NEWRI chosen as the choice partner for its research activities in Singapore (Toray, 2009). The Singapore-Delft Water Alliance (SDWA), now known as the Sustainable Development & Water Alliance, was also launched in 2007 as a multinational, interdisciplinary research centre of excellence for water knowledge involving PUB, NUS and Dutch research institute Deltares through an initiative of NRF. The alliance leverages on both Singapore and the Netherlands' complementary strengths in water management which has spawned a series of water-related research programmes. Building on the achievements of SDWA, a new alliance NUSDeltares was established in 2012, with the intention to bring scientific impact to address societal needs in Southeast Asia (NUS, 2013).

5 Conclusions

5.1 Broad assessment

Singapore's approach to environmental management, including its treatment of water, can be summed up by the 5Es of Engineering, Economics, Education, Enforcement and Engagement (Tan *et al.*, 2009):

- *Engineering*: Singapore has invested significantly in environmental infrastructure such as water production capacity, as well as in technology improvements and the development of human capital.
- *Economics*: This ensures the efficient production of services and the appropriate pricing of limited environmental resources, without encouraging over-consumption of these resources or services by providing subsidies.
- *Education*: This relates to helping people change their behaviour and mindsets towards environmental conservation for the common good. For instance, NEWater recently won the 'Water for Life' United Nations Water (UN-Water) Best Practices 2014 award in the participatory, communication, awareness-raising and education practices category.
- *Enforcement*: This reinforces education by ensuring that consequences are put in place for irresponsible behaviour.
- *Engagement*: Finally, Singapore recognises that a sustainable approach to environmental protection requires the engagement of all stakeholders to take ownership of their environment; the responsibility does not and cannot lie with the government alone.

In turn, Singapore's specific success in the area of water management has enabled it to export its expertise overseas. The city-state's effective addressing of a national crisis has providently placed it on the forefront of an arena that is growing in demand worldwide, well positioning it to enjoy sustained competitive advantage.

5.2 Critical success factors

There have been some critical factors that have led to Singapore's success in its water management strategy as well as leveraging on it for economic value added for the nation, which can be largely attributed to its policymakers and the close collaboration among various government agencies and beyond:

- *Singapore's holistic approach to water management*, that is pragmatic (a matter of survival) but also looks further to using its success and lessons learnt in creating economic value and impacting the city brand by positioning itself as a global player in the water industry.
- *Long-sighted integrated planning and the strong coordination between government agencies* have made the 'Four National Taps' and ABC Waters programmes possible. This is linked to a strong common vision and ownership of the role and importance of water to the nation.
- *The government paving the way in innovation through new water technologies* (e.g. NEWater) by bearing the initial demand and technological risks, and subsequently easing the transition to privatisation through sharing knowledge and expertise with the private sector.
- *'Putting money where the mouth is'* – sending the signal of strong support in the water sector through heavy investments in R&D and technology and supporting initiatives.

All this has been underpinned by a strong political will on the part of the top national leaders, perhaps driven more by necessity than economic benefit. This however may be a unique feature of being a city-state where setting goals and managing resources at different local, regional and national levels, as would be the case for most other cities, becomes a non-issue for Singapore. This is further magnified when considering the current lack of a real political opposition party in Singapore (though this may change in the future), which allows the ruling party to develop and enforce long-term strategies, in the context of a still hierarchical culture that more easily tends to follow rather than oppose those at the top. Nevertheless, other cities around the world can learn from Singapore's thinking and planning with a multi-faceted and long-term perspective, having the foresight to facilitate and even drive the development of unconventional and new technologies and emerging industries.

Notes

1 Singapore is also home to two R&D science and knowledge 'locations', namely the Biopolis (which deals with biomedical sciences) and the Fusionpolis (which covers information and communication technologies (ICT), media, physical sciences and engineering). It is also in the midst of developing a third hub called Mediapolis, which, as the name suggests, focuses on media.

2 This 'lack of silo' approach to planning is thus manifested in integrated urban solutions such as Marina Bay, which is envisioned as an environmentally friendly mixed-use precinct that also serves as a freshwater reservoir that will add to local water supply by 10 per cent.

3 One of the first fruits which has been borne from this initiative is the collaboration between PUB and Toray Industries, Inc., which will involve both parties co-developing water treatment-related technologies and products that will aid in providing solutions to the anticipated global water shortage.

4 PUB (previously known as the Public Utilities Board) is a statutory board under the Ministry of the Environment and Water Resources. It is the water agency that manages Singapore's water supply, water catchment and used water in an integrated way.

5 These companies span the entire water value chain from upstream component players like membrane and pumps' suppliers, midstream Original Equipment Manufacturers (OEMs) and system integrators, to downstream Engineering, Procurement and Construction (EPC), Design-Build-Operate (DBO) and Build-Own-Transfer (BOT) players.

References

Adam, S. (2010). Singapore economy expanded 17.9 per cent in first half, growth May ease. Retrieved 2014 from Bloomberg: http://www.bloomberg.com/news/articles/2010-08-08/singapore-s-economy-expanded-17-9-in-first-half-growth-to-cool-lee-says

Chen, S. (2013). Singapore growth quickened in 2013 as Lee pursues economic shift. Retrieved 2014 from Bloomberg: http://www.bloomberg.com/news/articles/2013-12-31/singapore-s-economy-grew-3-7-in-2013-prime-minister-lee-says

Economic Development Board (2012). Monthly manufacturing performance November 2012. Retrieved 2014 from: https://www.edb.gov.sg/content/edb/en/news-and-events/news/news-archives/2012/monthly-manufacturing-performance-november-2012.html

Economic Development Board (2014). Home strategy. Retrieved 2014 from: https://www.edb.gov.sg/content/edb/en/why-singapore/about-singapore/strategy/home-for-innovation.html

Economic Strategies Committee (2010). Retrieved 2014 from Key Recommendations to the Singapore Government: http://app.mof.gov.sg/data/cmsresource/ESC per cent20Report/ESC per cent20Full per cent20Report.pdf

Hawksford (n.d.). Singapore economy – a brief introduction. Retrieved 2014 from: http://www.guidemesingapore.com/incorporation/introduction/singapores-economy

Ho, P. (2014). Preparing for the Future. Retrieved 2014 from Civil Service College Singapore: https://www.cscollege.gov.sg/Style per cent20Library/Ethos per cent20Previous per cent20Issues/05Jul/04Prepare.pdf

Lee, H. L. (2013). Summit Keynote Address. Retrieved 2014 from Singapore Summit Conference 2013: http://www.singaporesummit.sg/system/application/views/website/pdf/Singapore per cent20Summit per cent202013 per cent20Post-Summit per cent20Report.pdf

Lee, K. Y. (2000). *From Third World to First: The Singapore Story: 1965–2000*. New York: HarperCollins.

Magnanti, T. (2012). President's Message – Establishing the Foundation of a New University. Retrieved 2014 from Singapore University of Technology and Design: http://www.sutd.edu.sg/presidents_message.aspx

Ministry of National Development (2002). From Garden City to City in a Garden. Retrieved 2014 from: http://www.mnd.gov.sg/MNDAPPImages/About per cent20Us/From per cent20Garden per cent20City per cent20to per cent20City per cent20in per cent20a per cent20Garden.pdf

NERI – NUS Environmental Research Institute (2014). NERI and GE launched visiting scientist scheme. Retrieved 2014 from National University of Singapore Environmental Research Institute: http://www.nus.edu.sg/neri/Urban_Water_Management.html

NUS – National University of Singapore (2013). NUS emerges top in world water research. Retrieved 2014 from: http://newshub.nus.edu.sg/headlines/1305/water_14May13.php

OECD – Organisation for Economic Co-operation and Development (2010). *Strong Performers and Successful Reformers in Education: Lessons from PISA for the United States*. Paris: OECD.

PUB (2014). The Singapore Water Story. Retrieved 2014 from: http://www.pub.gov.sg/water/Pages/singaporewaterstory.aspx

PUB, Atelier Dreiseitl Asia, Black & Veatch, CH2M HILL and CPG Consultants Pte Ltd (2013). 'ABC WATERS' entry – for the Global Water Awards 2013. Retrieved 2014 from: http://www.siww.com.sg/sites/default/files/ABC_Waters_entry_forGWA2013.pdf

Ranasinghe, D. (2014). In Singapore? Odds are, you're looking for a new job. Retrieved 2014 from CNBC: http://www.cnbc.com/id/101314663

Reuters (2013). Singapore's water companies aim to quench China's $850 billion thirst. Retrieved 2014 from: http://www.reuters.com/article/2013/06/02/singapore-water-idUSL3N0EA13020130602

Sedghi, A. (2012). The world's top 100 universities 2012 – how the Times Higher Education ranks them. Retrieved 2014 from *Guardian*: http://www.theguardian.com/news/datablog/2012/oct/03/world-top-100-universities-2012

Tan, P. and Seah, H. (n.d.). Impact of NEWater as feedwater for the production of ultra-high-purity deionised water and manufacturing process. *Future Fab International 16 – Process Gases, Chemicals, and Materials*.

Tan, Y. S., Lee, T. J. and Tan, K. (2009). *Clean, Green and Blue: Singapore's Journey Towards Environmental and Water Sustainability*. Singapore: ISEAS Publishing.

Toh, K. (2012). Singapore best in Asia for higher education. Retrieved 2014 from *Jakarta Post*: http://www.thejakartapost.com/news/2012/05/15/singapore-best-asia-higher-education.html

Toray (2009). Toray to establish Water Treatment R&D Center in Singapore – NTU is chosen as its first research partner in Singapore. Retrieved 2014 from http://www.toray.com/news/rd/nr090624.html

UNESCO – United Nations Educational, Scientific and Cultural Organization (2002). *1st United Nations World Water Development Report*. UNESCO.

World Economic Forum. (2014). *Global Risks 2014 Report*. Retrieved 2014 from http://www.weforum.org/reports/global-risks-2014-report

12 World Design Capital (Helsinki)

Alexander Otgaar and Luís Carvalho

1 Introduction

In this chapter we discuss the strategy of Helsinki to become a design-driven city, using the World Design Capital (WDC) as a catalyst for change. The key element of this strategy is the concept of an 'open city' that facilitates the participation and involvement of designers and end-users in the co-creation of products and services that make cities better places to live.

The strategy responds to several trends in urban policies and regional innovation strategies. First of all, we observe a growing interest in the use of (global) events for strategic purposes. Cities increasingly recognise the potential mobilising power of major sports, cultural and business events. The aim of hosting such events is not only to attract visitors and to get media attention, but also to secure post-event legacy such as long-lasting improvements in infrastructure and socio-economic conditions. WDC is not only an event, but also a 'city promotion' tool that highlights the achievements and progress of a city in this particular field. Obviously the event not only promotes the city but also the industry, in a similar way as the Barcelona's World Mobile Capital event (see Chapter 6).

Second, the concept of an open, design-driven city meets the growing demand for new forms of urban governance, planning and development in which hierarchical, top-down structures are at least partly replaced by bottom-up initiatives of, and networks between governments, businesses, knowledge institutions and citizens (end-users). It relates to popular concepts such as participatory planning (involving the community in designing places), living labs (involving end-users in designing products or services), open innovation (firms involving other external actors in innovation) and open data (sharing data to promote participation and to stimulate innovation, e.g. in the design of products or services).

The idea of a design-driven city is relatively new. As we will explain in the present chapter, Helsinki introduced this concept in its bid for WDC 2012. The International Council of Societies of Industrial Design (ICSID), which leads the WDC programme, has adopted the concept. For example, in December 2012 ICSID invited 'design-driven cities to bid for the World Design Capital (WDC) 2016 designation'. A quick search on the internet shows that several cities are described as 'design-driven' – including Seattle, Seoul, Cape Town (WDC 2014), Milan, Portland, Copenhagen and Montreal – however without a clear definition of what a design-driven city is. In this chapter we use Helsinki's definition, which is more or less 'using design for every aspect of urban life'. This implies that the demands and desires of the end-user have to be taken into consideration, not only in designing physical places, but also in designing services. As executive director of WDC Helsinki 2012 Pekka Timonen said in an interview: 'a design-driven city is a user-centric city' (Focus Finland, 2012, p.10).

Before we discuss Helsinki's strategy to become a design-driven city using WDC Helsinki 2012 as a catalyst we first explain the context in which this strategy has been developed. In Section 2 we assess the city's performance and identify contextual factors that influence this performance also considering recent changes. We particularly focus on context variables that are considered relevant for the rise of the design-driven city: not only changes in the economic and demographic structure, but also changes in governance and how the city and the region develop spatially. In Section 3 we present the city's vision and strategy, and we explain what actions relate to design. In Section 4 we arrive at the very essence of this case study answering the questions what the project is all about, how it has been or will be implemented and what were/will be the enablers (Section 5) and what direct and indirect results have been realised or can be expected (Section 6). The lessons and challenges ahead conclude the case study in Section 7.

2 City profile: capitals for sustainable competitiveness

2.1 Performance

Helsinki is the largest city, capital and undisputed economic centre of Finland. The metropolitan area, which consists of 14 municipalities, accommodates 1.4 million people, almost one-fourth of the country's population. Over the entire year of 2012, the population climbed by 17,200 people (City of Helsinki Urban Facts, 2013). This region represents one-third of the Finnish economy. With a share of 40 per cent in investments in R&D, Helsinki is also the knowledge and innovation champion of Finland. Expenditures for R&D accumulate to 4.5 per cent of the regional GDP, which is higher than the 3.7 per cent on a national level (City of Helsinki Urban Facts, 2013).

Comparing Helsinki with other metropolitan areas in Europe, growth has been stronger than average (Laakso and Kostiainen, 2011). Between 1980 and 2009 the number of jobs increased by 1.1 per cent per annum compared to an average 0.6 per cent per annum for all areas. In the same period output grew at an annual rate of 3.7 per cent compared to 2.0 per cent. During this period of steady growth, Helsinki experienced ups and downs more intensely than most other cities. After the severe crisis in the early 1990s (mainly caused by the disintegration of the Soviet Union, Finland's main trade partner at that time), the city entered a stage of rapid growth spurred by the ICT revolution (with Nokia as leading firm). With the burst of the ICT bubble in the first years of the new millennium Helsinki was again hit by a strong recession.

The 2008 credit crunch affected Helsinki too, but not as severe (yet) as other recessions in the past. In 2009, Finnish exports declined dramatically resulting in an 8 per cent drop in output. Because the Helsinki region is strongly oriented towards services for business and households – and less to manufacturing and construction – employment declined less than 2 per cent only. Production grew from 2010 until the second quarter of 2012 but declined slightly at the end of that year.

In terms of population Helsinki has experienced growth since the 1980s and further growth is expected. Since the year 2000, the population has been growing at an average rate of 1 per cent, partly due to a positive migration balance (explaining 50 per cent of the growth) and partly because of above-average natural growth. For the coming decades ageing will cause the dependency ratio[1] to rise from 0.40 in 2011 to 0.54 in 2030 (Manninen *et al.*, 2012).

2.2 Economic structure and knowledge

In comparison with other Finnish regions, the Helsinki region is specialised in the production and distribution of knowledge, arts and culture, financial services, scientific research,

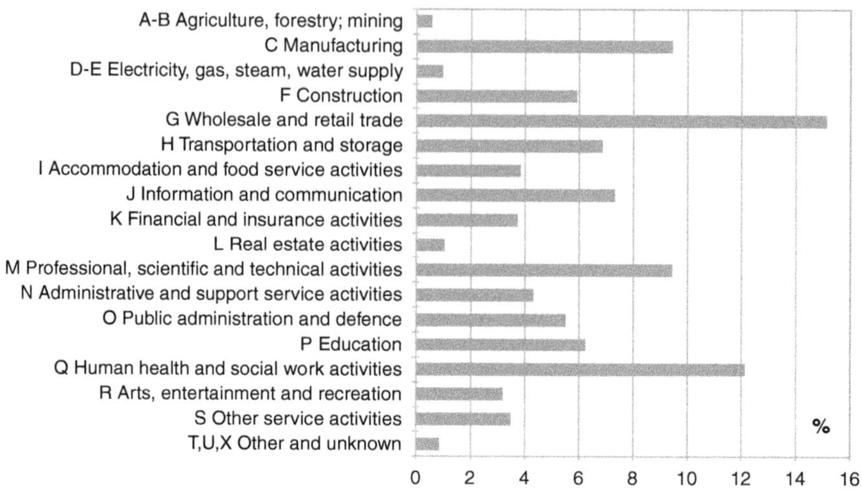

Figure 12.1 Employment in the Helsinki region (2010, NACE Rev2)

Source: Helsinki Urban Facts

wholesale trade and air transport. As indicated in Figure 12.1, the manufacturing industry provides less than 10 per cent of the jobs. In general the economic base can be characterised as knowledge-intensive and strongly oriented towards services. One of every ten workers is active in knowledge-intensive services (professional, scientific and technical activities); almost half of the national employment in this sector is concentrated in the Helsinki region. About 80 per cent of all jobs are in the service sector. It is also worth mentioning that 44 per cent of the workforce of Helsinki Region has completed a degree at tertiary level, and there are three universities in this region: the University of Helsinki, Aalto University and the University of the Arts Helsinki.

How many people work in design? Statistics give us some idea, as we can add the number of jobs in particular design-intensive sectors such as architecture, advertising, interior design, media production, fashion, urban planning and industrial design. In 2007, this core of design-industry professionals counted more than 10,000 members (see Table 12.1). In some of the sectors we can observe a strong growth in employment between 2003 and 2007,

Table 12.1 Design professionals in the Helsinki region

Sector	*Number of jobs 2007*	*Growth 2003–7*
Architectural services	2,151	+ 22.6%
Advertising agencies	3,747	+ 22.7%
Applied arts and fashion design	705	+ 46.3%
Urban/community planning	1,397	+ 30.2%
Industrial design	235	N/A
Interior design	288	N/A
AV production (film, television, video)	1,745	N/A
TOTAL	10,268	N/A

Source: Open Helsinki, WDC 2012 Application (2009)

up to 46.3 per cent for applied arts and fashion design, 30.2 per cent for urban/community planning, 22.7 per cent for advertising agencies and 22.6 per cent for architectural services. In its bid book for World Design Capital the city indicates that the actual number of people working in design-intensive industries is much higher (up to 200,000), using Richard Florida's definition of the creative class (Florida, 2002). Although this definition is debatable it is fair to say that Helsinki is a 'creative city'. In Finland, about one-third of all jobs in the creative industries are to be found in the capital region.

2.3 Social conditions

In comparison with other cities the social and living conditions in Helsinki are very good. In the egalitarian society of Finland income inequality is relatively low and almost all citizens have access to basic services such as affordable housing, healthcare and education. In Helsinki there are no neighbourhoods with high levels of deprivation. The main reason is that the City owns all the land and that housing policies – on a local and national level – provide strict guidelines for developers, enforcing the development of mixed neighbourhoods (rental and owner-occupied houses). Local policies state that in the development of new residential areas, 20 per cent of the housing must be state-subsidised rental housing, while an additional 40 per cent is regulated under HITAS: a special programme for affordable owner-occupied houses with regulated prices and qualities.[2] Another relevant factor is the City's policy to invest more money in the social infrastructure of neighbourhoods that score low in social rankings (positive discrimination).

Despite all this, policymakers are concerned about social segregation. Their main worry is that a growing number of citizens are not able to catch up with the emerging knowledge economy. The transition to a knowledge-based economy in Helsinki has led to increasing unemployment and income differences between higher and lower educated people (Vaattovaara, 2002; Vaattovaara *et al.*, 2010). We can also observe growing health inequalities: higher-educated people live between four and eight years longer than lower-educated people (Health Centre, 2007). Socio-economic problems tend to concentrate in particular areas, notably in eastern and north-western parts of the city.

The population density of the Helsinki region is relatively low compared to other major cities. The population is growing rapidly, for example in Helsinki the population increased by 8,600 persons in 2012. In the last decades the city has been confronted with urban sprawl and an expansion of the commuter region. Recently, however, the city has decided to redevelop old harbour areas – strategically located in or near the city centre – into residential areas, which will cause the population density to grow.

2.4 Institutional conditions

The political system of Finland has only two layers of government with elected representatives: the state and the local governments. In between there are regional councils whose main task is to promote inter-municipal cooperation. Municipalities are highly autonomous as they have the right to levy taxes (on income and corporate profits among others) and the right to develop and implement spatial plans (e.g. concerning residential areas). Municipalities use the income from taxes to deliver community services such as education, healthcare, social welfare, cultural, environmental and technical infrastructure and water supply. Local governments take care of about 30 per cent of all public expenditures in Finland. They may choose to deliver the services on their own or in cooperation with other local authorities and/or service providers.

The City of Helsinki is governed by the City Council whose members (85 in total) are elected by the population every four years. The daily management of the city is in the hands of the City Board that counts 15 members elected by the City Council. The Mayor and the four Deputy Mayors are responsible for the functioning of several departments and committees, such as the Economic and Planning Centre, the City Planning Committee, the Social Services and Health Department, the Education Department, the Cultural Office and Urban Facts.[3] In total the City of Helsinki employs 39,000 people, making it the largest employer of Finland.

As many issues go beyond the administrative boundaries of the municipality (e.g. transport, planning, the environment, specialised health services, marketing, housing, etc.), the City of Helsinki is involved in various inter-municipal partnerships. Examples are the Helsinki Region Transport Authority (HSL), the Helsinki Region Environmental Services Authority (HSY), the Hospital District of Helsinki and Uusimaa (HUS) and Greater Helsinki Promotion (GHP). In 1995 the Regional Council of Uusimaa and the cities of Helsinki, Espoo and Vantaa created Culminatum Innovation Oy Ltd: a platform for triple-helix cooperation involving knowledge institutions and businesses. From the very beginning the aim of this development corporation has been to promote the international competitiveness of the region. One of its main tasks is to implement the national Centre of Expertise Programme for the Helsinki region, targeting nine clusters: 'digibusiness', environmental technology, food processing development, health, 'living business', nanotechnology, tourism, ubiquitous computing and well-being. Design is not explicitly mentioned, but plays an implicit role in many of these clusters. It is particularly relevant in the living business cluster, which focuses on 'user-friendly living solutions' considering both the exterior and interior of urban houses. Many activities of Culminatum Innovation are financed through Tekes, the Finnish Funding Agency for Technology and Innovation.

Another relevant platform for triple helix cooperation is Forum Virium Helsinki Ltd. This platform was set up in 2005 by several anchor companies (Nokia being the most well known) in cooperation with 'partner companies' (including global firm such as IBM, CGI and Siemens) and public authorities (the City of Helsinki, but also national innovation funds such as Tekes and Sitra). The shares of Forum Virium Helsinki are 100 per cent owned by the City of Helsinki, but in contrast with Culminatum Innovation, it is a membership organisation. Private-sector members pay an annual fee (€20,000), which gives them a seat on the board. Its mission is to develop new digital services in a user-centric approach, involving the end-user in the design of these services, particularly by means of 'living labs'. Projects typically involve several companies willing to cooperate in pre-competitive prototyping. With a staff of 30 persons, Forum Virium Helsinki generates an annual turnover of around €3.5 million. The City of Helsinki subsidises the platform with a yearly €800,000 contribution from the Helsinki Innovation Fund. Forum Virium Helsinki is an active member of the European Network of Living Labs, which was founded in 2005 on the occasion of Finland's EU presidency.

2.5 *Conclusions: main issues*

In the last two decades Helsinki grew faster than many other European cities, in terms of output but also in terms of population. The post-2008 recession is not as deep as previous recessions from which the city seems to have recovered relatively fast. After the ICT crisis the city has successfully started the implementation of an economic diversification strategy through the action of well-coordinated innovation platforms and partnerships. Not only on a local level but also on a national level, the design industry – in its broadest definition – has been identified as a potential source of future job growth.

Expectations are that the population of Helsinki will grow considerably in the years to come, due to a combination of natural growth and migration. The new residential areas in the former harbours of the city will accommodate a large part of this growth. This massive redevelopment programme provides an opportunity for the city to design and develop new types of neighbourhoods and houses, responding better to the demands and needs of the users. Other important socio-demographic developments are the changing age structure and the growing concerns about social segregation. Helsinki has used World Design Capital Helsinki 2012 not only to promote the design industry, but also raise awareness about the use of design in all aspects of urban development.

3 City-wide vision and strategy

3.1 Vision

The City of Helsinki has formulated the city-wide vision in its Strategy Programme 2009–2012 as follows:

> As the capital city and the focus of the area Helsinki is an evolving international business and innovation centre relying on science, art, creativity and learning as well as good services. Helsinki's success will benefit the residents' wellbeing and the whole country. The Metropolitan Area will be developed as a uniformly operating area surrounded by nature, i.e. a good place to study, work and do business. (City of Helsinki, 2009)

The municipal vision articulates the objectives to be realised during the council term 2009–2012. In view of our case study it is particularly interesting to see how this comprehensive vision has been translated into strategic objectives related to design and user-involvement. One of the aims is, for example, to provide user-oriented municipal services and to strengthen residents' participation and responsibility. Other objectives are to make the city 'attractive, functional and enjoyable', to create 'interesting, beautiful and functional urban residential areas and environments' and to provide educational services that people need (City of Helsinki, 2009).

The City's strategy is well aligned with the metropolitan area's competitiveness strategy that presents the vision of a prosperous metropolis (Culminatum Innovation, 2009). The common vision for the region is nearly equal to the municipal vision; the only difference is that it doesn't identify Helsinki as the focus of the region. One of the focal points is 'to strengthen user-driven innovation environments'. The strategy also highlights the need to 'build good quality of life and a pleasant and secure living environment', 'to improve top-quality education and know-how' and to 'connect the metropolitan area to international networks'. Prosperous metropolis can be seen as the follow-up of the 2005 innovation strategy (Culminatum Innovation, 2005), which recognises the need to combine technology with art, design and quality of life.

3.2 Strategies

Both on a local and regional level the strategic objectives formulated above have been translated into more specific policy intentions and 'action lines'. Below we identify the five action lines in the prosperous metropolis strategy (2009) we consider most relevant in view of the ambitions to create a design-driven city:

- Improving public transport connections between the university campuses and enhancing students' housing conditions with the aim to create an integrated campus network.
- Developing new highly attractive residential waterfront areas: Jätkäsaari, Kalasatama, Östersundom and Kruunuvuorenranta.
- Developing Helsinki city centre as a meeting place and university campuses as user-driven innovation environments.
- Encouraging demand- and user-driven innovation activities in public procurements (Culminatum Innovation runs a special programme on this issue).
- Promoting the region as a forerunner in open user-driven development environments (with GHP as leading actor).

As we will explain in the next section, the strategy to become a design-driven city has not yet been adopted (at the time of writing, 2013) as an official strategy of the city or the region. It is the ambition expressed by WDC Helsinki 2012. What we can see, however, is that the city-wide vision and strategy recognise some objectives that could be seen as part of such a strategy: the development of user-driven innovation environments, the creation of attractive urban areas and the provision of user-friendly services.[4]

4 The project: WDC 2012

4.1 Vision and content

Helsinki has used the World Design Capital (WDC) Helsinki 2012 event to promote the embedding of design in every aspect of urban development. This vision is clearly related to the concept of an 'open city'. As put in Helsinki, an open city implies that 'ideas, thoughts and people can move freely without unnecessary creativity-hampering obstacles' and 'open data' (sharing information) is part of this conceptual approach.[5] Important features of 'embedded design' are user orientation and co-creation, thus requiring the active involvement of experts and end users in the development of products and services. As we will explain later on, Helsinki considers design as a philosophy to develop a better city that responds more accurately to the demands and needs of its users.

The idea of hosting the event emerged in 2008 as an initiative of the Cultural Department with support of the Mayor, but it quickly developed into an umbrella for 580 projects residing under and coordinated by the WDC office and carried out by numerous actors in the cities (Helsinki, Espoo, Vantaa, Kauniainen and Lahti). Before we take a closer look at the features of WDC Helsinki 2012, it is essential to understand how and why this initiative emerged, also considering what happened before 2008.

4.2 Brief story of the initiative

World Design Capital Helsinki 2012 has to be seen as the result of a chain of previous strategies and events. The deep recession of the early 1990s stimulated leaders from municipalities, universities and businesses to redefine the future of the Helsinki region. Through strong local cooperation efforts and support of the national government (Tekes) Helsinki managed to turn itself one of Europe's leading regions in ICT during the second half of the 1990s (van den Berg *et al.*, 2005).

During this decade Helsinki also succeeded in becoming a more international city, partly due to Finland's EU membership (1995). As part of this 'internationalisation strategy' the City successfully applied for European Cultural Capital 2000, with several annual festivals as a long-lasting legacy. Although the number of events initially declined in the years after 2000 (the hangover effect), more events were organised some years later. Some say that the Cultural Capital has also motivated grassroots movements of citizens organising events by themselves, such as the increasingly popular Restaurant Day twice a year during which one-day restaurants 'pop up' everywhere in the city.[6]

After the burst of the ICT bubble in 2000/2001 key actors realised that overspecialisation (in ICT) was a serious threat and more diversification was needed. Discussions resulted in a new vision (in the year 2005, see Section 3) with the ambition to develop the Helsinki region into a world-class place for innovation, combining technology with art and design. This vision was translated into a regional innovation strategy (Culminatum Innovation, 2005) that specified the aim to create 'common development platforms': settings in which technology/service developers and end-users come together. This strategy clearly got its inspiration from the first successful experiences with 'living labs' in the district of Arabianranta, as well as the overall tradition of the city's industries to embed design in products and services, such as ICT, pottery, textiles, furniture, etc. (e.g. van Winden *et al.*, 2012). In 2005, a new platform for cooperation – Forum Virium Helsinki – was set up, to promote the participation of users in the design and development of digital services (e.g. through living labs). Also worth mentioning here is the foundation of the Aalto University, named after the renowned designer Alvar Aalto, in 2010.[7]

At the end of 2008 the director of the Cultural Department (Pekka Timonen) asked the Mayor of Helsinki (Jussi Pajunen) if the City was willing to support a bid for becoming World Design Capital in 2012. Turin hosted the first edition of this biennial event organised under the auspices of the International Council of Societies of Industrial Design (ICSID).[8] This non-governmental organisation nominates every other year one city in the world for the status of WDC as a city that promotes design as a 'social and cultural stimulator' and as a tool for economic growth. Local governments can use the event to promote and brand cities as places of innovation and creativity.

The Mayor of Helsinki gave full support to the bid and convinced the Mayors of neighbouring municipalities to see the benefits of the WDC nomination and convinced them to join the initiative. The municipalities prepared 'the winning bid' in only three months time with the help of consultancy and marketing agencies. Face-to-face interviews with more than 150 experts – policymakers, entrepreneurs, professors and designers – gave input to a strategic framework that evolved in a collaborative way. The consulted experts agreed that the design industry in Helsinki – and in fact Finland as a whole – was (and still is) a strong asset, but not enough to distinguish the city from its competitors for being WDC. Therefore, the bid should not only emphasise the rich design legacy of Helsinki – with its famous designers and products – but also present something new, which would also appeal to ICSID and help strengthen the WDC brand: 'Open city, embedding design in life'. With this concept Helsinki succeeded in convincing ICSID to choose Helsinki. By promoting design as a way to improve public services and the functioning of the public sector in general, Helsinki was able to 'tip the toe to an ocean of opportunities'.

4.3 Open city, embedding design in life: the concept

In the application for WDC 2012 Helsinki expresses its ambition to use the philosophy of design in every aspect of life. Clearly it builds on a broad definition of design as used by

ICSID: 'Design is a creative activity whose aim is to establish the multi-faceted qualities of objects, processes, services and their systems in whole life cycles'.[9] Design can be seen as a systematic approach or philosophy with responsiveness to end-users as key characteristic. It involves recurrent steps such as observing the behaviour of users and developing, testing and improving prototypes. A well-designed product or service combines usability (functional use, user-friendliness) with sustainability (in ecological, social and economic terms) and desirability (it has to be appealing) (e.g. Lidwell and Manacsa, 2011).

The ambition of WDC Helsinki 2012 is to embed this 'way of thinking' in all aspects of urban management and development: regarding not only the physical appearance of the city (infrastructure, buildings, landscape, parks, etc.), but also 'immaterial areas'. Examples of so-called extended applications are to be found in service delivery (education, health-care, etc.), but also in planning economic, social and cultural development. Design can be embedded, according to this view, by involving users and experts in 'creating new things' from the very beginning. In this interactive process design accelerates implementation by visualising desired outcomes for the future. The process of embedding design in urban development consists of three steps: awareness (actors acknowledge that design is relevant for society), knowledge (actors understand what design is about and are familiar with user-centred innovation methods) and action (actors are able to act with design, embedding design in life).

Co-creation relates to the concept of an open city, as we explained in the introduction to this section. WDC Helsinki 2012 promotes the exchange of data, thoughts and ideas to make cities better places and to improve urban management. The bid was presented as an 'open book' allowing citizens to suggest activities and events through an open tender. During the event, the concept of an open city is put in practice by (among other things) hosting workshops and experimental labs and organising dialogues between people from various sectors and with various disciplines. Openness also means that the city invites people from other cities to share their ideas and to get inspired.

4.4 The programme

The programme of WDC Helsinki 2012 consists of 551 projects and 2,800 events organised mainly in Helsinki, Espoo, Vantaa, Kauniainen and Lahti, implemented by 14,500 people from 290 organisations (Deloitte, 2013). The projects can be divided into six categories: transforming the city, rethinking design, year of events, exhibitions, encounters and communications. About 50 per cent of the programme relates to initiatives of the cities, the government and the universities that remain in place after the event, which illustrates the expected legacy for Helsinki and Finland. Long-term effects can be particularly expected from the 58 initiatives in the category 'transforming the city', such as:

- An open international ideas competition for the Helsinki South Harbour, organised by the City Planning Department; entrants may use the results of a survey among nearly 500 residents about the area and its future.
- Active cooperation and partnership during the construction period in Jätkäsaari, one of the new residential areas, resulting in special design solutions (combining traditional design and service design) to prevent nuisance (e.g. dust, noise and traffic blockages).
- Customer orientation and intelligent public transport, with active participation of citizens in the development and testing of services that promote more flexibility in travel with different modes of public transport.

But also most of the 112 initiatives in the category 'rethinking design' show the ambition to change the way cities are designed, from hardware to software. Some examples:

- Design-led Innovations for Active Ageing: through system analysis, data visualisation and a user perspective various actors cooperate to develop solutions that support the independent and active life of senior citizens.
- Helsinki Region Infoshare: makes public information freely available for everyone, thus promoting their understanding of, and involvement in the development of urban neighbourhoods (see Box 12.1).
- Smarter Cities Challenge: participation of Helsinki in a global IBM project which aims to 'integrate and visualise open information', using among others the data of Helsinki Region Infoshare.

Box 12.1 Helsinki Region Infoshare

The Helsinki Region Infoshare is not exclusively or directly associated with the WDC Helsinki 2012; however, its vision closely relates with the openness and user-driven development ambitions of Helsinki. In a nutshell, the initiative consists of preparing and making freely available online a number of datasets and information concerning Helsinki Region. The web service (http://www.hri.fi/en) was launched in 2011 and nowadays there are lots of datasets such as city statistics, planning and geo-located information of the built environment, transport, event agendas, etc. In its initial stages, a prime objective was to increase the general awareness of the benefits associated with the open release of data in society, so that three long-term objectives can be accomplished:

1. government efficiency – e.g. through the development of better/more efficient data organisation and releasing routines, as well as more fit solutions and provisions for citizens (e.g. transport, environment, services), reducing procurement costs;
2. transparency and democracy – e.g. through increased accountability, openness and proximity between public decision-making and the citizen;
3. business and innovation – e.g. through the development of new (exportable) solutions and technologies developed with the support of the available information, releasing creativity and the faster development of better products.

The pioneer Washington DC's open data catalogue has inspired the initiative. Helsinki Region Infoshare is designed and managed by Helsinki Urban Facts and Forum Virium Helsinki. It contributed to developing a network of data owners who maintain and release public data in a common platform, following common standards. Again, a prime objective is to involve the users in co-designing better services and solutions in the city, benefiting from data that was previously closed or was of difficult access. In order to steer this involvement and increase awareness, a number of 'app contests' have been developed, which have resulted already in a number of applications. Some examples are:

- an application called Blindsquare to help people with visual disabilities to move in the city (based on data on the urban environment); more information: http://blindsquare.com/

- an application about finding places for dumping waste, directly connected with the database of the municipal waste company;
- an application to directly pay for parking in the city, now also connected with the transport company ('single click' parking).

In June 2013 the European Commission awarded Helsinki Region Infoshare with the European Prize for Innovation in Public Administration. The Commission's prize was part of a public sector innovation competition and was awarded in the competition's section on services for the benefit of citizens. The prize is €100,000 which has been used to develop services further including access to public information about city decision-making.

Approximately 80 per cent of all initiatives resulted from the open tender: many projects have been initiated by actors outside and without any existing relations with the city administration. In this way the open tender has succeeded in mobilising actors who are not the 'usual suspects' thus anticipating possible criticism that open tenders are not really open to all. It was the concept of an open, design-driven city (a comprehensive, long-term vision) that made it easier to gain the support of many different actors in society: designers, companies, universities, different municipalities, etc. WDC Helsinki 2012 clearly has become an umbrella for numerous initiatives that combine design (a user-centric approach) with open data.

5 Implementation

The project WDC Helsinki 2012 ran from 2010 to 2013. It included the development of the programme in 2010 and 2011 and the actual implementation of the programme (hosting the event) in 2012. However, most discussion partners indicated that the implementation of 'embedding design in every aspect of urban management and development' needs, naturally, a lot more time. Clearly they use a broader definition of the project, expecting new initiatives after 2013 to secure the long-term legacy of the event.

5.1 Organisation and finance

A consortium of five municipalities implemented WDC Helsinki 2012: Helsinki, Espoo, Vantaa, Kauniainen and Lahti. As displayed in Table 12.2, the budget for organising, managing and marketing the project (2010–13) was €17.8 million, with contributions from the five cities (€6 million), the government (€5 million) and corporate partners and foundations (€6.8 million) (also Figure 12.2). This budget was spent on the programme (€7.2 million), communication and marketing (€6 million) and personnel, license fees and administration (€4.5 million) (Deloitte, 2013). What remains is a surplus of €100,000. The contributions from corporate partners and foundations exceeded expectations, explaining the difference with the original budget of €16 million.

With a relatively small budget WDC Helsinki 2012 has been able to act as a lever for the financing of innovative design-related projects in the city. The International Design Foundation, set up to manage the whole project, did not act as a financing body but as a facilitator for the events and projects under its umbrella. Only some main events were

Table 12.2 Budget of WDC Helsinki 2012

Income		*Expenditure*	
Municipalities	€6 million	Programme and content	€7.2 million
Government (State)	€5 million	Communication and marketing	€6 million
Corporate partners and foundations	€6.8 million	Organisation and other expenses	€4.5 million
Total	€17.8 million	Total	€17.7 million

Source: Adapted from Deloitte (2013)

financed partly from the budget of WDC Helsinki 2012. Other events and development projects – only 50 per cent of the programme is event-based – mainly benefited from WDC Helsinki 2012 through the communication and networking opportunities it provided. The Foundation took care of the legal arrangements with ICSID concerning the use of logos and was very active in promoting the event internationally, e.g. by using diplomacy in order to get international press coverage. This all implies that the actual financial impact of WDC 2012 is much larger than €17.8 million.

5.2 Leadership

In the early stage of implementation – between 2008 and 2009 – leadership played a critical role in launching the bid and the project WDC Helsinki 2012. One of the first people to think of applying for the WDC title was Pekka Timonen who was at that time Director of the Cultural Department at the City of Helsinki. He got in contact with Eero Holstila, a key advisor to the Mayor who had already occupied many important roles in the city administration and was Director of the City of Helsinki Economic Development Office. In 2008, they presented their idea to the Mayor, Jussi Pajunen. With his support the bid process was formally

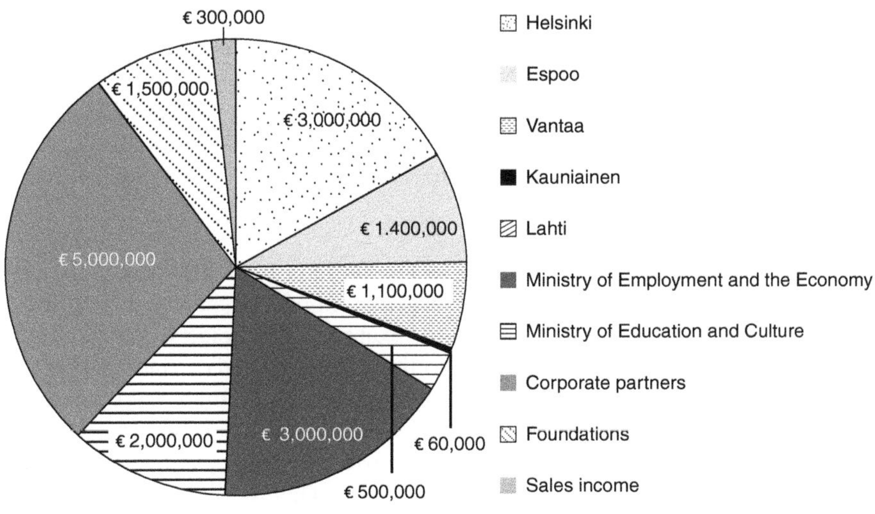

Figure 12.2 Budget of WDC 2012

Source: Adapted from Deloitte (2013)

started. Together with consultants, the Mayor and his advisors started to draft a vision for the application. The Mayor invited other municipalities, universities and firms in the metropolitan region to join the bid, which was, in retrospective, essential for winning the title.

In a relatively short period of time, consultants organised interviews and workshops with more than 150 people, thus involving the 'users' in this process. One of them was Yrjö Sotamaa: designer, former director of the University of Arts and Design, the (founding) father of many design associations, inspirer of Arabianranta and the brain and driver beyond the merger of Aalto University. During a brainstorming session in 2009 he made a critical contribution to the definition of the WDC Helsinki 2012 strategy by 'connecting the dots' and formulating a vision in such a way that others said 'this is what we were looking for'. Another important person in the development of the strategy was Jarmo Eskelinen, CEO of Forum Virium Helsinki and presently President of the European Network of Living Labs.

6 Results

In March 2013, WDC Helsinki 2012 presented the official evaluation report, prepared by Deloitte (2013). The positive effects and goals that have not been met are summarised in Table 12.3. An important finding is that 52 per cent of the projects would not have been implemented without the programme, or would have been implemented in a different manner. The authors of the report estimate that 37 per cent of the projects are either likely or certain to produce results after 2012, for example service factories continue in Aalto University and the Oiva Academy which trains City of Helsinki personnel has incorporated service design as part of their training.

Some other highlights:

- The project has featured in almost 8,000 international articles published in nearly 100 countries and another 7,000 in Finland.
- Over 80 international seminars, conferences and events were organised during WDC Helsinki 2012, which brought more than 2,000 experts worldwide to Finland.
- Between 2010 and 2012, more than 200 international events in 44 different countries have been organised within the framework of WDC Helsinki 2012.
- There is an effect on the development of tourism. Helsinki and surrounding municipalities report growth rates. According to research carried out by IBM analysts, the special effect of World Design Capital 2012 on Helsinki's tourism is estimated at 2.5 per cent. Without the event, tourism would have decreased in 2012 with 1.2 per cent instead of the actual increase of 1.3 per cent.

More facts and figures can be found in a document on the legacy published by WDC Helsinki 2012,[10] which presents ten statements on the long-term impact of the event (see Box 12.2). Each statement is supported by evidence, with examples of projects and (intermediate) results from the impact assessment and other studies that have been carried out. The 'legacy document' can be seen as an intermediate report with some preliminary facts and figures on the impact of WDC, mainly focusing on the positive effects of the event.

WDC Helsinki 2012 activated citizens and supported the participating cities in creating new types of practices that involve and empower citizens. Deloitte estimates that 37 per cent of the projects are either likely or certain to produce results after 2012. Many of the development projects were launched by the end of 2012, and they will continue far into the future. The total impact of the WDC Helsinki 2012 project can be assessed and reviewed from many

Table 12.3 Results of WDC Helsinki 2012

Positive effects	*Goals that have not been met*
The citizens became active. WDC mobilised citizens and supported in creating a rich and active city culture.	The general public's understanding of design did not change due to the WDC year.
WDC project's core actors' understanding of design has increased and broadened.	WDC failed in its objective to support the growth of the number of companies that have internalised design thinking. Design's importance to companies did not significantly change during the WDC year.
WDC strengthened the understanding of design in City organisations, as well as in universities.	
WDC increased citizens' interest in developing their own city.	The execution of the ideas created in the projects of WDC remains partially unclear. Too many projects remain on planning level.
WDC generated projects related to design and user-friendliness in public and private sectors with significant immediate effects.	Common key projects for the City organisations were not created.
The cities feel that WDC has helped them create new ways of operating that encourage citizen participation. The design education key projects created tools that will be utilised in design education also in the future.	WDC failed to support the cities in developing public services from a productivity and cost-effectiveness point of view with projects focusing on this perspective missing altogether from the programme.
WDC had some effect on the use of design in the private sector. The share of companies that have invested significantly in design during the last two years increased.	WDC poorly supported creation of company projects that utilise design in product and service development.
WDC strengthened (broad) cooperation in the metropolitan area, through its programme structure. WDC generated an unfortunately small number of new forms of cooperation, but supported the formation of new contacts.	Long-term effects would be more likely if WDC had succeeded in engaging the stakeholders more strongly in the programme and in creating a tighter network around the programme.
WDC strengthened the image of Finland as a design country.	

Source: Adapted from Deloitte (2013)

Box 12.2 The legacy of WDC Helsinki 2012

1. Design became a matter of social interest and a topic of public discussion.
2. City residents gained a deeper insight into the importance of design and its impact on their daily life.
3. The user's perspective in design was highlighted.
4. Design was used to solve problems in different environments.
5. Design education for children and youth was increased.
6. The design world came together in the design capital.
7. Helsinki raised its profile as a design destination.
8. The business community gained a better understanding of design as a competitive edge.
9. New forms of collaboration emerged in and between different sectors.
10. New methods and places for doing things together sprung up around the city.

angles, and assessments will change with time. For example, the results of development projects will only be seen over the course of time.

Discussion partners for this study are in general positive about WDC Helsinki 2012, but they also admit that there is still a long way ahead in creating a design-driven city or what some call a 'design democracy'. Apart from the many initiatives under 'the official flag' of WDC, there are some initiatives that emerged parallel to this event. While ICSID is very keen on the proper use of the WDC brand, the City considers this development of side events and projects as an important result in the process towards a design-driven city. It fits in with the idea of an 'open bid book'.

An important result of WDC Helsinki 2012 has been the development of new forms of cooperation in and between different sectors (see Box 12.2, statement 9). WDC can be seen as one of the many events to align actors within and outside the city administration, but also as a unique opportunity to promote design and user involvement among various actors. Expectations are that these new partnerships will continue to exist – these ideas have been included for example in the new strategy 2013–16 of the City of Helsinki (City of Helsinki, 2013).

In our interviews with representatives of the city administration we got the impression that, indeed, design has become a matter of social interest and a topic of public discussion (see Box 12.2, statement 1) also within the municipality. Clearly we observe awareness about the role of design in many aspects of urban development and management. At the same time, however, we also noticed that there is not yet full understanding (knowledge) among all our interviewees of what design actually means: they have their own interpretations. This is neither a surprise nor a problem: if design is a new way of thinking (as suggested by WDC Helsinki 2012), considerable time is needed to change the minds and attitudes of people. Some interviewees also find it hard to give examples of actions that would fit in a design-driven approach. The impact of WDC Helsinki 2012 on the design of residential areas (and planning in general) has not materialised yet (see Box 12.3); in comparison with other cities urban development is highly regulated and planned. The event's main influence has been that people have started talking about it.

Box 12.3 User-involvement in the development of new residential areas

Recently the City of Helsinki has adopted a new approach for the development of residential areas. In the past the City first developed a detailed plan and then invited market actors, asking them to implement the plan. It often appeared that the detailed plan was not realistic. Therefore, the City now cooperates more closely with the project developers. The two lines of designing an area and tendering for its development now run parallel. In the open tender the City formulates quality requirements and asks developers to formulate their vision on the design and how much they are willing to pay. This new approach can be seen as a first step towards more user involvement, although it is mainly about the involvement of developers and not so much of end-users in the design of an area. Citizens and other end-users are involved, but in a more traditional way: the City has appointed three officials to organise city hearings allowing end-users to express their views on draft and final plans. Also they sometimes organise workshops to ask people about their needs and wants. In Kalasatama, for example, they organised a brunch session that resulted in 3,000 ideas on how to develop this area.

For example, the event has created advocacy for trying out new types of housing that better fit specific users such as students. Architects but also other experts have been invited to discuss new ideas for student housing, for example in Jätkäsaari, one of the new residential areas. Although regulations are still in place, we also observe an increasing willingness to experiment with new ideas, taking a somewhat more flexible bottom-up approach. The universities play an important role in this transition, being aware of the need to further develop Helsinki as a 'student city' of international standing. They consider the current student housing policies too restrictive and are in favour of more variety in choices. More involvement of students will probably lead to more variety in the housing offer, including possibilities for couch surfing and other less conventional solutions.

One discussion partner expressed the view that the ambition of the City to adopt principles of design and user-centredness can be mainly observed at Culminatum Innovation and Forum Virium Helsinki, and not yet so much in other parts of the municipality, such as healthcare, social services, administration and planning. In practice, however, there are many interesting user-centric approaches going on, for example Laituri which is a meeting place of the City Planning Department in the centre of Helsinki that provides information about how Helsinki will develop now and in the future. Other good examples are 'democracy pilot projects' planned and implemented by citizens through which the City of Helsinki enhances democracy and participation. Furthermore, there are several other democracy initiatives like participatory budgeting, Ruuti programme for youth, online services for electronic feedback and new open data interfaces (Open Ahjo, Open Service Maps).

Despite these initiatives, the City of Helsinki is still a large and essentially bureaucratic organisation, which cannot yet be characterised as user-centric. WDC Helsinki 2012 has raised awareness about the importance of design and a user-centric approach, but more investments and efforts are needed to actually change the culture.

7 Conclusions

7.1 Broad assessment

The concept of a design-driven city was introduced by Helsinki and four neighbouring municipalities in their joint bid for the third edition of World Design Capital (2012). The concept responds to the growing need for participatory forms of urban governance and development, involving businesses, knowledge institutions and citizens in their role as end-users. As they won the bid, the case of Helsinki provides a first assessment of how this concept works in practice. The event WDC Helsinki 2012 ('the project' we analysed) – funded by the five cities, the state, corporate partners and foundations – has to be seen within this broader ambition of developing an open, design-driven city, functioning as a catalyst for the envisaged transition.

The idea of using an event to promote the transition to a design-driven city builds on various local assets and capitals: 1) Finland's reputation as a country of designers; 2) strong organising capacity with relatively well-developed relations between governments, businesses and knowledge institutions (multi-level governance and triple helix cooperation); and 3) a policy environment that seems increasingly susceptible to bottom-up, user-centred initiatives, despite a tradition of government control and top-down planning. The introduction of design principles to all aspects of urban development anticipates fundamental trends such as the increasing competition between regions and the need to be attractive for international knowledge workers, but also considering changes in the demographic structure of the city (e.g. a declining dependency ratio).

Although the concept of a design-driven city is not explicitly mentioned in the City's policies, it does fit in the city-wide vision and strategy, on the level of both the municipality and the metropolitan region. This particularly applies to the ambition to promote open, user-driven innovation environments, which could be relevant for applying design principles in the public realm as well (besides the more obvious applications in the private realm). This all demonstrates that WDC Helsinki 2012 cannot be isolated from previous policies such as the metropolitan area's innovation strategy that was launched in 2005, while it has also influenced other strategies such as the Competitiveness Strategy for the Helsinki Metropolitan Area launched in 2009. The Competitiveness Strategy was renewed and expanded consisting of the whole Helsinki Region in 2012.

Overall the evaluation of WDC Helsinki 2012 is positive. The event attracted many visitors and resulted in positive media attention for Helsinki and Finland, strengthening their image. Regarding the transition to a design-driven city, we conclude that people have become more aware of the role of design in many aspects of urban development. This particularly applies to people who were directly involved in one of the many projects under the WDC 2012 flag. However, the general public's knowledge about the role of design did not change significantly during the event. Also from our interviews we got the impression that there is still a long way to go: design principles are not yet fully integrated in urban management. It will be interesting to see if the experimental projects of WDC Helsinki 2012 – promoting stronger involvement of citizens – will become common practice in the years to come.

7.2 What have been critical factors?

Although the transition to a design-driven city has not been fully realised yet, we can identify some critical factors that explain the progress so far: the successful bid for WDC 2012, the development of the event programme and the impact it had on the city.

First, it is the inspirational vision of a design-driven city that mobilised actors in the region, but it also appeared to be crucial for convincing ICSID that Helsinki should become World Design Capital. For such an inspiring vision the City successfully consulted more than 150 experts, thus taking advantage of the existing Finnish knowledge and expertise on design and user involvement.

Second, citizens were invited to participate in the development of the programme from the very beginning. The open tender and the concept of an open bid book provoked initiatives from 'new kids on the block' with fresh ideas on how to apply design in everyday life. More than half of the projects would not have been realised without WDC Helsinki 2012, or at least not in the same manner.

Third, we identified leadership as an important factor. Several individuals from different sectors played a key role in moving the initiative forward: 1) the Director of the Cultural Department and the Mayor and his advisor (all from the public sector); 2) the former Director of the University of Arts and Design (a knowledge institution) and the CEO of Forum Virium Helsinki (more or less representing the private sector). The University of Helsinki also played a key role in this context. It launched two interactive meeting places (Think Corner and Minervatori) in the middle of Helsinki for everyone but especially for students, and in addition opened up a new library for students and researchers. This new Helsinki University Library is called Kaisa House.

Interestingly it seems that Helsinki has not only introduced a model for its own future development, but also for the development of other cities. ICSID seems to have taken over the concept of a design-driven city and it is very like that future World Design Capitals will be able to use the event for promoting the application of design and user involvement in various fields. But also for cities without the ambition to become World Design Capital the idea of a design-driven city may be a source of inspiration.

Notes

1 The dependency ratio is usually defined as the ratio between the population younger than 15 plus people older than 65 divided by the population between 15 and 64 years old.
2 The only other city in Finland with a HITAS scheme is Oulu. But while Oulu only guarantees the price for the first sell, in Helsinki it is also controlled for second sells (and so on), thus preventing individual owners from making a profit out of such a system.
3 An overview of all departments and committees can be found at http://www.hel.fi
4 In April 2013 the Helsinki City Council endorsed a new Strategy Programme for the years 2013–16 and the 14 municipalities of the Helsinki Region launched a new Competitiveness Strategy for the Helsinki Region. Both these strategies consist of very similar elements as there were in the previous strategies.
5 http://wdchelsinki2012.fi/en/open-helsinki-open-city
6 More examples can be found in the book *Helsinki Beyond Dreams*, edited by Hella Hernberg (2012).
7 Aalto University is the result of a merger between the School of Economics, the University of Technology and the University of Art and Design (TaiK) and has the ambition to stimulate interdisciplinary research and education.
8 Seoul was World Design Capital in 2010 and Cape Town hosted the event in 2014.
9 http://www.icsid.org/about/about/articles31.htm
10 http://wdchelsinki2012.fi/sites/default/files/121127_perinto_en.pdf

References

City of Helsinki (2009). *Strategy Programme 2009–2012*. Helsinki: City of Helsinki.

City of Helsinki (2013). *Strategy Programme 2013–2016*. Helsinki: City of Helsinki.

City of Helsinki Urban Facts (2013). *Helsinki Region Trends 2013*. Helsinki: City of Helsinki Urban Facts.

Culminatum Innovation (2005). *Innovation Strategy Helsinki Metropolitan Area*. Helsinki: Culminatum Innovation.

Culminatum Innovation (2009). *Prosperous metropolis; Competitiveness Strategy for the Helsinki Metropolitan Area*. Helsinki: Culminatum Innovation.

Deloitte (2013). *World Design Capital 2012: Embedding Design in Life*. Helsinki: Deloitte.

Florida, R. L. (2002). *The Rise of the Creative Class: And How It's Transforming Work, Leisure, Community and Everyday Life*. New York, NY: Basic Books.

Focus Finland (2012). *How to Design a City?* Helsinki: Focus Finland.

Health Centre (2007). *Annual Report 2007: Health for the Residents of Helsinki*. Helsinki: Health Centre.

Hernberg, H. (2012). *Helsinki Beyond Dreams*. Helsinki: Urban Dream Management.

Laakso, S. and Kostiainen, E. (2011). European Metropolises. Recession and Recovery. In *Statistics*, Vol. 19. Helsinki: City of Helsinki Urban Facts.

Lidwell, W. and Manacsa, G. (2011). *Deconstructing Product Design: Exploring the Form, Function, Usability, Sustainability, and Commercial Success of 100 Amazing Products*. Beverly, MA: Rockport Publishers.

Manninen, A., Vallittu A., Karvinen, M., Ranto, S., Askelo, S. and Lahti, T. (2012). *Helsinki*. In L. van den Berg, J. van den Meer and A. Otgaar (eds), *Cities as Engines of Sustainable Competitiveness: Looking Back at Two Decades of Urban Policies*. Rotterdam: Euricur.

Vaattovaara, M. (2002). Future Developments of Residential Differentiation in the Helsinki Metropolitan Area: Are We Following the European Model? *Yearbook of Population Research in Finland*, 38, 107–23.

Vaattovaara, M., Vilkama, K., Yousfi, S., Dhalmann, H. and Kauppinen, T. (2010). Contextualising ethnic residential segregation in Finland: migration flows, policies and settlement patterns. In R. Andersson (ed.), *Immigration, Housing and Segregation in the Nordic Welfare States*. Helsinki: University of Helsinki.

Van den Berg, L., Pol., P. and van Winden, W. (2005). *European Cities in the Knowledge Economy: The Cases of Amsterdam, Dortmund, Eindhoven, Helsinki, Manchester, Munich, Münster, Rotterdam and Zaragoza*. Aldershot: Ashgate.

Van Winden, W., Carvalho, L., van Tuijl, E. and van den Berg, L. (2012). *Creating Knowledge Locations in Cities*. Abingdon: Routledge.

World Design Capital 2012 Application (2009). *Open Helsinki: Embedding Design in Life*. Helsinki: City of Helsinki.

13 Strategies, dilemmas and capabilities for urban managers

Luís Carvalho, Leo van den Berg, Hazem Galal and Peter Teunisse

1 Introduction

This chapter brings together a number of insights derived from the previous case studies. Section 2 considers how the different types of urban strategies under analysis have been impacting on the cities' structural assets and capitals for sustainable competitiveness: *knowledge and economic capital, social conditions, environment and quality of life, image* and *institutional conditions*. To do so, key insights from the case studies are grouped across a typology of six key themes or domains: *sharing data in the digital city, using events for new urban legacies, regenerating the city, climate proofing and adaptation, promoting new cluster development* and *enhancing urban security*. Section 3 looks into the changing organising capacities and enablers analysed in the case studies, and how they relate to early results and pathways of urban change, namely: *strategic networks and collaboration, leadership, communication, finance, prioritisation and piloting* and *agile delivery*. Section 4 concludes with a number of conceptual and policy reflections, deriving a summary and dilemmas and capabilities to manage cities under a sustainable competitiveness framework.

2 Nurturing sustainable (urban) competitiveness: six contemporary themes and strategies

The case studies in this book analyse how contemporary urban development strategies are contributing to sustainable (urban) competitiveness, defined as 'the ability of cities to keep growing and developing over time while fostering social cohesion and environmental quality' (see also van den Berg *et al.*, 2014, p.4). The ambition was to better understand how and to what extent cities have been moving beyond economic 'boosterism' strategies toward combining economic competitiveness with higher order sustainability objectives.

Most of the case studies in this book touch upon more than one dimension of a city's sustainable competitiveness – economic, social and environmental – and thus also contribute to strengthen more than one city structural asset or 'capital': knowledge and economic structure; social conditions; environment and quality of life, accessibility; image; and institutional conditions. This section brings together a number of key trends and insights, organised under holistic themes, namely: 1) sharing data in the digital city; 2) using events for new urban legacies; 3) regenerating the city; 4) climate proofing and water management; 5) promoting new cluster development; and 6) enhancing urban security. Each theme is explored below and provides a link with a city's sustainable competitiveness and new organising capacities. As synthesised in Table 13.1, many of the projects impact on more than one domain, although each can be ascribed to a 'dominant' theme (see the bold 'X's in Table 13.1).

Table 13.1 Themes covered in the case studies

		Themes					
		Sharing data in the digital city	*Using events for new urban legacies*	*Regenerating the city*	*Climate proofing and adaptation*	*Promoting new cluster development*	*Enhancing urban security*
Digital Strategy	Manchester	**X**		X		X	
Dublinked	Dublin	**X**				X	
Integrated Housing Development	Addis Ababa			**X**			
Kasongan Pottery Cluster	Yogyakarta					**X**	
World Mobile Capital	Barcelona	X	**X**			X	
Pacifying Police Units	Rio de Janeiro		X				**X**
Park Railway North	Antwerp			**X**			
Phoenix	Dortmund			**X**			
RDM Campus	Rotterdam			**X**	X	X	
Water Management Strategy	Singapore				**X**	X	
World Design Capital	Helsinki	X	**X**				

Source: Own elaboration

Note: a bold 'X' identifies the dominant theme/challenge being tackled by each strategy or project.

2.1 Sharing data in the digital city

The convergence of digital technologies – for example, diffusion of smart personal devices, data-sharing platforms, ubiquitous and cloud computing – is opening up new possibilities for the delivery of urban services while creating economic and innovation opportunities in cities. Thus, most cities are progressively embarking on different sorts of digital and 'smart city' strategies to reap benefits from the new possibilities opened by ICT (information and communication technologies) (Glasmeier and Christopherson, 2015). In this book, the cases of Dublin, Manchester, Helsinki and Barcelona provide an insight into the ways such strategies are being devised.

Manchester has put in place a very comprehensive 'digital strategy', encompassing multiple dimensions (business creation, skill development and brokerage, technological experimentation, social inclusion, modernisation of the public sector, etc.) and distributing its leadership among many stakeholders in the city. In Barcelona, the city is using a large international event (the Mobile World Capital) to catalyse and hopefully anchor 'smart city'-related innovation locally. To that effect, new incubation, piloting and technology transfer initiatives are being devised to enhance the profile of Barcelona as a global frontrunner in ICT innovation.

Dublin and Helsinki have been working on a subset of digital-smart city initiatives. Their city councils are championing the open data movement, i.e. making local data on services (e.g. transport, planning, waste collection) freely available to wider audiences, such as businesses, researchers, entrepreneurs and citizens at large. By doing so, they expect to contribute to sustainable competitiveness in multiple ways, namely:

- *by increasing democratic participation, accountability and transparency.* With more (and more transparent) information about local government actions and provisions, public decision-making may become more accountable and citizens have enhanced possibilities (and incentives) to be actively involved in urban affairs;
- *by spurring innovation and new business opportunities.* The wide access to city data makes it possible to entice companies, researchers and entrepreneurs to more closely engage with – and hopefully solve – concrete city problems. New solutions such as urban 'apps' are being developed with open city data, leading to new businesses and local economic diversification (although this claim still needs more evidence);
- *by improving a city's service provision.* Ultimately, open data – and the new solutions developed from it – can contribute to challenge old provision models and the ways urban services (e.g. water, electricity, public transport) are delivered towards more inclusive and fit solutions.

The urban open data movement – and the respective city strategies – is still in its infancy, but is already delivering results. For example, in Helsinki, The Helsinki Region Infoshare (HRI) was launched in 2011 and has been primarily managed by the City of Helsinki (Department of 'Urban Facts') and Forum Virium (a public-private platform involving several private companies). The HRI has been nurturing a network of data owners who maintain and release public data in a common platform, following common standards; after an experimental period, it became part of the City Council's business-as-usual operations. All the data and information produced by the City Council became, by default, open data and should be easy to access online. This has been contributing to increased transparency and excellence in city governance, but also to involve users in co-creating better urban solutions. The city's vibrant community of software developers has been using the data to come up with new ways to improve city living (e.g. to facilitate parking, route planners, urban mobility by the visually impaired, and many others). In June 2013 the European Commission awarded HRI with the European Prize for Innovation in Public Administration.

A similar process has been unfolding in Dublin. The idea to develop an open data initiative in Dublin (Dublinked) dates back from 2009, when the city was strongly feeling the impact of the 2008 financial crisis. Beyond transparency, Dublinked was conceived to create new local business opportunities and to nurture a real-life test-bed for transnational companies interested in urban technology. Dublinked combines a number of features: a traditional datastore plus manifold community building and network-enhancing initiatives (e.g. data visualisation events, data workshops and 'app' contests). Beyond 'easy' datasets, Dublinked has been actively experimenting with the release of large, real-time flows of data (e.g. traffic and water flows). So far, among the development of manifold apps for smartphones, the city's ICT developer communities have been strengthened and the City Council has developed better data-releasing procedures. Apart from four city councils of Dublin's region, Dublinked has been closely supported by a local university and a major technology provider (which has recently opened a 'smart city R&D centre' in Dublin). This organisation model has ensured wide and equitable access to the data, the agility to carry out the initiative through the administration's bureaucracy while ensuring political checks-and-balances.

The cases in this book show how most (if not all) of the city capitals are being impacted from the unfolding 'digital skin' of cities (Rabari and Storper, 2015). Yet, they also suggest that new organising capacities are central to the early success of smart city and open data initiatives, namely when they envision moving beyond 'technology pushes' to embrace transparency, participation and the bottom-up involvement of local communities. As for the latter, distributed leadership and the balanced involvement of private companies, grassroots movements and agile brokers within the municipal administration are becoming pivotal. In this book's case studies, these factors have proved important in many ways, namely to 1) challenge entrenched routines and vested interests in city data ownership; 2) nudging city policymakers to move beyond their comfort zone; and 3) encourage the involvement and buy-in of companies, entrepreneurs and citizens in the process.

2.2 *Using events for new urban legacies*

Hosting large-scale sports and cultural events has long been used by cities as 'catalysts' of urban transformation. The quintessential example is the 1992 Summer Olympics in Barcelona, which provided the necessary impulse to implement the city vision of the Mayor Pasqual Maragall. The event speeded up a number of public and private investments (e.g. transport infrastructure, hotels, new urban areas), propelling growth and enhancing the image and attractiveness of the city. Nevertheless, to each success story there are many more examples of over investment in 'white elephants' such as big stadia and oversized concert halls. Moreover, the long-term legacy of such events have been questioned on many fronts, e.g. for displacing former poor communities in favour of the well offs, for overheating the urban and national economy, etc. All in all, there is nowadays a considerable consensus in the literature that boosting a city's economy through large events, at least in isolation, is not a consistent road to sustainable competitiveness (e.g. Richards and Palmer, 2012).

Over the last years and following the footsteps of Barcelona, Rio de Janeiro placed a strong bet hosting large sports events – World Cup 2014, Olympics 2016 – to strengthen its 'global city' status. The local, state and federal governments saw these events as catalysts to urban competitiveness, namely by improving the city's lagging transport infrastructure and enhancing the safety of its most problematic slums. These improvements may be crucial for the sustainable competitiveness of a large and crowded metropolis like Rio. The outcomes of Rio's strategy are still to be observed, but its planning stage already shows the increasing criticism and scrutiny of the planned legacies by the local and international communities (e.g. concerning the balance between public and private investment; community displacement and human rights issues, etc.).

Beyond mega-event boosterism, other cities in this study are working to reap softer, innovative legacies from other international events, seeing sustainable competitiveness in a different way. For example, in Helsinki, the opportunity to host the World Design Capital in 2012 was used by the city as instrumental to nurture and embed new participatory forms of urban governance, involving businesses, knowledge institutions and citizens through 'design-thinking' philosophies and tools. For such an inspiring vision, the City Council successfully consulted with more than 150 experts, thus taking advantage of the existing Finnish knowledge and expertise on design and user involvement. Moreover, citizens were actively invited to participate in the development of the one-year event's programme from the very beginning. The open tender and the concept of an open bid book provoked initiatives from 'unusual suspects' with fresh ideas on how to apply design in everyday life (e. g. home care for the elderly and new ways of public participation).

The case from Barcelona presented in this book skips the city's well-known 'Olympic narrative' and analyses the legacy planning for a completely different type of event: the Mobile World Capital. Barcelona gained the right to be Mobile World Capital (MWC) for seven consecutive years (2012–18) from GSMA, the association that represents mobile operators worldwide. Through MWC, Barcelona envisions connecting local ICT companies and SMEs with global ICT providers in order to improve the city's innovation ecosystem. A specific programme to do this, connected with the MWC, is the Mobile World Hub, to be based in the well-known 22@district. The Mobile World Hub programme is giving rise to different competence centres around different ICT and mobile solutions. One example is the 'mSchools' programme. This programme has been developed in close cooperation with GSMA and the Catalan government and aims to lower dropout rates and ultimately prepare students to deal with, and create, businesses in today's digital world.

All in all, as the planned event legacies become less focused on heavy infrastructure, the cases of Helsinki and Barcelona suggest that combining good communication (of the less visible yet long-term benefits of events) and bottom-up approaches with distributed sources of leadership is fundamental: it contributes to give clear guidance while empowering unusual suspects, tapping into local (sometimes hidden) sources of knowledge and imagination. Naturally, the long-lasting impacts and urban legacies of such events in the city are still to be assessed in more depth. Yet, the cases already demonstrate that it is possible to devise more socially inclusive and hopefully enduring urban legacies that go beyond oversized infrastructure, seasonal job creation and income trickle-down.

2.3 Regenerating the city

In order to attract and retain knowledge-intensive activity, people and skills, many cities have been working hard to improve the quality of the living environment. As manufacturing has declined in many cities (particularly in Europe), many urban competitiveness strategies focused on the regeneration of old and deprived city districts (city centres, waterfronts, railway and industrial estates), creating better access, cleaning the physical environment and providing the conditions for private investments in new quality housing. Hence, the rationale has often been to tackle sustainable competitiveness by providing better infrastructure, amenities and nudging image change (van den Berg *et al.*, 2014).

The cases of Dortmund's Phoenix Project and Antwerp's Park Railway North are illustrative of these developments. Contrary to sectoral initiatives, a key feature of such projects is that they envision integrated approaches – linking economic, social and environmental dimensions all together in more holistic solutions. These features place strong claims on a city's organising capacity: they often require combining agile delivery vehicles with enhanced citizen participation and communication strategies to keep up the momentum during long implementation periods and to gain social support.

However, the evidence shows that while generally successful in enhancing the physical/ environmental conditions – for example bringing old city areas back to life and attracting new private real estate investments – such initiatives find it hard to contribute to economic diversification and the improvement of overall social conditions in a city. In Antwerp, despite the green upgrading and the enhanced amenities around the new park, the neighbourhoods around the new Park Railway North remain amongst the most deprived. And in Dortmund, while the redevelopments around Phoenix East (targeting higher-class population segments) had been progressing fast, the revitalisation of the more socially deprived Hörde district has been very slow. A relatively recent trend in the revitalisation of old neighbourhoods is to also make space for technological incubators, creative factories and other sorts of knowledge locations. Yet, as the case of Dortmund suggests, their net effect in nurturing economic diversification has been limited, as most companies simply relocate from other areas of the region.

The cases of Rotterdam and Addis Ababa in this book provide new insights into the relations between urban regeneration initiatives and sustainable competitiveness, for different reasons. In Rotterdam, the on-going re-development of the RDM Campus, in formerly derelict port areas, is a world frontrunner in a new generation of regeneration initiatives. It envisions dealing with the long-term transition of former port sites into new urban and industrial functions until 2040. Together with business-as-usual aspects (e.g. development of new office and industrial space, functional mix, etc.), an underlying planning philosophy consists of involving 'unusual suspects' (e.g. new entrepreneurs, researchers, students) and piloting new ideas that challenge the status quo on many fronts (e.g. the development and experimentation of living in floating houses). From a radically different socio-economic context, the case of Addis Ababa highlights the relevance of housing tenure and ownership as a key dimension of the city's sustainable development process, which is often taken as given in developed economies. In Addis Ababa, the use of condominium development for urban renewal and slum upgrading has shown the potential to be replicated in other areas of the city characterised by very poor and not upgradable housing. Moreover, the case shows considerable innovations in context, namely related to finance and organisation improvements that were necessary to implement the scheme.

In summary, the cases in our study are illustrative of the increasing attention paid to the (hard-to-reach) link between greening, environmental quality, economic development and sustainable competitiveness. As the societal demands behind urban regeneration grow, the cases in this book also bring to the fore a number of challenges for city officials and planners, namely 1) the need to more convincingly deal with the social remits of the projects beyond the physical and economic payoffs; 2) the need to cater for flexibility and temporary uses in the area during the long period of project development; and 3) the project's finance, namely curbing speculation while developing more flexible financial solutions that can cater for market swings during the project development period.

2.4 Climate Proofing and Adaptation

Climate change has been an important trend in local policy agendas over recent years (Bulkeley *et al.*, 2011). Urban strategies have been shifting from pure environmental agendas towards embracing 'climate adaptation' and 'climate proofing' strategies. Climate change is naturally a global issue, but many of the most acute challenges and some solutions can be better visualised in cities (e.g. Carvalho *et al.*, 2012). Acting on climate change is a long-term agenda for cities, but one that will largely influence their sustainable competitiveness.

The links between climate adaptation agendas and sustainable competitiveness in cities are manifold. Many (if not most) urban climate agendas tend to have a utilitarian perspective on the issue (see Chapter 1). On the one hand, they see climate change as endangering a city's environment and quality of life (e.g. through heightened pollution, flooding and other natural hazards) and thus jeopardising the city's functioning and attractiveness for businesses and citizens. On the other hand, many urban policymakers see a link between climate proofing their cities, industrial innovation and new businesses associated with exportable 'cleantech' (e.g. to retrofit the built environment, deal with renewable energy sources, mobility solutions, etc.).

The story of Singapore's water management strategy is exemplary in the latter respect. The city's authorities proudly stress how the city-state was able to turn a major vulnerability – restricted water access in an unfavourable geographical situation – into a major opportunity, that is, the development of a world-leading water technology industry. One of the distinctive factors of Singapore's water strategy has been its integrative dimension, aiming to simultaneously achieve water self-sufficiency, engaging citizens in more sustainable consumption behaviours and develop economic value out of that. Many leading world companies and a strong knowledge base was nurtured that keeps generating new water-related innovation. Naturally, the pivotal role played by Singapore's economic developmental state in supporting the industry is not easy to replicate in other contexts; nevertheless, the case is illustrative of what the link between climate adaptation and sustainable competitiveness can render.

The case of Rotterdam illustrates a relation between urban regeneration, climate proofing and sustainable competitiveness. The city has been active in developing a number of 'green deals' and 'climate adaptation' agendas. One of the drivers is to deal with the risk of rising sea levels, but this is seen as going hand in hand with the development of new industries and competences. In the Stadshavens area (former 'city ports'), the vision is to create space for the development of new breakthrough climate innovations, some of them challenging the existing status quo. The area defined a number of key innovation fields – namely 'moving' (mobility), 'building' (construction) and 'powering' (application of 'sustainable energy') – with the aim to experiment solutions that seem utopic today, but in which the city wants to excel in the future. One example is the development of floating houses.

Due to the novelty of such solutions, it becomes critical to embrace uncertainty and to take small steps at a time. The link between nurturing climate change solutions and cities is increasingly recognised, but will require a number of new planning and governance modes (Loorbach, 2010). As has long been stressed by socio-technical studies of innovation (e.g. Kemp *et al.*, 1998), and also suggested by this study, urban climate proofing and innovation strategies will tend to require experimentation, piloting and the involvement of 'unusual suspects' beyond established companies and government bureaucrats.

2.5 Promoting cluster development

As globalisation intensified, many cities and regions worldwide have embraced the challenge of fostering clusters, i.e. agglomerations of economic activities that benefit from each other's presence, e.g. for sharing resources, matching competences or exchanging knowledge (Belleflamme *et al.*, 2000). Clusters have long been recognised as key drivers of the competitive advantage of cities and nations (Porter, 1990). From a sustainable competitiveness perspective, supporting cluster development can be an important strategy to strengthen and/or diversify a city's knowledge and economic base.

Cluster policies are well known in the western world (yet with varying degrees of success). Yet, there is less literature on such strategies and their implementation processes in emerging economies, which often target low-tech industries. The case of Kasongan pottery cluster in Yogyakarta is an example. The Department of Trade and Industry in Yogyakarta has brought together a bottom-up cluster programme of the local government with a top-down programme from the state in order to strengthen the competitiveness of the Kasongan pottery cluster. As such, it is rather different from other cluster policies in Indonesia, which are mainly developed by a top-down approach. The programme's focus is on training of pottery makers (including design, marketing and the use of ICT), access to fairs and exhibitions, technology transfer (e.g. new heating systems) and new production techniques. Interviewees reported that the institutional structure around the cluster has improved considerably (new networks, trust, shared resources, etc.), being a major exception to the norm in Indonesia, which is still challenged with bureaucracy and corruption.

Other cases in this book illustrate how urban strategies in apparently unrelated domains become combined with cluster-supportive policies. For example, the enhanced democracy and transparency goals of Helsinki's City Council have been leading to the development of new apps and companies based on the city's open data. Nowadays, beyond transparency, the council pursues several initiatives to steer new communities of innovators around open city data. Naturally, those communities of ICT developers were there long before, but the council's policies contribute to strengthen them further. Finally, another trend in urban competitiveness strategy is to link cluster development with the development of new solutions to societal challenges. Instead of cherry-picking concrete sectors to support, the idea is to set a common local challenge to which many related industries can contribute. One example is the link between climate adaptation and cleantech innovation, as illustrated above by the cases of Rotterdam and Singapore.

2.6 Enhancing urban security

Cities are places of great economic prosperity, but also of sharp social imbalances and wide gaps between the haves and have-nots (van den Berg *et al.*, 2006). In the large cities of emerging economies, this situation has been amplified as a result of massive rural-urban migrations. Also in Europe and North America, many cities have witnessed events and situations of social unrest. It is commonly accepted that tackling these issues requires integrated and long-term policies (combining social, economic and physical interventions), but the ways to make it happen are much more controversial. Safety from crime and the perception of security are widely acknowledged as important locational factors. They impact on the investments and location decisions of people and companies. The image of many cities is also closely linked with the perception of safety (or the lack of it), affecting their sustainable competitiveness.

This book analysed one recent flagship, security-enhancing strategy unfolding in the city of Rio de Janeiro: the innovations behind Pacifying Police Units (UPP), operating in different slums in the city. Rio's pacifying approach contrasts with the traditional 'slum clearance programmes' and with other 'new public management' efficiency measures to streamline police operations (Carneiro, 2012). Introduced in 2009, the UPP programme has been hailed as transformational in reducing violence, cutting crime and increasing security in Rio's slums. The UPP, delivered by the state government of Rio de Janeiro, aims at reclaiming territories controlled by gangs of drug dealers and private militias. UPP Social, a follow-up programme jointly delivered by the municipal government, seeks to re-integrate residents of pacified favelas back into the city's economic and social fabric through coordination of, and improved access to, public and private services.

To date, the UPP programme has generally been regarded as successful, not only due to evident crime reduction, but also due to the spill over socioeconomic development effects (e.g. education rate improvements, new business attraction and access to public services). The UPP programme has attracted millions of dollars in donations and there are plans to expand the programme beyond the city of Rio de Janeiro to other cities. However, it also has been showing a number of hindrances and challenges. More recently, many commentators and interviewees question the real social inclusion potential of the UPPs. Moreover, there is evidence of displacement effects: crime has been moving away from the most problematic slums around the city centre, but increasing in other sub-urban areas of the metropolis.

3 Organising capacity and key enablers

This section, in line with Chapter 1, discusses the organisational capabilities and critical enablers involved in the implementation of the urban strategies analysed in this book. How have they been implemented 'on the ground'? What do they reveal concerning the organising capacities of cities, and which challenges and lessons can be derived?

Overall, the more demanding the urban strategy's objectives – i.e. with new and more complex combinations between economic, social and environmental elements – the more the enablers and organising capacities needed to bridge the implementation gap. In this study, we paid attention to six critical types of enablers linking visions with results: *strategic networks and collaboration, leadership, communication, finance, prioritisation and piloting,* and *agile delivery.*

3.1 Strategic networks and collaboration

As expected, the formation of networks and partnerships proved to be a critical enabler in all the case studies, influencing the quality of the other five enablers. Throughout the case studies, it was possible to identify several types of networks and collaborations between local governments and three other main types of stakeholders:

- private sector (see Section 3.1.1);
- universities and knowledge institutes (see Section 3.1.2);
- grassroots organisations and citizens (see Section 3.1.3).

Networks and partnerships varied in depth, breadth and quality. In some of the more demanding projects and strategies, all the types of stakeholders had to come together, bringing additional knowledge and expertise, financial resources and legitimacy. Moreover, most of the partnerships in the analysed projects did not start overnight. The most successful ones tend to involve stakeholders that knew each other before and are grounded in trust and mutual understanding.

Partnerships bring the skills and resources needed to target sustainable competitiveness objectives in cities. Yet a key challenge is to engage the right number of stakeholders while keeping projects agile enough to cope with unexpected change.

3.1.1 Private involvement

In the urban studies literature, there is strong scepticism about the involvement of corporations and private parties in urban development affairs (see e.g. van Winden, 2013, for a review). The arguments are manifold: 1) private sector organisations tend to have shorter-term horizons and higher pressure for immediate results, which can be incompatible with sustainable urban development strategies; 2) evidence shows that local industrial elites and 'old boys' networks can steer policy agendas in their own interest, and appropriate public resources in the process; and 3) private sector parties tend to cherry pick their involvement only in the most profitable urban issues.

The aforementioned arguments and threats should be taken seriously. The types of private involvement they refer to are often closely linked with the limitations behind many large urban (re-)development strategies over the last decades. However, it should also not be taken at face value. Some case studies in this book found that the involvement of private companies has been pivotal for the success of many of the projects under analysis, in many domains and in developed and emerging economies. In fact, many companies and transnational corporations involved in this book's case studies are increasingly looking at cities and local communities as part of their long-term competitive environment and are willing to invest in them (e.g. van den Berg *et al.*, 2004). This is more the case with companies that embrace 'shared value' approaches, i.e. 'policies and operating practices that enhance the competitiveness of a company while simultaneously advancing the economic and social conditions in the communities in which it operates' (Porter and Kramer, 2011, p.66). As suggested by Porter and Kramer (2011), creating shared value is not about the peripheral concerns of the company (e.g. as with corporate social responsibility), but moves directly to the core of its strategy, being integral to competing and making profits.

Two examples from the case studies are illustrative. In Dublin, a leading transnational ICT company strongly contributed to the launch and implementation of Dublinked, not only with a 'self-interested eye' to the data for their own R&D but also to diffuse the value of open data in Dublin's society. This involved early cooperation with the City Council to signal which data could have higher relevance (water, energy and transport) while Dublin's City Council made sure that there would be no privileged treatment, making data available on an open platform. Moreover, support to Dublinked is provided (with time and expertise rather than cash) on technical issues such as publishing standards or the requirements for publishing dynamic data. Since open data is such a new field in cities, both the city and corporation are jointly learning about how to take it forward. In the process, the company is learning how to adjust business models and the city council is learning how to devise more innovative and equitable procurement modes.

A rather different type of private involvement was seen in Rio de Janeiro. Beyond donations, tokenism and traditional corporate social responsibility schemes, a number of private companies are contributing to the city slum pacifying scheme. One example comes from the electricity distribution system operator, who is championing an initiative to exchange recyclable waste for cheaper/free energy to households in pacified slums. The company is interested in decreasing energy theft and increasing the citizen's engagement with the company, but in the process the local community gains better energy access and protects the environment, amplifying the effects of the pacification initiatives. The same company is deploying a large number of social initiatives there, such as training people to save energy, do their finance, find jobs, etc. These are seen as beneficial initiatives for the people and also, ultimately, for the company's profits.

Despite the need for caution when closely involving private companies in urban development affairs, there are growing opportunities for cooperation between City and enterprise. Many companies are increasingly ready and willing to invest in the urban environment(s) in which they operate. However, the involvement of private companies in urban development strategies does not (and should not necessarily) lead to shrinking or privatising public services. On the contrary, the fruitful cooperation with private parties in urban development will require more, not less competences and skills from City managers and their teams.

3.1.2 Universities' and knowledge institutes' involvement

Besides their key educational mission, universities are often perceived as 'ivory towers', the opposite of the needs of many cities that have to deal with fast change. However, this image is not entirely true – over the last decade, many universities, educational and research institutes have been closely engaging with local and regional authorities to change their urban and regional economies (Goddard and Vallance, 2013). Many universities increasingly consider 'societal engagement' as part of their mission. Moreover, universities are important sources of knowledge and often can act as independent brokers, mediators and facilitators in urban development projects. Universities can provide the social legitimation for more innovative or controversial initiatives.

For example, in Rotterdam, the concept of the RDM Campus as a place to link education, research and business was initiated and is managed by a university of applied sciences and a school on the vocational level. These schools are the first tenants on the former shipyard and among the leaders in the redevelopment process, together with the Port Authority. This is a sizable departure from their original core missions, but at the same time, pivotal for the research and experimental ambitions of the area. In Dublin, although the City Council initiated Dublinked (the City's open data initiative), its management was shared with a local university. This was important for many reasons. First, it helped to get the right expertise to manage such an innovative endeavour, to which there was little knowledge at the City Council; second, their involvement was relevant to pool up the necessary finance to support its experimental stages; third, it was pivotal to nudge the engagement of other partners at the metropolitan level, as the university was seen as a neutral and trustworthy partner. A similar phenomenon happened in Yogyakarta, where a university professor played the role of broker, inspiration and driving force behind the successful pottery cluster initiative.

The role of universities in the sustainable competitiveness of cities is becoming more important in many fields. Besides their core educational and research missions, universities and knowledge institutes are, increasingly, key partners (if not engines) of urban development projects. Their expertise and respected role in society are invaluable to support experimentation and legitimise some of the most innovative approaches.

3.1.3 Grassroots and citizen involvement

The role of citizen participation in urban affairs has long been emphasised but is frequently forgotten. In some instances it is confused with tokenism, for example informing or consulting citizens about decisions already taken to avoid seeing initiatives blocked by popular protest (Arnstein, 1969). A higher degree of citizen involvement assumes deeper levels of participation, for example through effective partnerships, decision making, implementation responsibilities and power delegation. Such partnerships are usually relevant because they empower unusual and unexpected 'suspects', bring novelty and the viewpoints of different users, and identify previously hidden problems and opportunities. How did it play out to influence the outcomes of the case studies in this book?

For example, in Helsinki, the vision for the World Design Capital Helsinki emerged out of several discussions with the civil society (and not as a top-down vision within the City Council). Moreover, the initiatives of the World Design Capital were all implemented by users and emerged from a fully open bidding process. Citizens and the civil society had the chance and the power to propose and organise the types of initiatives to be carried out. An organising committee curated the proposals, but the overwhelming majority was accepted.

In Antwerp, citizen participation has been a vital factor for the realisation of the park project. Residents – represented by a small but very influential team – were very outspoken about a new function for the abandoned railway yard: a green lung and nothing else. During the implementation period participation became more refined on details of the project. Several parts were changed to meet the citizens' desires (like conservation of railway buildings and providing water elements).

In Manchester, the involvement of ICT community groups and grassroots organisations has been essential to the implementation of the digital strategy (for example, in the fields of open data and citizen involvement). Many new developments, innovations and policy directions are emerging in these groups and not in the offices of big companies or the City Council. One example is the 'Mad Lab', a grassroots organisation that hosts meetings of several local groups of 'nerds', hobbyists and innovators. The City of Manchester, through its arm's length unit to deal with digital issues (Manchester Digital Development Agency) has been closing engaging and supporting the activities of Mad Lab (not with finance, but with public endorsement and cooperation in many events).

Some of the cases in this book suggest that beyond achieving 'societal support' for large (and sometimes controversial) projects, the sustainable competitiveness of cities requires that new players and 'unusual suspects' become involved and truly empowered in urban policymaking platforms. Citizen involvement can contribute to enhance the quality of the policies and challenge crystallised policymaking routines. City officials dealing with economic development need to be apt to 'go on the street' to identify new trends and developments.

3.2 Leadership

As suggested in Chapter 1, leadership has been a critical enabler throughout the case studies, but to the extent that it has been distributed among urban stakeholders (e.g. Beer and Clower, 2014). In this study, the most remarkable examples of leadership involve the capacity of the ones in charge to effectively distribute their power and sometimes even 'let it go'. In those situations, it becomes difficult to differentiate a single leader as leadership is spread among individuals who exert different types of power (e.g. institutional, technical, boundary spanning and financial) in different ways but organically aligned to a common purpose.

For example, in Dublin, Dublinked's leadership is closely associated with Dublin City Council's City Manager and his power to champion the (innovative and risky) project within the administration at a time of budget constraints. His leadership was important, ensuring 'connectedness', both outside and inside the City Council, and re-shaping the partnership as difficulties emerged (e.g. legal challenges). His leadership was also central to legitimising such an 'out-of-the-box' initiative within the administration, e.g. by making sure that different departments would share their data for a fast implementation. Since the project was backed at such a high level, it could overcome the capture of bureaucratic departments and responsibilities were delegated to a newly created experimental unit with less crystallised routines and interests. This way a group of executive managers (with the right technical and communication skills) could emerge and act as central facilitators within the organisation, driving the open data process forward.

In Manchester, the Digital Strategy relies not on one but on multiple leaders and orchestrators. First, it has been strongly supported by the leader of the political cabinet and the Council's Chief Executive. Second, it is currently spearheaded by a newly appointed Assistant Chief Executive for 'ICT and Communications', giving it the status of a true 'IT Master Plan'. Third, and most importantly, the Digital Strategy is coordinated and led on a daily basis by Manchester Digital Development Agency, whose (now former) Director has been involved in the development of ICT strategies in Manchester since the early beginning (late 1980s). His role was essential to assure stability, while simultaneously allowing for smooth transitions and for new variety to permeate the Digital Agenda over time. Moreover, his role was pivotal as a boundary spanner who could connect, bring together and understand the 'languages' of different 'worlds' (public administration, universities, companies, community groups, IT activists, among others).

The leadership required to carry out complex urban development strategies requires multiple skills and new types of urban managers. It relies less on single people but on their capacity to connect and distribute power to other stakeholders, within and outside the public administration. Within a sustainable competitiveness framework, leaders should be able to inspire, connect stakeholders and challenge the status quo while keeping up enthusiasm.

3.3 Communication

Throughout the case studies, communication was closely linked with the capacity to mobilise partnerships and involve citizens in urban projects. It was more important in cases in which the outcomes were hard to visualise beforehand, contributing to gain advocacy and societal support.

For example, in Dortmund, the Phoenix site was previously separated from the rest of the city fabric and considered a 'forbidden area'. Therefore, communication with society was crucial, not only to create awareness, but also to get support for the redevelopment plans. For example, the idea to create a big lake was initially perceived as 'crazy and unrealistic', with negative media attention. Beyond conventional information centres and brochures, important communication measures to tackle the scepticism involved: 1) the organisation of temporary events to bring people to the site during the regeneration process (e.g. theatre festival, culture nights, arts and industrial history exhibitions); 2) the promotion of the area's industrial heritage and landmarks; and 3) the organisation of field trips for citizens, in order to brand and promote the location and raise awareness about new uses.

In the case of Rotterdam's RDM Campus, after a number of tensions in the past (e.g. concerning the location of new industrial activities), new modes of communication with inhabitants have been set up, through the so-called 'Heijplaat Vitaal' meeting platform. Here, the key actors meet inhabitants on a monthly basis, not only informing them but also leaving space for feedback, and as such inhabitants are involved in the development process.

This new platform strategy has turned out to be a success and is also used by the Port Authority in other port areas that are close to housing areas.

Overall, good communication is moving to the core of urban development initiatives. Beyond slogans and brochures, there is an increasing need to professionalise communications functions and develop new innovative ways to communicate change in the city.

3.4 Prioritisation and piloting

By touching upon multiple dimensions simultaneously, sustainable competitiveness-oriented projects often require prioritisation and piloting. This is because, as suggested, their ultimate objectives are too complex to be attained with single and short-term initiatives. Yet, despite their comprehensive and often long-term visions, they require concrete action to start and create momentum and constituency. Pilots and prototypes allow for early visualisation of solutions to see what works and what does not, and to create new coalitions and visions about the impacts of projects in new urban futures. Prototyping and experimentation is moving to the core of urban strategies, namely when they target complex transitions and societal challenges, such as climate change (e.g. Bulkeley *et al.*, 2011). The Stadshavens and RDM Campus strategy in Rotterdam is exemplary in this respect: prototyping, experimentation and flexibility are key dimensions of the whole concept.

In Dublin, the launch and early implementation of Dublinked started with the collection and release of less complex datasets, with limited privacy issues, which were technically easy to deal with and which were almost exclusively from the public sector. This helped to generate early enthusiasm and give a good dimension to the data repository, while testing the user's feedback and paving the ground for more demanding stages. Starting with the most complex and controversial pieces would slow down the process and create many blockages from the early beginning. Moreover, it allowed for the early visualisation of what a fully-fledged open data city would look like and contributed to engaging new stakeholders in the process, for example the transport authority for the release of additional datasets.

In Rio de Janeiro, the slum pacification scheme (UPP) deliberately prioritised a limited number of interventions to start with, out of the hundreds of slums in the city. The first slum to be pacified was Santa Marta, a smaller one close to the city centre that could act as a pilot and demonstrator of the scheme's potential. Subsequently, UPPs were carried out in Cidade de Deus, one of the biggest in Rio, in order to see how the scheme would work in a rather different setting. After getting a sense of the challenges 'on-the-ground', for different contexts, the scheme could be fine-tuned, gaining traction and advocacy.

Prioritisation and piloting allowed for creating momentum, as well as visualising solutions, engaging stakeholders early on and showing the business case for demanding, sustainable competitiveness solutions. As urban competitiveness projects and strategies target more demanding and long-term objectives, they require cities to incubate new ideas and accelerate their impact through trial-and-error and rapid prototyping.

3.5 Finance

The projects and strategies in our study, different as they are, range in size from many millions (e.g. the redevelopment of the RDM Campus or the Phoenix project) to hundreds of thousands of dollars or euro (e.g. the development of Dublinked open data initiative). One conclusion that can be drawn, perhaps evident and considering the obvious differences in dimension, is that good projects do not necessarily require heavy financing. Moreover, it could also be said that good financial solutions are a necessary but not sufficient condition for success. Having said this, budgets are under strain and finding new ways of financing growth is a very topical issue in cities of both developed and emerging economies.

For example, in Manchester, where 'doing better for less' has become an important mantra in urban management affairs, the City's ambitious Digital Strategy is currently anchored in various funding sources. It combines European and national funding with small seed money from the City Council (to leverage innovative projects), volunteer work and the own internal resources of the different stakeholders involved (including private companies). This combination of sources has been key to leverage different projects in a context of growth financing constraints.

In a completely different context, the Municipality of Addis Ababa, after exclusively using its own budget during the first years of the IHDP – Integrated Housing Development Programme (almost 50 per cent of the capital expenditure in 2004–5) and with no international donor funding, started selling bonds to the Commercial Bank of Ethiopia to finance the programme, including construction costs. The bank also came to an agreement with the Housing Development Corporation to provide a loan service to all condominium beneficiaries, whereby the bank paid 80 per cent of the unit price on behalf of the beneficiary at the handover of the property (the beneficiaries paid the 20 per cent down payment themselves) and the beneficiary entered into a loan agreement with the bank to pay back this amount subject to interest.

In Rotterdam's RDM Campus, the funding model has been adapted from the Port Authority's business model, which is rather different from the ways conventional urban redevelopment projects have been funded. As explained in the case study, the funding model is based on long-term contracts with large or core tenants. As such, there are limited financial risks, and income generated by long-term contracts can be used to cross-fund other parts of the area, including the incubator centre in the Innovation Dock that works with flexible contracts. This is a new approach for urban area-based redevelopment, which normally works with a strict master plan that has been pre-financed with large investments, while the Port Authority's approach is more flexible and includes less financial risk.

By combining multiple players, sustainable competitiveness projects open up multiple funding opportunities, but there are also many challenges in coordinating and pooling finance from different sources. Sustainable competitiveness requires sustainable and responsible finance, flexible enough to cope with changing circumstances.

3.6 Agile delivery

Many of the projects under analysis required substantial changes in the traditional ways of local government organisation. The silo-oriented mentality of many administrations is challenged by sustainable competitiveness objectives. Many departments within city administrations had to move beyond their comfort zone and entail new (and risky) endeavours. In many cases, heavy and bureaucratic structures had to be complemented or replaced by new agile delivery vehicles within the city administration, with the ability to pursue the right partnerships and with the lean structures that allow for fast change and adaptation. This proved particularly important when the project was very innovative and with uncertain results and when multiple partnerships had to be pursued, at multiple levels.

For example, in Manchester, Manchester Digital Development Agency (MDDA) is the City Council's arm's length unit in charge of coordinating the Digital Strategy. The unit is tasked, among other things, with organising training initiatives and raising digital awareness, and supporting digital-related actions with other stakeholders and innovative ICT pilots in the city. For less 'institutional' groups outside the City Council (e.g. social media activists and grassroots organisations), MDDA acts a route to access the City Council and raise awareness for new developments of the digital and IT world. MDDA supports new groups by facilitating their own services but also by providing them with legitimacy and visibility within policy spheres.

In Dublin, the Council established 'The Studio', a new experimental taskforce under the responsibility of the City Manager. It was responsible for generating and prototyping new solutions and for working at engaging staff and customers using design-thinking tools. A multidisciplinary team of designers, planners and architects, among others, staffed it. The Studio started as a way to internalise the recently acquired knowledge gained by seven city officers in two nine-month collaborative projects called 'Designing Dublin: Learning to Learn and Love the City'. A key role is the creation of new ideas to solve old city problems. It was instrumental in launching Dublin's open data initiative, as the initiative required thinking 'out-of-the-box' and challenging pre-existing routines and vested interests of other departments.

All in all, enhancing agility is a challenge for city administrations, heightened by sustainable competitiveness challenges. New delivery models and organisations can deliver better services and rebuild trust between the city administration and its many stakeholders. Yet, a key challenge is to increase agility and empower new arm's length organisations while keeping democratic checks and balances on the City's strategies.

4 Conclusion and final reflections

4.1 Sustainable competitiveness and urban strategies: old wine in new bottles?

This book started with the ambition to provide new insight into how cities have been influencing their competitive position over time, namely through innovative types of urban development strategies. Urban development objectives have overall been shifting from pure wealth creation to combine social and environmental dimensions in a more thoughtful manner, while avoiding economic booms and busts in cities. In this context, a key assumption is that cities – through their actors and stakeholders – are not passive receivers of external trends, but often have some degree of leeway to influence their own development 'on the ground'. Therefore, how innovative are the projects under analysis, and to what extent do they pave the ground for new ways of planning urban economic development and sustainable competitiveness?

To be fair, in some of the projects and strategies under analysis, despite their integrative vision, it is still possible to find evidence of traditional ways of planning urban economic development. For example, despite their achievements in regenerating old city areas and attracting private investment, some urban redevelopment initiatives under analysis have shown considerable limitations in tackling pressing social issues and equate the environment with high-level amenities. In other cities, while aware of the need to build long-term legacies, big event attraction is still seen as a catalyst to concentrate several large investments in a short period of time. And the design of some cluster-oriented policies could not overcome the usual difficulties promoting crossovers with other industries and economic diversification.

Having said that, the book also shows many flagship examples of cities and projects that are embracing new local economic development approaches and following new roads in the pursuit of sustainable competitiveness. This has been happening for different reasons, often in combination. First, some cities have been gaining considerable experience in urban economic strategies and their actors have been able to see what works and what does not, avoiding the most common pitfalls of the past. Second, cities have been reacting to new trends and societal challenges, which require the development of radically new types of urban development initiatives (e.g. related to digitalisation, the open data movement, climate adaptation, etc.). Third, new actors and unusual suspects became progressively involved in urban economic affairs, influencing and contributing to the initiatives' visions in different ways (e.g. grassroots movements, private companies, etc.).

Beyond dealing with new urban development challenges, many of the strategies under analysis have been incorporating new concepts and insights from different scientific disciplines, departing considerably from previous concepts. There are many examples. One is the relevance of connecting local clusters of innovative activities to external knowledge sources or 'pipelines' (Bathelt *et al.*, 2004), which is explicitly recognised in many strategies, such as in Yogyakarta (linking local pottery producers with advanced buyers), Singapore (linking local water-technology champions with global value chains) or Barcelona (explicitly using an international event to anchor new companies and ICT-related experimentation in the city). Another example is the consideration of open innovation and the formation of communities of practice (e.g. Chesbrough, 2006; Amin and Roberts, 2008) as pivotal in the redevelopment of urban knowledge locations (e.g. in Rotterdam's RDM Campus). Finally, insights from design thinking (Brown, 2008), complexity theory (Teisman and Klijn, 2002) and sustainability transitions (e.g. Loorbach, 2010) made inroads into the planning and implementation of new urban development strategies, as observed in Dublin and Rotterdam.

Simultaneously, it is also possible to witness the decline of former conceptual approaches grounding competitiveness strategies in cities. One example concerns 'cherry-picking' sector policies and the support of 'winning' sectors. While still popular (and effective) in countries with a strong developmental state – as the water strategy in Singapore illustrates – this approach has been proving its limitations and has been giving way to new innovation policies that focus not on sectors but on domains of related activities linked with specific local challenges (e.g. climate proofing, see the case of Rotterdam) (e.g. Cooke, 2008). And in Rio de Janeiro, the new police-pacifying scheme (UPP) departs considerably from previously failed 'new public management approaches' (Carneiro, 2012) to embrace new hybrid and pragmatic solutions to complex challenges of the city's slums.

Notwithstanding the limitations analysed throughout the chapters, the case studies in this book provide early encouraging evidence of the results of these approaches in their cities' sustainable competitiveness. Some of these concepts are being applied in urban development policies and practices for the first time and, as such, the coming years will tell more about their ultimate contributions. Yet, at this time it is possible to conclude that urban sustainable competitiveness projects and strategies should avoid one-size-fits-all solutions and should be seen as space and time specific. For example, the urban development challenges witnessed in Helsinki are at sharp contrast with the ones that motivate new urban development strategies in Addis Ababa, Rio de Janeiro or Rotterdam. Moreover, the differences in institutional conditions across cities suggest urban development strategies should be carefully tailored to a city's context (Rodriguez-Pose, 2013). Under this reality, simple policy mobility across cities (e.g. McCann, 2011) will always be incomplete, at best, or harmful, at worst.

4.2 New dimensions of urban organising capacity

As proposed in Chapter 1, the strategies and projects under analysis – their vision, ambitions and design – result from the interplay between external trends and the city 'capitals' (their structure and dynamics). Yet, a number of organising capacities and enablers contributed to explain how well the project managed to bridge the implementation gap between a vision and tangible results. In this study, we paid attention to six critical types of enablers: strategic networks and collaboration, leadership, communication, finance, prioritisation and piloting, and agile delivery.

A general recognition from the case studies is that the more demanding a project's content – i.e. with new and more complex combinations between economic, social and environmental objectives – the more the enablers that become needed to implement it. Strategic networks and collaboration proved to be the critical enabler, influencing the quality of the other five. Without collaboration through good partnerships, it becomes difficult to 'enable the other enablers'. Yet, the analysed enablers and the overall organising capacity involved in a city's strategies should be seen not as a sufficient but as a necessary condition for a project's capacity to deliver.

As synthesised in the previous sections, strategic networks and collaboration are the glue that brings sustainable competitiveness projects together. More and more, city councils are not exclusive players but the facilitators of broader collaborations. In earlier formulations of organising capacity (van den Berg *et al.*, 1997), networks and collaboration were already seen as critical elements in a City's organising capacity; however, the role of concrete actors and stakeholders was not explicitly fleshed out. The case studies in this book make it more explicit, bringing to the fore new trends and dimensions of private sector involvement, universities and research institutes, not-for-profit and grassroots organisations. Good collaboration under a sustainable competitiveness framework requires trust, mutual understanding and flexibility/capacity to try out new approaches, but, as previously stressed, it is strong and continuous communication that makes collaboration and networks tick.

The cases in this book bring to the fore three new dimensions of the organising capacity of cities that were not explicitly considered before as such: finance, prioritisation and piloting, and agile delivery. To a large extent, they reflect the growing need to react to shrinking budgets (namely in the aftermath of the financial crisis of 2008), as well as to the need to experiment and prototype new urban solutions before their wider rollout. The new demands put on cities to tackle societal challenges (such as climate change) while fostering new urban economies will put strong demands on these types of enablers. Yet, our cases suggest that these dimensions and enablers can only be enacted not by the action of heroic urban leaders (as suggested in the earlier formulations of organising capacity), but through distributed leadership. In all the analysed case studies, the implementation of concrete visions required connecting and distributing power among the key stakeholders both within and outside the public administration, and inspiring them to action.

4.3 Sustainable competitiveness: dilemmas and capabilities

The initiatives analysed in this study show that there are few single recipes for sustainable competitiveness. Moreover, the context of each city plays a pivotal role for the success of projects and strategies. As said, even if there are common challenges and learning possibilities across cities, initiatives that made sense for a city in certain moments of time could be unfit for others. All in all, embracing a sustainable competitiveness framework requires a new type of urban manager, who can connect capabilities and resources and distribute power, but also who can understand nuances in context and make well-informed choices in time and place. Even if there is not a single blueprint, this final section wraps-up and distils a number of key choices and dilemmas, and prescribes some actions and capabilities deemed important for urban managers and economic development officials dealing with the sustainable competitiveness of their cities.

4.3.1 Choices and dilemmas

- *Transformational or incremental initiatives?* The sustainable competitiveness of cities doesn't necessarily rely on how big or small the projects are. Large transformational initiatives are often acclaimed as game-changers but are not a guarantee of high (or balanced) impact on a city's capitals. Smaller, path-breaking initiatives can be more effective in the medium or long run, namely if they became scalable and pave the way towards new ways of doing things in cities. This is not to say that the sustainable competitiveness of cities does not require heavy investment. However, in many circumstances, a series of small, community-owned projects can be more effective than 'big-bang' turnkey solutions.
- *Broader, multi-domain or specialised initiatives?* In many occasions, choices have to be made between tackling multiple domains simultaneously or concentrating efforts on a few initiatives and domains. Sustainable competitiveness requires that different capitals are harnessed and evolve in harmony. However, it can be difficult to tackle multiple domains at once. It is important not to lose sight of the big picture and make sure that different capitals are progressing, even if not catalysed within a single initiative. Broad, umbrella-like strategies have advantages (e.g. generating momentum and visibility) but prioritisation of efforts is often necessary.
- *Status quo or regulation-busting projects?* On many occasions, urban managers are confronted with the challenge of 'how far' they want (and are ready) to go with their projects and strategies. While some projects can largely run within established cultural-legal structures, others tend to challenge vested interests, legal frameworks and organisational cultures, and propose radically new ways of doing things (e.g. sharing data and user involvement). Risk-return trade-offs need to be equated, bearing in mind that bolder reforms often require challenging established routines.
- *Municipal borders or functional urban regions?* The projects and strategies in this book vary in spatial scale. Some are implemented within cities (and their administrative borders); others take a clear (and purposeful) supra-municipal dimension. While some sustainable urban competitiveness issues call for more contained scales, others operate and make more sense at the level of functional urban regions. Moreover, some initiatives can be very successful at the neighbourhood or city level but create unsustainable situations in metropolitan regions (e.g. crime displacement). However, choosing the best scale is not only a question of context but of political sensitivity, and there are trade-offs involved. While projects at the level of functional urban regions can make it easier to involve the right stakeholders and look at problems from a more integrated perspective, they may also create political and institutional tensions that can also ultimately hamper the initiatives' success.
- *Diversifying partnerships or arms length body?* Involving stakeholders proved pivotal in augmenting sustainable competitiveness projects and in accessing knowledge, finance and support. However, there are also trade-offs associated. The involvement of too many stakeholders can give rise to rigid structures, making it difficult to keep projects agile and adaptable to changing contexts over time. Moreover, involving too many stakeholders can also generate visions that are too broad, hampering prioritisation and concentration of efforts.

4.3.2 Capabilities and actions

- Conduct a throughout assessment of the emergent, sometimes hidden dynamics of the city. Different cities are impacted by the same challenge in different ways, and the visions and rationales for similar initiatives can vary widely. Even if cities can learn from each other, there is no one single path to sustainable competitiveness.
- Strive to move single-domain projects towards multi-domain initiatives (e.g. from pure economic towards social and environmental aspects). Not every dimension can be linked with one another and there are latent conflicts, but there are many bridges and complementarities to be explored. And long-term development projects should be flexible enough to cope with change while keeping the direction recognisable.
- Identify potential leaders within and outside your organisation to which leadership can be distributed. Sustainable competitiveness projects require strong and resilient leadership at different levels, in different organisations, both technical and political. At a closer look, single powerful and omnipotent leaders are very rare.
- Involve unusual suspects. Beyond the business and policy elite, citizens, new entrepreneurs, not-for-profit organisations and other 'new kids on the block' are often central to get fresh ideas and to better recognise the user's perspective.
- Think about the most appropriate spatial level for the intervention (e.g. neighbourhood, city, metropolitan area). Functional, not administrative, areas are often more appropriate to tackle sustainable urban competitiveness challenges and make it easier to involve the stakeholders that matter.
- Clearly identify the capabilities (technical, expertise, networking, political, institutional) that are likely to be necessary in the medium term. Consider involving other stakeholders from the onset that can add to and complement the core capabilities of your organisation. Universities and private companies often have complementary resources *vis-à-vis* city councils and are increasingly ready to invest in the cities where they operate.

References

Amin, A. and Roberts, J. (eds). (2008). *Community, Economic Creativity, and Organization*. Oxford: Oxford University Press.

Arnstein, S. R. (1969). A ladder of citizen participation. *Journal of the American Institute of Planners*, 35(4), 216–24.

Bathelt, H., Malmberg, A. and Maskell, P. (2004). Clusters and knowledge: local buzz, global pipelines and the process of knowledge creation. *Progress in Human Geography*, 28(1), 31–56.

Beer, A. and Clower, T. (2014). Mobilizing leadership in cities and regions. *Regional Studies, Regional Science*, 1(1), 5–20.

Belleflamme, P., Picard, P. and Thisse, J.-F. (2000). An economic theory of regional clusters. *Journal of Urban Economics*, 48(1): 158–84.

Brown, T. (2008). Design thinking. *Harvard Business Review*, 86(6), 1–10.

Bulkeley, H., Broto, V. C., Hodson, M. and Marvin, S. (eds) (2011). *Cities and Low Carbon Transitions*. Abingdon: Routledge.

Carneiro, L. P. (2012). A politica de pacificação do Rio de Janeiro: um estudo sobre liderança e inovação. In A. Pinheiro and F. Veloso (eds). *Rio de Janeiro: um estado em transição*. Rio de Janeiro: Fundação Getúlio Vargas.

Carvalho, L., Mingardo, G. and van Haaren, J. (2012). Green urban transport policies and cleantech innovations: Evidence from Curitiba, Goteborg and Hamburg. *European Planning Studies*, 20(3), 375–96.

Chesbrough, H. W. (2006). *Open Innovation: The New Imperative for Creating and Profiting from Technology*. Boston, MA: Harvard Business Press.

Cooke, P. (2008). Regional Innovation Systems, Clean Technology & Jacobian Cluster Platform Policies. *Regional Science Policy & Practice*, 1(1), 23–45.

Glasmeier, A. and Christopherson, S. (2015). Thinking about smart cities. *Cambridge Journal of Regions, Economy and Society*, 8(1), 3–12.

Goddard, J. and Vallance, P. (2013). *The University and the City*. Abingdon: Routledge.

Kemp, R., Schot, J. and Hoogma, R. (1998). Regime shifts to sustainability through processes of niche formation: the approach of strategic niche management. *Technology Analysis & Strategic Management*, 10(2), 175–98.

Loorbach, D. (2010). Transition management for sustainable development: a prescriptive, complexity-based governance framework. *Governance*, 23(1), 161–83.

McCann, E. (2011). Urban policy mobilities and global circuits of knowledge: toward a research agenda. *Annals of the Association of American Geographers*, 101(1), 107–30.

Porter, M. (1990). *The Comparative Advantage of Nations*. New York, NY: Free Press.

Porter, M. E. and Kramer, M. R. (2011). Creating shared value. *Harvard Business Review*, 89(1/2), 62–77.

Rabari, C. and Storper, M. (2015). The digital skin of cities: urban theory and research in the age of the sensored and metered city, ubiquitous computing and big data. *Cambridge Journal of Regions, Economy and Society*, 8(1), 27–42.

Richards, G. and Palmer, R. (2012). *Eventful Cities*. Abingdon: Routledge.

Rodríguez-Pose, A. (2013). Do institutions matter for regional development? *Regional Studies*, 47(7), 1034–47.

Teisman, G. R. and Klijn, E. H. (2002). Partnership arrangements: governmental rhetoric or governance scheme? *Public Administration Review*, 62(2), 197–205.

Van den Berg, L., Braun, E. and van der Meer, J. (1997). *Metropolitan Organising Capacity: Experiences with Organising Major Projects in European Cities*. Aldershot: Ashgate.

Van den Berg, L., Braun, E. and Otgaar, A. (2004). Corporate community involvement in European and US cities. *Environment and Planning C*, 22, 475–94.

Van den Berg, L., Pol, P., Mingardo, G. and Speller, C. (2006). *The Safe City: Safety and Urban Development in European Cities*. Aldershot: Ashgate.

Van den Berg, L., van der Meer, J. and Carvalho, L. (eds) (2014). *Cities as Engines of Sustainable Competitiveness: European Urban Policy in Practice*. Aldershot: Ashgate.

Van Winden, W. (2013). Delivering sustainable urban solutions: a new chapter of corporate involvement in urban management. In M. P. van Dijk, J. van der Meer and J. van den Borg (eds), *From Urban Systems to Sustainable Competitive Metropolitan Regions: Essays in Honour of Leo van den Berg*. Rotterdam: Erasmus University Rotterdam.

Index

For Product Safety Concerns and Information please contact our EU representative GPSR@taylorandfrancis.com
Taylor & Francis Verlag GmbH, Kaufingerstraße 24, 80331 München, Germany

www.ingramcontent.com/pod-product-compliance
Ingram Content Group UK Ltd.
Pitfield, Milton Keynes, MK11 3LW, UK
UKHW050754180926
475701UK00007B/1049
9780367668310